AF707110

Becoming a Platform in Europe

On the Governance of the Collaborative Economy

Maurizio Teli and Chiara Bassetti
(Editors)

This article/publication is based upon work from COST Action CA16121, supported by COST (European Cooperation in Science and Technology).

COST (European Cooperation in Science and Technology) is a funding agency for research and innovation networks. Our Actions help connect research initiatives across Europe and enable scientists to grow their ideas by sharing them with their peers. This boosts their research, career and innovation.

www.cost.eu

EUROPEAN COOPERATION IN SCIENCE & TECHNOLOGY

Funded by the Horizon 2020 Framework Programme of the European Union

now the essence of knowledge

Published, sold and distributed by:
now Publishers Inc.
PO Box 1024
Hanover, MA 02339
United States
Tel. +1-781-985-4510
www.nowpublishers.com
sales@nowpublishers.com

Outside North America:
now Publishers Inc.
PO Box 179
2600 AD Delft
The Netherlands
Tel. +31-6-51115274

ISBN: 978-1-68083-840-4
E-ISBN: 978-1-68083-841-1
DOI: 10.1561/9781680838411

Copyright © 2021 Maurizio Teli and Chiara Bassetti

Suggested citation: Maurizio Teli and Chiara Bassetti (eds.). (2021). *Becoming a Platform in Europe – On the Governance of the Collaborative Economy*. Boston–Delft: Now Publishers

The work will be available online open access and governed by the Creative Commons "Attribution-Non Commercial" License (CC BY-NC), according to https://creativecommons.org/licenses/by-nc/4.0/

Table of Contents

DOI: 10.1561/9781680838411.ch1

Chapter 1

Introduction: Questioning the Collaborative Economy

By Maurizio Teli and Chiara Bassetti

1.1 From Sharing to Caring

This book is one of the main results of the Working Group 4, "Mechanisms to activate and support the collaborative economy", of the COST (European Cooperation in Science and Technology) Action 16121 "From Sharing to Caring: Examining Socio-Technical Aspects of the Collaborative Economy", that started in the spring of 2017 with the overarching goal of developing "a European network of actors focusing on the development of collaborative economy models and platforms and on social and on technological implications of the collaborative economy through a practice-focused approach" (CA 16121, 2016; Avram *et al.*, 2017).

One of the relevant aspects of the Action has been to question the collaborative economy in its various instances, from bottom-up peer to peer solidarity (Bassetti *et al.*, 2019) to corporate owned platforms (Rossitto and Lampinen, 2018), in relation to what are known as the European social values, that is the respect for human dignity and human rights, freedom, democracy, equality and the rule of law. Looking at forms of economic organization, as the ones collected under the label "collaborative economy", with the lenses of the European social values brings immediately to the question of the governance of such economic activities and how they organize work and social life.

Engaging in understanding the governance of the collaborative economy, brings with it three potential ways of looking at how collaborative economy platforms organize labor and sociality. Within a *processual perspective*, it is possible to question how the genealogy of specific platforms and the technical and organizational choices coming with it have brought to certain outcomes, in terms of technological features (Bødker *et al.*, 2020), collaborative models and practices (Avram *et al.*, 2019), organizational structures, values and contradictions (e.g., Barbu *et al.*, 2018). Within a *comparative perspective*, comparing different platforms or their use in different contexts (e.g., Clausen and Velázquez García, 2017) allows mapping and navigating the complexities surrounding the platforms under scrutiny, along with their diverse relational qualities, such as: the local/global dimension, the cross- and intra-industries differences and similarities, the forms of ownership, the profit/not-for-profit motive, and the various relations with existing institutions, ranging from governments to trade unions, from municipal actors to social movements. Within a *narrative perspective*, one has the opportunity to investigate the manifold elements that build up the platform self-presentation, such as general social goals and dynamics — from the United Nations Sustainable Development Goals to community dynamics, passing through rhetorics of innovation and jobs creation —, the re-articulation of legal aspects (labour regulations, data management, or welfare protections), and technological features aimed at supporting sharing and trust — like privacy protection, or rating and reputation systems (e.g., Richardson, 2015).

Through the collection of various contributions, this book takes a comprehensive approach able to highlight the processual, comparative, and narrative dimensions of the collaborative economy, helping us to address a variety of questions, such as: How do platforms re-articulate, describe, and implement power structures (Lampinen *et al.*, 2018)? Are they innovating in a way that is based on caring social relations or promoting exploitative practices (Light, 2019)? How are economic value, on the one hand, and social and cultural values, on the other hand, produced, circulated, and transformed by platform initiatives (Bassetti *et al.*, 2018; Light *et al.*, 2017)? How the production of goods and services, collaborative subjects, and collective narratives is legally, socially, and technically organized in platform initiatives

(Lampinen and Brown, 2017)? How existing institutions support, favour, or create obstacles to caring and/or exploitative platforms (Cibin *et al.*, 2019; Teli *et al.*, 2020)? Before proceeding with the presentation of how the book addresses those issues, we need to take a step back and introduce some definitions on what we call "collaborative economy".

1.2 Collaborative and Sharing Economy: A Plethora of Practices and Definitions

Businesses and initiatives that today go under the label "collaborative" or "sharing" economy "range from the small, grassroots-funded variety featured in Shareable to the big and venture-backed, many of which are online platforms" (Balaram, 2016). The domain of activity — accommodation, mobility, food, delivery, etc. — and the geographical scale — local, national, supranational — are similarly varied. Moreover, such initiatives can be carried out with or without any mediation between providers and consumers; when an intermediary is involved, this generally happens via online platforms, which is why they have been defined as "collaborative" platforms. Finally, the practices involved in this kind of economic activities are manifold, including barter, swap and rental; loan, crowdfunding and crowdsourcing; collective purchase, joint ownership and co-creation (cf. Bardhi and Eckhardt, 2012; Botsman, 2013, 2015; Botsman and Rogers, 2011; Frenken, 2017). The panorama is therefore highly varied, we offer an illustrative snapshot in Table 1.1.

Such a variety is accompanied by variability, as the considered activities are rapidly evolving and the involved actors (both individual and collective ones) change at a fast pace. This favoured the flourishing of a plethora of definitions and labels over the last ten years, ranging from "peer-to-peer" to "gig", "crowd", "on-demand", or "access" economy. As Balaram (2016) wrote half a decade ago, taking stock of what happened until that moment,

> In 2009, Airbnb, TaskRabbit, and Uber were fledgling start-ups [...] In 2010, writer and social entrepreneur Rachel Botsman began popularising the ideas underpinning these start-ups under the banner of "collaborative consumption". [...] By 2011, collaborative consumption gave way to the more intuitive, media-friendly term the "sharing economy". [...] the sharing economy is conflated with the "collaborative economy", which emphasises the role that internet technologies play in making connections between distributed groups of people, or with the "access economy" because of the focus on reducing the need for ownership [...]. The "gig economy" and the "on-demand economy" are the most recent additions to our vocabulary, [...] especially when referring to labour of TaskRabbit or Uber's nature.

Table 1.1. The panorama of collaborative and sharing economy initiatives.

	Mobility	Accommodation	Food	Work	Other Goods
On-demand	Ride hailing (e.g. Uber, Lyft, Bolt)	e.g. Airbnb, Fairbnb	e.g. Foodora, Just Eat, Glovo	Gig work (e.g. TaskRabbit)	
Sharing	Ride sharing (e.g. Blabla car), bike sharing, park sharing (e.g. Just Park)	Couch surfing	Meal sharing (e.g. Meal Sharing), also ethical purchasing groups (e.g. GASs, Alveari)	Time banks	Tool libraries and the like (e.g. Pumpipumpe), recovered factories (e.g. Rimaflow)
Second-hand	Repair and re-use (e.g. Ciclonauti)	/	Food sharing and recovery of wasted food	/	Purchase (e.g. Etsy, Ebay), barter
Product-service	Vehicle rental from a company	Hotel, apartment rental from a company	/	Co-working spaces	e.g. Clothes rental from a shop/company

At the same point in history, the European Commission felt the need to provide a definition, which remained an overarching, "umbrella" one, and explicitly open to change:

> For the purposes of this Communication, the term "collaborative economy" [7] refers to business models where activities are facilitated by collaborative platforms that create an open marketplace for the temporary usage of goods or services often provided by private individuals.
>
> [7] The term collaborative economy is often interchangeably used with the term 'sharing economy'. Collaborative economy is a rapid evolving phenomenon and its definition may evolve accordingly.
>
> (COM 2016, 356, p. 3)

Once more in 2016, Juliet Shor tried instead to restrict the definition of sharing economy by maintaining that "sharing refers to predominantly private, and often non-commercial transactions". Building on Frenken *et al.* (2015) understanding of sharing as "consumers granting each other temporary access to under-utilized physical assets ('idle capacity'), possibly for money", Frenken (2017) further attempted a definition of the sharing economy with the aim to distinguish it from the terms on-demand economy (also called gig economy), second-hand economy, and product-service economy (e.g., renting). In their view, sharing is characterized by the combination of:

- *access* to, rather than ownership of resources, or temporary rather than permanent access (a feature also shared by product-service and on-demand economies);
- *peer-to-peer* exchange (characterizing second-hand and, possibly, on-demand economy too);
- access to *goods*, more specifically "shareable goods", rather than services (featuring also in second-hand and product-service economies).

The issues at stake, however, are broader. It is not only a matter of more or less specific definitions, analytical distinctions and categorization. As it is always the case with practices, it is also a matter of tacit values, ideals and Weltanschauung — of culture, to put it shortly.

> The movement began with locally-based, grassroots-funded initiatives such as tool libraries and timebanks, but now seems to be led by global, venture-backed corporations. [...] Early proponents of the sharing economy were advocating for peer-to-peer exchange [...] as rooted in the commons, which encourages shared ownership over, or access to, resources [...] sustainability, openness, and solidarity.
>
> (Balaram, 2016)

What is at stake, therefore, is above all the values orienting economic activity. And this is true not only for the strictly defined sharing economy, but also — as the emergence of Fairbnb, in contrast with Airbnb shows — for the so-called collaborative and sharing economy (CSE) at large, including also on-demand economy. This is the terrain covered in this book, with a particular attention to question the relation between the collaborative economy and the European social values in a way that is open to the complexity of the task. To be able to design both policies and technologies with values in mind, is vital for the future of our societies. To understand those issues, the process bringing to this book has been, in itself, collaborative and complex.

1.3 Collaborative Thinking, Writing and Editing: How the Book Came to Be

Starting in Spring 2019, and leveraging on the work pursued during the previous two years within the Working Group 4 (WG4) of the COST Action "From Sharing to Caring", members started to discuss key issues for *a European collaborative economy that cares*: from media representation of platforms, to the role of institutions in relation to platform-based initiatives, passing through the relation with social movements and the legal framework. A series of *brainstorming sessions* (see e.g., Fig. 1.1) was held during several face-to-face meetings: on March 15, 2109 in Zagreb, Croatia; on April 11, 2019 in Rome, Italy; on May 20, 2019 in Vilnius,

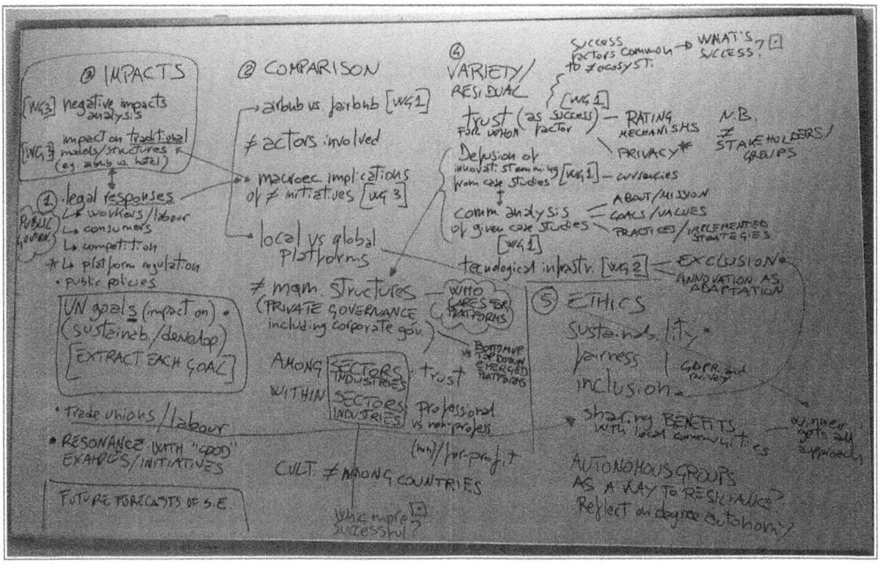

Figure 1.1. Whiteboard of one of the WG4 Brainstorming Sessions.

Lithuania. The overarching questions addressed were the following:

- Which values we as European citizens want to foster?
- What do we mean by a caring collaborative economy? What would be a European platform for a caring economy and society?
- What's unique in a European perspective? What's distinctive about local initiatives and other forms of collaborative economy activities in Europe?
- Do we have "European" technologies? What does that mean?
- What would mean inclusion in European caring economy platforms?

These questions have driven the discussion on some sub-themes to be explored, like the relation between social values and market failures, the organization and governance of labour, which societal problems to tackle, for the benefit of whom, and fostering which values. This discussion brought to a reflection on the space for institutional action around the collaborative economy in Europe, starting with a focus on the relation between local needs and global technologies.

A further meeting was held on October 24, 2019 in Edinburgh, Scotland. On such an occasion, the book call for chapters was shared and some participants presented ideas for potential chapters, looking for cross-country and multi-disciplinary collaborations in the spirit of the COST networking ambitions and the call for chapters guidelines. In the following months, we facilitated such a collaboration by further disseminating the call, collecting proposals, and providing potential authors with an online space where to share abstracts, express interest in given topics, search for writing partners able to provide empirical data and/or further disciplinary/national perspectives on such a topic, or offer one's data/perspective to other participants. *Co-authorship* is therefore a tenet of every chapter of this book.

First drafts of chapter proposals were then collected by Spring 2020 and a process of *mutual review* was established among prospective authors. Alongside the editors, at least two out of the pool of authors reviewed each chapter three times, and all took part in two mutual review meetings in June and in October 2020 (first version and second version review, respectively). Third versions were evaluated only in written form, by the same pool of reviewers, in late Autumn 2020, and final versions were collected in January 2021. Therefore, although editorship responsibility remains ours, editing too may be seen as a collaborative and interdisciplinary effort.

1.4 The Chapters: From Individual Motivations to the Future

As said, the process that has brought this book to life has been extremely collaborative, both in the elaboration of its topic and concerning the single chapters.

While moving from the brainstorming sessions to the first chapter ideas we, as editors, felt the need of writing a call for chapters capable of attracting authors also beyond the ones who participated in the brainstorming sessions, as well as to summarize the main points coming out of those sessions. The three perspectives mentioned at the outset – processual, comparative, and narrative – emerged out of this work. Together with the call for chapters, we needed to elaborate on a potential title for the book. We came to "Becoming a Platform in Europe", for a series of reasons. First, the verb *becoming* connects to the processual perspective, conveying the meaning of an ongoing process of platformization both behind us and in front of us. Second, adding *a platform* brings the processual perspective together with the narrative perspective, since platforms as things are defined and described, and platforms as organizations —companies or grassroots initiatives— tell stories about themselves. Finally, the reference *in Europe* both delimits the geography under scrutiny and allows for the comparative perspective to emerge.

In structuring the book, we have decided to start with the chapters including comparisons among countries or cases, to later present the chapters that dig deeper on specific processual or narrative aspects. In this way, we aim at providing the readers with some instruments to frame the specific aspects unders scrutiny in the light of a comparative dimension. Among those potential instruments, we begin with trying to understand the *individual choices in participating* in the collaborative economy. Majetic and Vega (Chapter 1), as well as Angelovska, Čeh Časni, and Lutz (Chapter 2), examine comparatively the influences on participation of a variety of factors, mainly distinguishing between economic, technological, and non-economic elements. The picture that those chapters offer is a multi-faceted one, providing a differentiated understanding of participation in the collaborative economy.

Moving beyond the understanding of individual motivations, the chapter by Diogo, Sanna, Bernat, and Vaiciukynaitë (Chapter 3) and the one of Rossitto, Lampinen, Light, Diogo, Bernat, and Travlou (Chapter 4) investigate *the use of platforms at the local level.* In the first case, the attention is on the use of one of the classical examples of the collaborative economy, the sharing of bikes and e-scooters for urban mobility, that provides meaningful insights on commonalities and differences among the provision of services and their uses in four European capitals, Budapest, Lisbon, Rome and Vilnius. Chapter 4, on the other hand, discusses "the platform paradox in community initiatives", investigating why and how grassroots initiatives of solidarity in several countries end up relying on Facebook as an infrastructure, although in many ways their values are in opposition to the social media giant.

The analysis of the relation between *platform design and political dimensions* is the key contribution of the chapter by Cruciani and Lewkowicz (Chapter 5) and the one of Goyens and Huybrechts (Chapter 6). Whereas the former allows understanding how grassroots initiatives can be supported by ways of designing capable of motivating participation in voluntary, time-consuming activities, the latter focuses on how the collaborative economy itself can be read as a political phenomenon, changing local relations. Moving beyond the local level, Koka, Kruja and Hysa (Chapter 7) discuss how research on the use of collaborative economy platforms, specifically AirBnB in Albania, points to the need of policy interventions. On a specular side, through two situated case studies, Larner (Chapter 8) shows how it is possible to think not only of policy intervention but to imagine alternative business models that reflect solidarity and collaboration while trying to ensure economic and financial viability. Dumančić, Naèinović Braje, and Aleksić (Chapter 9) remind us that thinking about the economy means to think about the way work is regulated at the legal level, with a focus on the differences in the worker-employer relation between traditional jobs and platform-mediated ones.

All the chapters mentioned so far point to the complexity of understanding the collaborative economy—a condition which does not favour effective policymaking, especially if values-oriented and aimed at medium-to-long-term outcomes. Sanna and Michelini (Chapter 10) offer a supporting tool by focusing on the methodologies to assess the *impact of the collaborative economy*, discussing also their policy implications. The last two contributions presented in this book further move discussion towards the future. More specifically, Crombie, Kollegala, and Zehle (Chapter 11) question recent technological developments, proposing to imagine a new design stack, an ensemble of technologies that can support solidarity and cooperation, while Subaşı, Fedosov, and Bates (Chapter 12) report on an experience of imagination of future cooperatives in the domain of the collaborative economy, rethinking basics of contemporary economy like currency and data. These last chapters are crucial in the overall picture provided by this book, as they highlight one fundamental aspect of "Becoming a Platform in Europe", the need to turn the understanding of the European social values into actionable processes flowing through policy, technology, and organizing.

References

Avram, G., Choi, J. H., De Paoli, S., Light, A., Lyle, P., and Teli, M. (2017). Collaborative Economies: From Sharing to Caring. *Proceedings of the 8th International Conference on Communities and Technologies*, 305–307. https://doi.org/10.1145/3083671.3083712

Avram, G., Choi, J. H., Paoli, S. D., Light, A., Lyle, P., and Teli, M. (2019). Repositioning CoDesign in the age of platform capitalism: From sharing to caring. *Co Design*, *15*(3), 185–191. https://doi.org/10.1080/15710882.2019.1638063

Balaram, B. (2016). Fair share – reclaiming power in the sharing economy, Royal Society for the encouragement of Arts, Manufactures and Commerce. https://www.thersa.org/reports/fair-share-reclaiming-power-in-the-sharing-economy

Barbu, C. M., Laurentiu Florea, D., Ogarca, R. F., and Barbu, M. C. R. (2018). From Ownership to Access: How the Sharing Economy Is Changing the Consumer Behavior. *Amfiteatru Economic*, *20*(48), 373–387. https://doi.org/10.24818/EA/2018/48/373

Bardhi, F. and Eckhardt, G. M. (2012). Access-Based Consumption: The Case of Car Sharing. *Journal of Consumer Research*, *39*(4), 881–898. https://doi.org/10.1086/666376

Bassetti, C., Botto, F., and Teli, M. (2018). The Commonfare Project. Designing to Support Grassroots Welfare Initiatives. *DigitCult – Scientific Journal on Digital Cultures*, *3*(1), 31–40. https://doi.org/10.4399/97888255159095

Bassetti, C., Sciannamblo, M., Lyle, P., Teli, M., Paoli, S. D., and De Angeli, A. (2019). Co-designing for common values: Creating hybrid spaces to nurture autonomous cooperation. *CoDesign*, *15*(3), 256–271. https://doi.org/10.1080/15710882.2019.1637897

Bødker, S., Lewkowicz, M., and Boden, A. (2020). What's in a word? Platforms Supporting the Platform Economy. *Proceedings of the 11th Nordic Conference on Human-Computer Interaction: Shaping Experiences, Shaping Society*, 1–10. https://doi.org/10.1145/3419249.3420167

Botsman, R. (2013). The sharing economy lacks a shared definition. Fast Company, 21. https://www.fastcompany.com/3022028/the-sharing-economy-lacks-a-shared-definition

Botsman, R. (2015). Defining the sharing economy: what is collaborative consumption–and what isn't. Fast Company, *27*(1). https://www.fastcompany.com/3046119/defining-the-sharing-economy-what-is-collaborative-consumption-and-what-isnt

Botsman, R. and Rogers, R. (2011). What's Mine Is Yours: How Collaborative Consumption is Changing the Way We Live. HarperCollins, UK.

Cibin, R., Teli, M., and Robinson, S. (2019). Institutioning and Community Radio. A Comparative Perspective. *Proceedings of the 9th International Conference on Communities & Technologies – Transforming Communities*, 143–154. https://doi.org/10.1145/3328320.3328392

Clausen, H. B. and Velázquez García, M. A. (2017). Collaborative Economy in Tourism in Latin America: The Case of Argentina, Colombia, Chile and Mexico. In Dredge, D. and Gyimóthy, S. (eds.), *Collaborative Economy and Tourism: Perspectives, Politics, Policies and Prospects* (pp. 271–284). Springer International Publishing. https://doi.org/10.1007/978-3-319-51799-5_16

Commissione Europea. (2016). Communication From the Commission to the European Parliament, the Council, the European Economic and Social Committee and the Committee of the Regions: A European Agenda for the Collaborative Economy. COM(2016) 356.

CA 16121 (COST Action "From Sharing to Caring"). (2016). Memorandum of Understanding. https://e-services.cost.eu/files/domain_files/CA/Action_CA16121/mou/CA16121-e.pdf

Frenken, K., Meelen, T., Arets, M., Van de Glind, P. (2015). Smarter Regulation for the Sharing Economy. The Guardian (May 20, https://www.theguardian.com/science/political-science/2015/may/20/smarter-regulation-for-the-sharing-economy, Retrieved October 23, 2016.)

Frenken, K., Schor, J. (2017). Putting the sharing economy into perspective. *Environmental Innovation and Societal Transitions*, *23*, 3–10.

Lampinen, A., Lutz, C., Newlands, G., Light, A., and Immorlica, N. (2018). *Power Struggles in the Digital Economy: Platforms, Workers, and Markets*. 417–423. Scopus. https://doi.org/10.1145/3272973.3273004

Lampinen, A. and Brown, B. (2017). Market Design for HCI: Successes and Failures of Peer-to-Peer Exchange Platforms. *Proceedings of the 2017 CHI Conference on Human Factors in Computing Systems*, 4331–4343. https://doi.org/10.1145/3025453.3025515

Light, A. and Briggs, J. (2017). Crowdfunding Platforms and the Design of Paying Publics. *Proceedings of the 2017 CHI Conference on Human Factors in Computing Systems*, 797–809. https://doi.org/10.1145/3025453.3025979

Light, A. and Miskelly, C. (2019). Platforms, Scales and Networks: Meshing a Local Sustainable Sharing Economy. *Computer Supported Cooperative Work (CSCW)*, *28*(3), 591–626. https://doi.org/10.1007/s10606-019-09352-1

Richardson, L. (2015). Performing the sharing economy. *Geoforum*, *67*, 121–129. https://doi.org/10.1016/j.geoforum.2015.11.004

Rossitto, C. and Lampinen, A. (2018). Co-Creating the Workplace: Participatory Efforts to Enable Individual Work at the Hoffice. *Computer Supported Cooperative Work (CSCW)*, *27*(3), 947–982. https://doi.org/10.1007/s10606-018-9319-z

Schor, J. (2016). Debating the sharing economy. *Journal of Self-Governance and Management Economics*, *4*(3), 7–22.

Teli, M., Foth, M., Sciannamblo, M., Anastasiu, I., and Lyle, P. (2020). Tales of Institutioning and Commoning: Participatory Design Processes with a Strategic and Tactical Perspective. *Proceedings of the 16th Participatory Design Conference 2020 – Participation(s) Otherwise – Volume 1*, 159–171. https://doi.org/10.1145/3385010.3385020

DOI: 10.1561/9781680838411.ch2

Chapter 2

Socio-environmental Determinants of Willingness to Participate in the Collaborative Economy

By Filip Majetić and Rodrigo Perez-Vega

This chapter explores the role of socio-environmental determinants in the willingness to participate in the collaborative economy (CE). The CE refers to remunerated and non-remunerated peer-to-peer sharing of underused resources via online (collaborative) platforms. Socio-environmental determinants represent a key distinctive feature of the CE and yet the field of empirical studies is far from being saturated. The present study used a non-representative sample of 363 EU-based respondents. The set of explored variables included the respondents' demographic characteristics and socio-environmental determinants of sociability, materialistic orientation, consumers' need for uniqueness (NFU), and environmental concern. The main findings revealed women being more willing to participate in the CE as well as environmental concern and propensity to make unpopular choices (a dimension of NFU) having a positive effect on the participation willingness. The proposed model accounted for 9% in variability of the willingness to participate in the

CE. This indicates that the socio-environmental determinants of participation in the CE are most likely (heavily) outweighed by the economic and/or technological ones.

2.1 Introduction

Considering the lack of consensus on the collaborative economy (CE) conceptualization (Dredge and Gyimóthy, 2015; Murillo *et al.*, 2017), we begin with defining the research field. In this chapter, the CE refers to peer-to-peer sharing of underused resources via online (collaborative) platforms. It embraces sharing of, i.e. providing/acquiring temporary access to, resources among acquaintances and strangers, as well as remunerated and non-remunerated forms of sharing (Frenken and Schor, 2017; Benoit *et al.*, 2017). Although peer-to-peer sharing itself has a long-standing tradition, the key novelty is that transactions are conducted through online platforms and that the scope of sharing is extended to strangers, who were made reachable by the introduction of platforms themselves (Belk, 2014; Schor, 2014; Kathan *et al.*, 2016). Since the collaboration is realized through the activity of sharing, and the shared resources are collaboratively consumed, the terms collaborative economy, sharing economy, and collaborative consumption have commonly been used to label the same phenomenon (e.g., Frenken and Schor, 2017; Leoni and Parker, 2019, Introduction, this volume). Since the European Union officially uses the "collaborative economy" term (European Commission, 2018), we opt for the same label.

This type of service production and consumption has attracted increased attention of businesses, politicians, researchers, customers, advocacy groups, and the media. The global CE scene has been rapidly growing (see European Commission, 2018; Eurostat, 2019), traditional industries and everyday lives have been disrupted (e.g., Uber has influenced the flexibilization of the taxi industry and Airbnb has influenced the touristification of residential neighborhoods), countries have introduced new legal frameworks to regulate the CE, and researchers have investigated a wide range of topics, in particular the determinants of participation (e.g., Prieto *et al.*, 2017), business models (e.g., Gyimóthy, 2017), and economic, sociocultural, and environmental impacts of the CE (e.g., Zervas *et al.*, 2017; Guo *et al.*, 2018; Jiang *et al.*, 2018; Piscicelli *et al.*, 2015, Sanna & Michelini, this volume). Additionally, and more importantly for our research rationale, the increased attention has been based on a common understanding that the CE has the potential to foster, and is underpinned by the values of socializing, socialization, community building, sustainable consumption, and environmental protection (Botsman and Rogers, 2010; Binninger *et al.*, 2015; Hamari *et al.*, 2016).

In relation to this widespread understanding, empirical research has indeed identified both economic and non-economic determinants of attitudes towards the CE, of the intention to participate, and of CE participation itself. Within the non-economic group, socio-environmental and technological determinants seem to be the most represented. For instance, Owyang *et al.* (2013) identified social (e.g., desire for community, drive for sustainability) and technological determinants of participation (e.g., the increase of social networking sites and mobile devices) in addition to economic ones (e.g., monetize excessive inventory). In a similar vein, Tussyadiah (2015) reported elements of economic, social, and technological determinants to drive participation. Lack of trust, lack of efficacy with regards to technology, and lack of economic benefits could hinder the collaborative consumption while societal aspects of sustainability and community as well as economic benefits could favour it. Furthermore, Hamari *et al.* (2016) reported that both economic and socio-environmental elements (e.g., sustainability, enjoyment) influence the attitudes and behavioral intentions towards engaging in the CE. Their findings suggest that economic determinants are more closely related to the intention to use, while the social determinants have a greater impact on the attitudes towards the CE (but not necessarily on the intentions to engage). In comparison to monetary and moral motives (sharing as an act of moral integrity), social-hedonistic motives had a larger impact on CE attitudes in Bucher *et al.* (2016) study too.

Unlike economic and technological determinants of CE participation that are inherent also to the entire body of digitally based mainstream economies represented by Amazon, eBay, and HousingAnywhere (Kenney and Zysman, 2016), the socio-environmental, i.e. not-for-economic-profit, determinants are a key distinctive feature of the CE.[1] In other words, in this context, socio-environmental determinants make the CE an alternative, prosocial and less money-driven business activity.

However, the actual impact of socio-environmental determinants has not received significant research attention. Although the most comprehensive studies on CE determinants such as Bucher *et al.* (2016) and Hawlitschek *et al.* (2016) explored elements from the not-for-economic-profit category, they did so in the context of attitudes (not participation) and did not include CE "distinctiveness/alternativeness".

---

1. Placing the sign of equality between the socio-environmental and not-for-economic-profit determinants is in line with (Belk, 2014, p. 1597) definition of collaborative consumption/economy which embraces "people coordinating the acquisition and distribution of a resource for a fee or other compensation". Apart from barter trading and time banking practices, the "other compensation" might also refer to non-economic compensation such as increased subjective well-being when socializing, protecting the environment or consuming sustainably (see e.g., Guillen-Royo, 2019). Of course, the latter does not imply that we socialize, protect the environment or consume sustainably solely for egoistic purposes.

Therefore, to address the research gap, this study will investigate the role of *socio-environmental determinants* in the willingness of EU-based population to *participate* in the collaborative economy. The study is based on primary quantitative data analyzed using univariate and multivariate statistics.

The relevance of the study emerges from its practical application too. Namely, better understanding the determinants of (current and potential) users' willingness to use collaborative platforms can help their creators to improve the design, promotion, and on-going operation of the platforms. This might be particularly important in the context of socio-environmental determinants since they tend to be more complex and difficult to analyze than the economic ones.

In what follows, to introduce the set of explored CE determinants and set up the hypotheses, we begin with presenting previous research findings on demographic characteristics and not-for-economic-profit motives relevant in the context of both CE participation and intention/willingness to participate. The literature review is followed by the methodology section and results of statistical analyses. The final section includes discussion and concluding observations.

2.2 Literature Review

2.2.1 Demographic Predictors

The most frequently explored demographic predictors of CE use are gender/sex,[2] age, individual purchasing power (e.g., monthly net income, employment status) and education. An inconclusiveness of the overall results is partially attributed to different industries the studies are related to (Alonso-Almeida *et al.*, 2020). For instance, Alonso-Almeida *et al.* (2020) found gender as a significant predictor of the industry-unspecific CE participation – in favor of women. On the other hand, in the P2P accommodation and carsharing industries men were more prone to using collaborative platforms (Pesonen and Tussyadiah, 2017; Le Vine *et al.*, 2014).

Furthermore, younger age groups were found more willing to engage in industry-unspecific CE (Owyang *et al.*, 2014), P2P accommodation services (Pesonen and Tussyadiah, 2017) and carsharing (Le Vine *et al.*, 2014). Morency *et al.* (2012) reported findings similar to Le Vine *et al.* (2014); people aged 35–44 were most likely to use carsharing services and people below the age of 35 were more likely to

2. We have decided to refer to gender/sex because previous literature is inconsistent in regard to this dimension, with some references looking at gender and some others at sex. It is outside the scope of the paper to engage in a substantive discussion on the use of gender or sex in quantitative studies, so we adopted a compromised solution, that we know carries with it limitations. We consider those limitations not particularly relevant for the overall goal of the chapter.

use them than people aged 45 and above. Using the rationale behind Tussyadiah (2015) and Stokes *et al.* (2014) findings, in comparison with the group of youngest people ($\leq$34), it might be that people aged 35–44 were more likely to use the services due to their higher purchasing power. On the other hand, Prieto *et al.* (2017) found that having a job had no effect on carsharing usage. Additionally, the level of income had no effect in the case of industry-unspecific CE participation (Alonso-Almeida *et al.*, 2020) and participation in P2P accommodation services (Pesonen and Tussyadiah, 2017).

Finally, Prieto *et al.* (2017) found a positive correlation between level of education and the usage of car sharing services and Tussyadiah (2015) reported that users of collaborative consumption in travel were more educated than non-users. No effect of education level on the willingness to participate in the ridesharing activities was found in Boateng *et al.* (2019) study.

Based on the above summarized findings, we hypothesize that gender/sex (H1a), age (H1b), and level of education (H1c) represent significant predictors of willingness to participate in the CE. In terms of directionality, women, younger people, and those who are more educated are expected to be more prone to CE participation.

2.2.2 Socio-environmental Determinants

Regarding the previous studies on socio-environmental determinants, the more individuals are willing to interact with other people, the higher their willingness to develop and maintain collaboration opportunities with others, including strangers (Wang *et al.*, 2015; Fox, 1984). Therefore, the desire to increase and empower social connections has been theorized as an important determinant to engage in the CE (Botsman and Rogers, 2010; Rowe, 2017). (Tussyadiah, 2015, p. 10) found that social motivations to participate in the CE were indeed important (although of less relative importance than the economic/monetary ones) and related to a desire to "get to know, interact and connect with local communities in a more meaningful way". In this context, it seems reasonable to assume that social motivation is related also to the desire for community building – in which communities get developed through the process of socialization (Gheitasy *et al.*, 2014), which in turn increases the level of social capital (e.g. trust in strangers, reciprocal interactions, Putnam, 2000). Furthermore, Bucher *et al.* (2016) reported a positive correlation between sociability and both moral and social-hedonistic motives to participate in the CE. Within this field of sociability, previous studies have emphasized the ability to gain reputation among the collaborators as an additional driver of participation, as well as the CE having the potential to boost online and offline socializing (Tussyadiah, 2015; Hamari *et al.*, 2016). Lack of trust in people, especially in strangers, was commonly mentioned as the key "social" barrier to CE participation (Owyang *et al.*, 2013; Tussyadiah, 2015).

Based on the presented findings, we hypothesize that sociability has a positive effect on willingness to participate in the CE (H2).

Second, the CE seems to have intensified a long-standing research interest in the relationship between materialism and consumers' willingness to share[3] (see e.g., Richins and Dawson, 1992). Namely, a materialistic orientation places acquiring and possession of material objects very high within an individual's hierarchy of values seeing them as an indicator of personal success and a source of happiness (Pilch and Górnik-Durose, 2016; Richins, 2004). Hence, Lindblom *et al.* (2018) study on Finnish respondents revealed that materialistic orientation was negatively related to consumers' attitudes towards the CE but positively related to intentions to get engaged in its activities. (Davidson *et al.*, 2018, p. 364) also reported materialistic orientation to be positively correlated with willingness to participate in the cases where users (not only consumers but also providers of services): (a) were searching for "transformative and hedonic experiences that will improve their self-image and wellbeing" (identified in the case of US respondents) and (b) where participation provides them with an "increased perceived utility" operationalized through flexibility, convenience, and availability (identified in the case of Indian respondents). Bucher *et al.* (2016) found a significant effect of materialism on monetary motives to participate in the CE, but not on socio-hedonistic and moral ones. Alonso-Almeida *et al.* (2020) study investigated the sharing economy within the model of new materialism, i.e. a "hybrid model in which property and the enjoyment of goods coexist with the enjoyment of experiences, which are becoming increasingly more important", and reported that CE participation is positively associated with both new materialism consumer awareness and new materialism social awareness. Akbar *et al.* (2016) researched the effect of possessiveness and non-generosity (dimensions of the Belk (1982) operationalization of materialistic orientation) on sharing intention: the possessiveness dimension was found to have a negative effect, while non-generosity had no effect. In the "materialism" → "sharing intention" → "sharing participation" relationship, materialism (i.e. the significant dimension of possessiveness) had a "highly significant" indirect effect on the sharing participation.

Based on the presented findings, we hypothesize that materialistic orientation has a positive effect on willingness to participate in the CE (H3).

Furthermore, (Akbar *et al.*, 2016, p. 4219) argued that this three-party relationship might be influenced by "certain consumer characteristics", in particular

---

3. In the context of this study, the "consumers" term refers to users of collaborative platforms i.e. both providers and consumers.

by consumers' need for unique consumer products (e.g., customized products). Need for uniqueness (NFU) refers to an individual's desire to be (perceived as) different from others (Lynn and Harris, 1997) and is often communicated via material objects that consumers display (Tian, 2001). The concept consists of the following three dimensions (Tian, 2001, pp. 52–53, 55). First, Creative Choice Counterconformity where consumers try to differentiate themselves "from most others" by making consumer choices "that are likely to be considered good choices by these others" (e.g., acquiring material objects to achieve a personal image that cannot be duplicated). Second, "Unpopular Choice Counterconformity" where the differentiation is made through consumer choices that "deviate from group norms" and might be socially disapproved (e.g., acquiring and wearing clothes that might offend other people). Third, "Avoidance of Similarity" i.e. "loss of interest in, or discontinued use of, possessions that become commonplace" (e.g. losing interest in brands once they become over-popular). Consumer research has studied the role of NFU consumption choices in the context of mainstream economy in depth (e.g., Latter *et al.*, 2010) and across various socio-cultural settings – including collectivist societies (e.g., Zhu *et al.*, 2015). However, in the context of CE, an alternative form of economy where users/individuals are predominantly considered to be prosumers (Botsman and Rogers, 2010), i.e. where NFU might have a simultaneous role in both production and consumption patterns, only a few papers have explored the topic. Hence, after taking the NFU moderator into the equation of "materialism" → "sharing intention" → "sharing participation", (Akbar *et al.*, 2016, pp. 4219–4221) revealed the following. First, "sharing intention decreases with high materialism particularly for those consumers with low desire for unique consumer products". Second, in comparison with the respondents who scored low on the NFU scale, those who scored high were "more likely to act (participate) according to their sharing intentions". Third, "the indirect effect of materialism on the sharing participation that is mediated by intention formation was much stronger" among those who reported weak and moderate NFU. Apart from the moderating role, the direct effects of NFU have also been explored – mostly within the CE fashion area. For instance, Lang and Armstrong (2018) found that consumers' NFU encourages intentions to swap items, probably as it would allow the consumer to display new unique objects, but there is a lesser inclination to rent products. In a similar vein, Becker-Leifhold and Iran (2018) and Matthews *et al.* (2019) posit that NFU, as part of hedonic motives, encourages users to engage in the fashion industry of CE. On the other hand, Hawlitschek *et al.* (2016) evaluated 17 motives for industry-unspecific CE participation and labeled the NFU as one of the insignificant ones.

Based on the presented findings, we hypothesize that consumers' need for uniqueness has a positive effect on willingness to participate in the CE (H4).

Finally, from the very beginning of today's collaborative economy, environmental protection and environmental long-term sustainability have been communicated as key elements (Botsman and Rogers, 2010; Pesonen and Tussyadiah, 2017). However, comprehensive empirical findings on this aspect are still in an early stage and somewhat inconclusive (Ertz *et al.*, 2018). For instance, Hamari *et al.* (2016) found a positive effect of perceived sustainability (sustainable consumption and ecological concerns) on CE attitudes but did not find a direct effect on behavioral intentions to get engaged. When CE attitudes were introduced as a mediator, only a small total effect of sustainability on behavioral intentions was reported. Therefore, the authors (2016, p. 2047) suggest that "sustainability might only be an important factor for those people for whom ecological consumption is important". Furthermore, Tussyadiah (2015) found that perceived sustainability (e.g., willingness to reduce the negative impacts of travel on the environment) does represent a driver of the CE, but the economic benefits were (again) a stronger motivator of participation. This quantitative finding on environmental motives being less prominent than economic/monetary ones was supported also by Binninger *et al.* (2015) qualitative study. On the other hand, contrary to the findings of Hamari *et al.* (2016) and Tussyadiah (2015), Pesonen and Tussyadiah (2017) found that "environmental friendliness" (e.g., personal consumption being reduced due to environmental reasons) did not affect the usage of P2P accommodation services.

Based on the presented findings, we hypothesize that environmental concern has a positive effect on willingness to participate in the CE (H5).

2.3 Methodology

2.3.1 Overview of the Procedure

Since the topic of our study is an ICT enabled phenomenon, the data was collected among internet users through an on-line questionnaire administered by Prolific Academic Ltd. in July 2020 (see Palan and Schitter, 2018). The data analysis consisted of descriptive statistics, principal component analyses (PCA) to assess the instruments' dimensionality, and hierarchical regression analysis to assess the predictive potential of the model(s). The analyses were conducted using IBM SPSS Statistics 25.

2.3.2 The Sample

The initial convenience sample consisted of 395 European Union based respondents. The final sample of 363 respondents was reached after excluding those who did not pass the attention check and the outliers. The outlier status was checked for all the variables using the Mahalanobis distance procedure with 45 degrees of freedom, i.e. critical Chi-square value of 80.08 at $\alpha = 0.001$ (Leys *et al.*, 2018).

According to the Kolmogorov-Smirnov test, values of all the variables deviate from normal distribution ($p < 0.001$). However, "the shape of the distribution may not be severely non-normal" because all the variables' absolute values of both skewness and kurtosis are within the acceptable range of ≤ 3.0 and ≤ 10.0, respectively (Kline, 2011, p. 77).

The overrepresentation of the respondents aged 18–29 (77.1%) as well as Polish (27.5%) and Portuguese (24.5%) respondents emerges from the "first come, first served" approach in data collection which Prolific.co commonly uses in the case of convenient i.e. non-representative samples (see Table 2.1).

Table 2.1. Basic demographic characteristics of the respondents.

Variable	Category	Frequency	%
Sex[4]	Female	127	35.0
	Male	233	64.2
	Missing data	3	0.8
Age	18–29	280	77.1
	30–39	51	14.0
	40–49	20	5.5
	50–59	9	2.5
	60–69	2	0.6
	70+	1	0.3
Country of residence	Austria	2	0.6
	Belgium	3	0.8
	Czech Republic	5	1.4
	Denmark	2	0.6
	Estonia	4	1.1
	Finland	3	0.8
	France	3	0.8
	Germany	2	0.6
	Greece	18	5.0
	Hungary	8	2.2

(Continued)

4. To simplify the data collection/analysis and avoid (over)simplifying the plurality of gender identities, the gender/sex-related questionnaire item explicitly referred to "sex". The response options were "Female", "Male", and "Prefer not to say" (Missing data).

Table 2.1. Continued

Variable	Category	Frequency	%
	Ireland	2	0.6
	Italy	33	9.1
	Latvia	2	0.6
	Luxembourg	1	0.3
	Netherlands	1	0.3
	Norway	2	0.6
	Poland	100	27.5
	Portugal	89	24.5
	Slovenia	4	1.1
	Spain	31	8.5
	Sweden	5	1.4
	Switzerland	2	0.6
	United Kingdom	41	11.3
Area of residence	Urban	242	66.7
	Semirural	91	25.1
	Rural	29	8.0
	Missing data	1	0.3
Highest level of education	Primary	8	2.2
	Secondary	133	36.6
	Post-secondary non-university	50	13.8
	Bachelor or equivalent	103	28.4
	Master or equivalent	68	18.7
	Doctorate or equivalent	1	0.3
Employment status	Student/Retired	158	43.5
	Employed	162	44.6
	Unemployed	43	11.8

2.3.3 The Instruments

The willingness to participate in the collaborative economy was explored through 4 items (see Table 2.2) using an adapted version of the scale elaborated by Balderjahn *et al.* (2013). Sociability was measured with 5 items (Table 2.3) using Goldberg *et al.* (2006) scale. Materialistic orientation was assessed with 6 items (Table 2.4) using the short form of Material Values Scale (MVS) developed by Richins (2004).

Table 2.2. A single component solution for "Willingness to participate in the collaborative economy" obtained using PCA.

Variables	Component 1
If you could afford to buy a product you need/want, to what extent would you be willing to:	
instead of buying, borrow it from strangers via online/collaborative platforms?	0.83
instead of buying, renting it from strangers via online/collaborative platforms?	0.80
If you owned a product you currently do not need, to what extent would you be willing to:	
temporarily share it with strangers via online/collaborative platforms?	0.77
temporarily renting it to strangers via online/collaborative platforms?	0.80
λ	2.55
%variance	63.83

K-M-O = 0.695; Bartlett Chi = 547.405; df = 6; $p = 0.000$.

Consumers' need for uniqueness was measured with 8 items (Table 2.5) using an adapted version of the short form of Customers' need for uniqueness scale developed by Ruvio *et al.* (2008). Environmental concern was measured with 6 items (Table 2.6) using Alzubaidi *et al.* (2021) scale. The respondents were asked to indicate their opinion on 5-point Likert scales. Apart from the NFU scale, all other scales (adapted versions) have previously been used in similar contexts.[5]

Regarding the instruments' dimensionality assessment, in all the cases, we retained only the components with eigenvalue (λ) of at least 1 and the items with factor loading of at least 0.5 (Hair *et al.*, 2009). If more than one component was extracted (eigenvalue ≥ 1), to enhance the results' interpretability, Varimax rotation with Kaiser normalization was employed. The scales' reliability was assessed using Chronbach's Alpha (α) with the least acceptable value set at 0.70 (Taber, 2018). The results of the principal component analysis are the following.

In line with the Balderjahn *et al.* (2013) approach, the analysis indicated that "willingness to participate in the collaborative economy" should be regarded as a unidimensional concept. The scale is of acceptable reliability ($\alpha = 0.811$).

In line with the Goldberg *et al.* (2006) approach, the analysis indicated that "sociability" should be regarded as a unidimensional concept. The scale is of acceptable reliability ($\alpha = 0.818$).

5. For instance, Davidson *et al.* (2018) and Lindblom *et al.* (2018) used the same Richins (2004) scale to explore the role of materialistic orientation in willingness to participate in the CE.

Table 2.3. A single component solution for "Sociability" obtained using PCA.

Variables	Component 1
I enjoy bringing people together.	0.79
I enjoy being part of a group.	0.81
I love to chat.	0.77
I love surprise parties.	0.66
I am interested in people.	0.82
λ	2.99
%variance	59.83

K-M-O = 0.842; Bartlett Chi = 635.894; df = 10; $p = 0.000$.

Table 2.4. A single component solution for "Materialistic orientation" obtained using PCA.

Variables	Component 1
I admire people who own expensive homes, cars, and clothes.	0.67
The things I own say a lot about how well I'm doing in life.	0.58
Buying things gives me a lot of pleasure.	0.72
I like a lot of luxury in my life.	0.75
My life would be better if I owned certain things I don't have.	0.66
I'd be happier if I could afford to buy more things.	0.72
λ	2.83
%variance	47.10

K-M-O = 0.750; Bartlett Chi = 563.277; df = 15; $p = 0.000$.

Unlike Richins' study (2004), where "materialistic orientation" was reported to be a three-dimensional concept (dimensions of "success", "-happiness", and "centrality" in acquiring and possessing goods) as well as Dević *et al.* (2015) and Müller *et al.*'s study (2013), where the scale was recognized as two-dimensional ("happiness" and "centrality"), our analysis indicated that it should be regarded as unidimensional. The scale is of acceptable reliability ($\alpha = 0.773$).

"Consumers' need for uniqueness" was assessed with items representing the dimensions of "Creative choice counter conformity" and "Unpopular choice counter conformity" (Ruvio *et al.*, 2008). The "Avoidance of similarity" dimension

Table 2.5. Two component solution for "Consumers' need for uniqueness" obtained using PCA.

Variables	Component 1	Component 2
I often combine possessions in such a way that I create a personal image that cannot be duplicated.	0.63	
I often try to find a more interesting version of run-of-the-mill products because I enjoy being original.	0.78	
I actively seek to develop my personal uniqueness by buying special products or brands.	0.83	
Having an eye for products that are interesting and unusual assists me in establishing a distinctive image.	0.84	
When it comes to the products I buy and the situations in which I use them, I have broken customs and rules.		0.65
I have often violated the understood rules of my social group regarding what to buy or own.		0.89
I have often gone against the understood rules of my social group regarding when and how certain products are properly used.		0.88
I enjoy challenging the prevailing taste of people I know by buying something they would not seem to accept.		0.65
λ	2.66	2.55
%variance	33.27	31.88

K-M-O = 0.833; Bartlett Chi = 1169.920; df = 28; $p = 0.000$.

was excluded because, unlike the counter conformity dimensions, it explores the achievement of uniqueness by mere avoiding to consume (over)conventional products and/or services i.e. it does not imply engagement in alternative consumption solutions.[6] In line with the Ruvio *et al.* (2008) approach, the analysis indicated that the concept should be regarded as two-dimensional (Varimax rotation converged in 3 iterations). Based on the extracted components (and following the same authors' labeling), the first component is labeled "Creative choices" and the second is labeled "Unpopular choices". Both scales are of acceptable reliability: $\alpha = 0.813$ and $\alpha = 0.816$, respectively.

6. (Ruvio *et al.*, 2008, p. 53) operationalized "Avoidance of similarity" through four items: (1) when a product I own becomes popular among the general population, I begin to use it less; (2) I often try to avoid products or brands that I know are bought by the general population; (3) as a rule, I dislike products or brands that are customarily bought by everyone; (4) the more commonplace a product or brand is among the general population, the less interested I am in buying it.

Table 2.6. A single component solution for "Environmental concern" obtained using PCA.

Variables	Component 1
I am concerned about the condition of the environment.	0.83
Humans are ruining the environment.	0.72
I would give up some economic goods for a cleaner environment.	0.71
The condition of the natural environment is getting worse every year.	0.67
I am concerned about natural resource shortage in the future.	0.74
We all need to change our behavior to protect the natural environment.	0.78
λ	3.33
%variance	55.48

K-M-O = 0.852; Bartlett Chi = 765.241; df = 15; $p = 0.000$.

In line with the Alzubaidi *et al.* (2021) approach, the analysis indicated that "environmental concern" should be regarded as a unidimensional concept. The scale is of acceptable reliability ($\alpha = 0.833$).

Based on the PCA, factor scores for each of the extracted components were calculated (Regression method) and these new variables were used in the regression analysis.

2.4 Results

The hierarchical regression analysis consisted of five blocks/models. The demographic characteristics of sex, age, area of residence, highest level of education, and employment status were used in the first block. Regardless of the fact the "Area of residence" variable (Urban/Semi-rural/Rural) was not commonly included in the previous topic-related studies (see e.g., Prieto *et al.*, 2017) and, consequently, was not represented in the hypotheses, it was added to the analysis to better depict the sample heterogeneity. The subsequent regression analysis blocks introduced socio-environmental determinants of "sociability", "materialistic orientation", "creative choices" and "unpopular choices" (representing the social determinants), and "environmental concern" (representing the environmental dimension).

All the assumptions of linear regression have been met: normality of the residuals, homoscedasticity, linearity (the normal P-P Plot of standardized residuals showed the points placed either on or close to the line, while the Scatterplot of standardized residuals showed a rectangular scatter of points densely populated in

Table 2.7. Hierarchical regression analysis of demographic characteristics, "Sociability", "Materialistic orientation", "Creative choices", "Unpopular choices", and "Environmental concern" on "Willingness to participate in the collaborative economy".

	Model				
Variables	**1**	**2**	**3**	**4**	**5**
Sex (Reference: Female)	−0.24**	−0.23**	−0.23**	−0.25**	**−0.22****
Age	−0.09	−0.09	−0.09	−0.09	−0.09
Area of residence	0.01	0.02	0.02	0.03	0.04
Level of education	0.04	0.03	0.03	0.02	0.01
Employment status	0.05	0.05	0.06	0.05	0.06
Sociability		0.07	0.08	0.08	0.04
Materialistic orientation			−0.06	−0.06	−0.06
Creative choices				−0.01	−0.01
Unpopular choices				0.15**	**0.15****
Environmental concern					**0.15****
R	0.25	0.26	0.26	0.30	0.33
R^2	0.06	0.07	0.07	0.09	0.11
ΔR^2	0.06**	0.01	0.00	0.02*	0.02**
R^2_{adj}	0.05	0.05	0.05	0.07	**0.09**

Note. $N = 363$. Model 1 – $F(5, 353) = 4.551$**; Model 2 – $F(6, 352) = 4.127$**; Model 3 – $F(7, 351) = 3.720$**; Model 4 – $F(9, 349) = 3.858$**; Model 5 – $F(10, 348) = 4.328$**.
*$p < 0.05$, **$p < 0.01$.

the central part), multicollinearity (Tolerance values = >0.8, VIF values = <1.3) and autocorrelation (Durbin-Watson value = 1.781).

In terms of interpreting Table 2.7, the respondents' demographic characteristics explained 5% of the variance in willingness to participate in the collaborative economy (R2 Adj. = 0.05). Sex was the only significant predictor in Model 1 ($\beta = -0.24$**) indicating that women expressed more willingness to engage in this type of economy (women were assigned the code 0). This finding is in line with Alonso-Almeida *et al.* (2020) conclusions and supports the gender/sex-related hypothesis (H1a).

In Model 2, "sociability" was introduced but, contrary to Tussyadiah (2015), it was not a significant contributor. The explained share of the phenomenon's variance remained the same which made the sociability hypothesis unsupported (H2).

In Model 3, in line with the Lindblom *et al.* (2018) findings and as it was the case with "sociability", introducing "materialistic orientation" did not increase the explained share of variance in the outcome. This finding means that our hypothesis relating to materialism orientation (H3) is rejected.

Model 4 included both "creative" and "unpopular choices" and this inclusion, controlling for the previously entered variables, accounted for an additional 2% ($\Delta R2 = 0.02^*$) in the outcome's variability. Unlike "creative choices", "unpopular choices" did significantly contribute to the phenomenon's explanation ($\beta = 0.15^{**}$), indicating that those who were more prone to making "unpopular choices" were more willing to engage in the CE. This finding is in line with Becker-Leifhold and Iran (2018) and Matthews *et al.* (2019) conclusions and has partially supported the consumers' need for uniqueness hypothesis (H4).

By introducing "environmental concern" (Model 5), controlling for the previously entered variables, an additional 2% ($\Delta R2 = 0.02^{**}$) was accounted for in the outcome's variability, supporting the environmental concern hypothesis (H5). In line with Hamari *et al.* (2016) and Tussyadiah (2015) study, this determinant ($\beta = 0.15^{**}$) indicated environmental concern being positively related to the willingness to take part in the CE. The entire model explained 9% of variance in the outcome variable (R2 Adj. = 0.09).

Regarding the final Model, the effect size can be labeled as small – Cohen's $f2 = 0.12$ – where 0.02 indicates small and 0.15 medium effect size (Cohen, 1988). While sex had the greatest contribution ($\beta = -0.22^{**}$), "environmental concern" and willingness to make "unpopular choices" made roughly the same contribution to the explanation of willingness to participate in the CE ($\beta = 0.15^{**}$).

2.5 Discussion and Conclusion

The factors that emerged as influencing willingness to participate in the collaborative economy are sex, the need to engage in unpopular choices, and environmental concerns. From these three factors, engaging in unpopular choices, as part of consumers' need for uniqueness, emerges as a novel driver of the CE. Previous literature has found that need for uniqueness is an important determinant of the CE, although it has been mainly associated with the dimensions of making creative consumer choices and avoiding similarity (Lang and Armstrong, 2018; Matthews *et al.*, 2019). However, our study showed that the dimension of creative choices might play a less significant role than previously thought, and instead *consumers proneness to make unpopular choices might be a greater factor* behind the willingness to participate. This can be explained due to unpopular choices becoming a manifestation of self-expression and identity (Sengupta and Sreejesh, 2017). From a managerial perspective, the determinant of "unpopular choices" suggests that CE organizations that are looking to increase participation from users might consider not only creating novel service encounters, but also delivering experiences that fall within the long tail of service encounters (Anderson, 2007) – as users might be looking for unpopular choices. Additionally, in the same context, adding social influence cues to the website to determine popularity, might not only be helpful to determine the most popular choices (Perez-Vega *et al.*, 2016), but also to identify those that are less seen/popular.

Furthermore, although the main aim of this study was not to explain the highest possible share of variability in willingness to participate in the CE, the final model, including the demographic characteristics and socio-environmental determinants, explained only 9% of the outcome's variability. Moreover, 5% (out of 9%) was explained by the demographic characteristics. The implications and key potential explanations of this finding are discussed below.

First, it might be that some other instruments for assessing the selected socio-environmental concepts would be more useful in depicting the reality – if it differs from the reality we have depicted. For instance, it might be that the materialistic orientation scale should be focused on the dimensions of possessiveness and non-generosity (Belk, 1982) instead of the happiness/centrality in acquiring and possessing goods. On the other hand, materialistic orientation might not even be an appropriate concept for elucidating the role of (attitudes towards) material possessions in the CE. In comparison with the materialism vs. anti-materialism, a more appropriate dichotomy might be favouring ownership vs. favouring access over ownership (Bardhi and Eckhardt, 2012). Namely, the CE relies on exchange of various material objects and on people who favour access over ownership, not on anti-materialists who voluntarily restrain themselves from these material objects. In any case, both the explored concepts and employed scales were selected based on previous research findings.

Second, the final result might be influenced by the research decision to employ the CE conceptualization more restrictively than it is usually the case – in line with, for example, Davidson *et al.* (2018) but not fully in line with, for example, Owyang *et al.* (2013) or Hamari *et al.* (2016).[7] As an illustration, sociability (assessed by the same scale we used) might turn out to be a significant contributor if the collaborative economy is vaguely conceptualized i.e. if blurry lines are set between it and its main surrounding concepts – for instance, on-demand (e.g., Uber), rental (e.g., Zipcar), and second-hand economy (e.g., Ebay) (see Frenken and Schor, 2017). In other words, in the context of sociability triggering CE participation, there might be a difference between, on one hand, renting cars from rental companies using online platforms and, on the other hand, sharing with strangers your own temporarily underused car, clothes, or house. Going one step further, based on this line

7. For instance, both studies include acquiring/providing permanent access to the shared resources i.e. change of ownership. The present study, as it was emphasized in the introduction, embraces solely the temporary access.

of reasoning, it also seems fair to question the "real" size of the CE – once CE actors get distinguished from actors of on-demand, rental, and second-hand economy.

Third, the outcome variable embraces willingness to both share your own and borrow/rent other peoples' resources and, based on the PCA results, it was explored as a unidimensional concept. Different findings might have been revealed if the willingness to produce and consume had been treated as separate outcome variables. For instance, in comparison with consumption, a higher score on the materialistic orientation scale might have a (significant) negative effect on willingness to produce i.e. share your own resources with strangers.

Finally, in spite of all the above-mentioned methodological considerations, the need to explore certain aspects of the topic more thoroughly (e.g., providers vs. receivers), and the fact the study was done using a non-representative sample, it seems safe to assume that the economic and/or technological determinants of willingness to participate in the CE most likely (heavily) outweigh the socio-environmental ones. In other words, our findings support Tussyadiah (2015), Binninger *et al.* (2015), and Hamari *et al.* (2016) findings who took into account both economic/monetary and socio-environmental determinants and reported the intentions and motives to participate in the CE to be related primarily to the economic i.e. non-socio-environmental reasons. From this point of view, the overall extensive discussion on a wide range of influential not-for-economic-profit motivations to participate – from Botsman and Rogers (2010) book onwards – seems to be nurtured primarily for marketing purposes i.e. showing that the collaborative economy is beyond mainstream solely monetary profit-driven digitally based economies.

References

Alonso-Almeida, M., Perramon, J., and Bagur-Femenías, L. (2020). Shedding light on sharing economy and new materialist consumption: An empirical approach. *Journal of Retailing and Consumer Services, 52*, 101900.

Alzubaidi, H., Slade, E.L., and Dwivedi, Y.K. (2021). Examining antecedents of consumers' pro-environmental behaviours: TPB extended with materialism and innovativeness. *Journal of Business Research, 122*, 685–699.

Anderson, C. (2007). *The Long Tail: How Endless Choice is Creating Unlimited Demand.* Random House.

Akbar, P., Mai, R., and Hoffmann, S. (2016). When do materialistic consumers join commercial sharing systems. *Journal of Business Research, 69*(10), 4215–4224.

Balderjahn, I., Buerke, A., Kirchgeorg, M., Peyer, M., Seegebarth, B., and Wiedmann, K.-P. (2013). Consciousness for sustainable consumption: scale development and new insights in the economic dimension of consumers' sustainability. *Academy of Marketing Science Review, 3*(4), 181–192.

Bardhi, F. and Eckhardt, G.M. (2012). Access-based consumption: the case of car sharing. *Journal of Consumer Research, 39*(4), 881–898.

Becker-Leifhold, C., and Iran, S. (2018). Collaborative fashion consumption-drivers, barriers and future pathways. *Journal of Fashion Marketing and Management: An International Journal, 22*(2), 189–208.

Benoit, S., Baker, T.L., Bolton, R.N., Gruber, T., and Kandampully, J. (2017). A triadic framework for collaborative consumption (CC): Motives, activities, and resources & capabilities of actors. *Journal of Business Research, 79*, 219–227.

Belk, R.W. (2014). You are what you can access: Sharing and collaborative consumption online. *Journal of Business Research, 67*(8), 1595–1600.

Belk, R.W. (1982). Acquiring, possessing, and collecting: fundamental processes in consumer behavior. In Bush, R.F. and Hunt, S.G. (eds.), *Marketing Theory: Philosophy of Science Perspectives* (pp. 185–190). American Marketing Association.

Binninger, A.S., Ourahmoune, N., and Robert, I. (2015). Collaborative consumption and sustainability: a discursive analysis of consumer representations and collaborative website narratives. *Journal of Applied Business Research (JABR), 31*(3), 969–986.

Boateng, H., Kosiba, J.P.B., and Okoe, A.F. (2019). Determinants of consumers' participation in the sharing economy. *International Journal of Contemporary Hospitality Management, 31*(2), 718–733.

Botsman, R. and Rogers, R. (2010). *What's Mine is Yours. The Rise of Collaborative Consumption.* HarperCollins Publishers.

Bucher, E., Fieseler, C., and Lutz, C. (2016). What's mine is yours (for a nominal fee) – Exploring the spectrum of utilitarian to altruistic motives for Internet-mediated sharing. *Computers in Human Behavior, 62*, 316–326.

Davidson, A., Habibi, M.R., and Laroche, M. (2018). Materialism and the sharing economy: A cross-cultural study of American and Indian consumers. *Journal of Business Research, 82*, 364–372.

Dević, I, Majetić, F., and Krnić, R. (2015). Vrijednosne preferencije hrvatskih građana kao odrednice materijalizma. *Druştvena Istraživanja, 24*(4), 555–576.

Dredge, D. and Gyimóthy, S. (2015). The collaborative economy and tourism: critical perspectives, questionable claims and silenced voices. *Tourism Recreation Research, 40*(3), 286–302.

Ertz, M., Durif, F., Lecompte, A., and Boivin, C. (2018). Does "sharing" mean "socially responsible consuming"? Exploration of the relationship between collaborative consumption and socially responsible consumption. *Journal of Consumer Marketing, 35*(4), 392–402.

European Commission (2018). *Study to Monitor the Economic Development of the Collaborative Economy in the EU – Final Report.* Publications Office of the European Union.

Eurostat (2019). *Are You Part of the Collaborative Economy?* Retrieved from https://ec.europa.eu/eurostat/web/products-eurostat-news/-/DDN-20190416-1

Fox, S. (1984). The sociability aspect of extraversion as a situation-specific dimension. *Social Behavior and Personality: An International Journal, 12*(1), 7–10.

Frenken, K. and Schor, J. (2017). Putting the sharing economy into perspective. *Environmental Innovation and Societal Transitions, 23*, 3–10.

Gheitasy, A., Abdelnour-Nocera, J., Nardi, B., and Rigas, D. (2014, June). Designing for online collaborative consumption: a study of sociotechnical gaps and social capital. In *International Conference on Human-Computer Interaction* (pp. 683–692). Springer.

Goldberg, L.R., Johnson, J.A., Eber, H.W., Hogan, R., Ashton, M.C., Cloninger, C.R., and Gough, H.C. (2006). The international personality item pool and the future of public-domain personality measures. *Journal of Research in Personality, 40*, 84–96.

Guillen-Royo, M. (2019). Sustainable consumption and wellbeing: does on-line shopping matter? *Journal of Cleaner Production, 229*, 1112–1124.

Guo, Y., Xin, F., Barnes, S.J., and Li, X. (2018). Opportunities or threats: the rise of online collaborative consumption (OCC) and its impact on new car sales. *Electronic Commerce Research and Applications, 29*, 133–141.

Gyimóthy, S. (2017). Business models of the collaborative economy. In *Collaborative Economy and Tourism. Tourism on the Verge* (pp. 31–39). Springer.

Hair, J.F., Jr., Black, W.C., Babin, B.J., and Anderson, R.E. (2009). *Multivariate Data Analysis (7th ed.).* Pearson Prentice Hall.

Hamari, J., Sjöklint, M., and Ukkonen, A. (2016). The sharing economy: Why people participate in collaborative consumption. *Journal of the Association for Information Science and Technology, 67*(9), 2047–2059.

Hawlitschek, F., Teubner, T., and Gimpel, H. (2016). Understanding the sharing economy – Drivers and impediments for participation in peer-to-peer rental. *49th Hawaii International Conference on System Sciences*, (pp. 4782–4791). IEEE.

Jiang, B. and Tian, L. (2018). Collaborative consumption: Strategic and economic implications of product sharing. *Management Science, 64*(3), 1171–1188.

Kathan, W., Matzler, K., and Veider, V. (2016). The sharing economy: Your business model's friend or foe? *Business Horizons, 59*, 663–672.

Kenney, M. and Zysman, J. (2016). The rise of the platform economy. *Issues in Science and Technology, 32*(3). Available at: https://issues.org/the-rise-of-the-platform-economy/ (Accessed 17 September 2020).

Kline, R.B. (2011). Principles and practice of structural equation modeling. Guilford.

Le Vine, S., Zolfaghari, A., and Polak, J. (2014). *Carsharing: Evolution, Challenges and Opportunities.* 22th ACEA Scientific Advisory Group Report. European Automobile Manufacturers Association.

Lang, C. and Armstrong, C.M.J. (2018). Collaborative consumption: The influence of fashion leadership, need for uniqueness, and materialism on female consumers' adoption of clothing renting and swapping. *Sustainable Production and Consumption, 13*, 37–47.

Latter, C., Phau, I., and Marchegiani, C. (2010). The roles of consumers need for uniqueness and status consumption in haute couture luxury brands. *Journal of Global Fashion Marketing, (1)*4, 206–214.

Leoni, G. and Parker, L.D. (2019). Governance and control of sharing economy platforms: Hosting on Airbnb. *The British Accounting Review, 51*(6), 100814.

Leys, C., Klein, O., Dominicy, Y., and Ley, C. (2018). Detecting multivariate outliers: Use a robust variant of the Mahalanobis distance. *Journal of Experimental Social Psychology, 74*, 150–156.

Lindblom, A., Lindblom, T., and Wechtler, H. (2018). Collaborative consumption as C2C trading: analyzing the effects of materialism and price consciousness. *Journal of Retailing and Consumer Services, 44*, 244–252.

Lynn, M. and Harris, J. (1997). The desire for unique consumer products: A new individual differences scale. *Psychology & Marketing, 14*(6), 601–616.

Matthews, D., Rothenberg, L., and Gopalakrishnan, S. (2019). The impact of mass customization on fashion-innovative students: an assessment of need for uniqueness, self-identity, and perceived performance risk. *International Journal of Fashion Design, Technology and Education, 12*(3), 293–300.

Morency, C., Habib, K.M.N., Grasset, V., and Islam, M.T. (2012). Understanding members' carsharing (activity) persistency by using econometric model. *Journal of Advanced Transportation, 46*(1), 26–38.

Murillo, D., Buckland, H., and Val, E. (2017). When the sharing economy becomes neoliberalism on steroids: unravelling the controversies. *Technological Forecasting & Social Change, 125*, 66–76.

Müller, A., Smits, D.M., Claes, L., Gefeller, O., Hinz, A., and Zwaan, M. (2013). The German version of the Material Values Scale. *GMS Psycho-Social-Medicine, 10*, 1–9.

Owyang, J., Tran, C., and Silva, C. (2013). *The Collaborative Economy: Products, Services, and Market Relationships Have Changed as Sharing Startups Impact Business Models. To Avoid Disruption, Companies Must Adopt the Collaborative Economy Value Chain.* Altimeter.

Owyang, J., Samuel, A., and Grenville, A. (2014). Sharing is the new buying [Website]. Available at: www.web-strategist.com (Accessed 17 September 2020).

Palan, S. and Schitter, C. (2018). Prolific.ac—A subject pool for online experiments. *Journal of Behavioral and Experimental Finance, 17*, 22–27.

Perez-Vega, R., Waite, K., and O'Gorman, K. (2016). Social impact theory: An examination of how immediacy operates as an influence upon social media interaction in Facebook fan pages. *The Marketing Review, 16*(3), 299–321.

Pesonen, J. and Tussyadiah, I. (2017). Peer-to-peer accommodation: drivers and user profiles. In: Dredge, D. and Gyimóthy, S. (eds.) *Collaborative Economy and Tourism. Tourism on the Verge*. Springer.

Pilch, I. and Górnik-Durose, M.E. (2016). Do we need "dark" traits to explain materialism? The incremental validity of the Dark Triad over the HEXACO domains in predicting materialistic orientation. *Personality and Individual Differences, 102*, 102–106.

Piscicelli, L., Cooper, T., and Fisher, T. (2015). The role of values in collaborative consumption: insights from a product-service system for lending and borrowing in the UK. *Journal of Cleaner Production, 97*, 21–29.

Prieto, M., Baltas, G., and Stan, V. (2017). Car sharing adoption intention in urban areas: what are the key sociodemographic drivers? *Transportation Research Part A, 101*, 218–227.

Putnam, R.D. (2000). *Bowling Alone: The Collapse and Revival of American Community*. Simon and Schuster.

Richins, M.L. (2004). The material values scale: measurement properties and development of a short form. *Journal of Consumer Research, 31*(1), 209–219.

Richins, M.L. and Dawson, S. (1992). A consumer values orientation for materialism and its measurement: scale development and validation. *Journal of Consumer Research, 19*(3), 303.

Rowe, P.C. (2017). Beyond Uber and Airbnb: the social economy of collaborative consumption. *Social Media+ Society, 3*(2), 2056305117706784.

Ruvio, A.A., Shoham, A., and Makovec Brenèić, M. (2008). Consumers' need for uniqueness: short-form scale development and cross-cultural validation. *International Marketing Review, 25*(1), 33–53.

Schor, J. (2014). *Debating the Sharing Economy*. Available at: www.greattransition.org (Accessed 17 September 2020).

Sengupta, A.S. and Sreejesh, S. (2017). Impact of other customers in high and low involvement services: moderating role of customer's need for uniqueness. *Journal of Indian Business Research, 9*(1), 41–58.

Stokes, K., Clarence, E, Anderson, L., and Rinne, A. (2014). *Making Sense of the UK Collaborative Economy*. Nesta Collaborative Lab.

Taber, K.S. (2018). The use of cronbach's alpha when developing and reporting research instruments in science education. *Research in Science Education, 48*, 1273–1296.

Tian, K.T., Bearden, W.O., and Hunter, G.L. (2001). Consumers' need for uniqueness: scale development and validation. *Journal of Consumer Research, 28*(1), 50–66.

Tussyadiah, I.P. (2015). An exploratory study on drivers and deterrents of collaborative consumption in travel. In *Information and Communication Technologies in Tourism 2015* (pp. 817–830). Springer.

Wang, Y., Benner, A.D., and Kim, S.Y. (2015). The cultural socialization scale: assessing family and peer socialization toward heritage and mainstream cultures. *Psychological Assessment, 27*, 1452–1462.

Zervas, G., Proserpio, D., and Byers, J.W. (2017). The rise of the sharing economy: estimating the impact of airbnb on the hotel industry. *Journal of Marketing Research, 54*(5), 687–705.

Zhu, J.L., Cui, C.C., Fordham, K. (2015). Consumers' need for uniqueness and customization behavior among chinese -consumers. In: Spotts, H. (eds.) *Revolution in Marketing: Market Driving Changes.* Developments in Marketing Science: Proceedings of the Academy of Marketing Science. Springer.

DOI: 10.1561/9781680838411.ch3

Chapter 3

The Influence of Demographics, Attitudinal and Behavioural Characteristics on Motives to Participate in the Sharing Economy and Expected Benefits of Participation

By Julijana Angelovska, Anita Čeh Časni and Christoph Lutz

The sharing economy is a relevant economic phenomenon of recent times and important for sustainable economic growth. This chapter considers the motivational factors that drive and hinder participation in the sharing economy. It investigates the impact of both economic or non-economic drivers and what role demographics, attitudinal and behavioural characteristics play as antecedents of those drivers. We rely on rich data from a 12-country survey to conduct our analysis, and we distinguish between three categories of respondents: providers, consumers and aware non-users. Trust, innovativeness and materialism are considered as important attitudinal antecedents, while volunteering is used as the key behavioural antecedent. We find that economic motives outperform non-economic motives overall. However, compared with providers and aware non-users, consumers are more strongly driven by economic motives, especially those who are more educated and trusting. Additionally, younger, more educated, more innovative, materialistic and volunteering respondents are driven more than others by non-economic motives. Finally, providers with lower household income, who are more educated and innovative are more likely to be driven by economic motives, while providers that have more trust in people and volunteer more frequently are more likely to be driven by non-economic motives. Overall, the chapter contributes to a more differentiated understanding of participation in the sharing economy in terms of motives and their antecedents. We discuss theoretical and practical implications of the findings.

3.1 Introduction

As discussed in the introduction to this volume (Introduction, this volume), the sharing economy is a broad concept that lacks a commonly accepted definition. It is sometimes referred to as collaborative consumption (Botsman and Rogers, 2011), access-based consumption (Bardhi and Eckhardt, 2012), or commercial sharing systems (Lamberton and Rose, 2012). The sharing economy has the potential to create substantial value, by promoting economic growth, technological innovation, environmental sustainability, and social inclusion; factors central to the United Nations Sustainable Development Goals (SDGs; Boar *et al.*, 2020). In this context, the sharing economy is of particular interest, because, in contrast to many other sustainable innovations, certain sharing economy sectors are scaling up rapidly.

This study aims to provide a comprehensive understanding of the motives for participation in the sharing economy. Synthesising previous studies, and in line with a holistic approach to the topic, both economic and non-economic motives are considered. Particularly, we understand non-economic motives broadly to include hedonic (fun), social (social interaction/meeting people) and social responsibility aspects. The chapter does not only investigate the key motivational factors for sharing economy participation in Europe but also the relative importance of demographics and selected attitudinal and behavioural characteristics in shaping motives. The analysis draws on data from a large survey conducted in 12 European countries on the state of the sharing economy (Andreotti *et al.*, 2017). Using univariate and multivariate statistical methods, we investigate the role of demographics, three relevant attitudinal constructs (trust, innovativeness, materialism) and one important behavioural correlate (volunteering). We study their influence on both economic and non-economic motivational factors among providers, consumers and aware non-users. The analysis reveals distinct differences between these three groups. Consumers tend to be driven mostly by economic motives, and this is particularly the case for consumers with high levels of trust and innovativeness. Providers, by contrast, are also motivated by non-economic factors. Trust and volunteering are identified as antecedents of non-economic factors. The findings allow for a holistic understanding of how social characteristics shape motives for participation in the sharing economy.

Studying antecedents of motives is important because it deepens our understanding of the dynamics of participation and how motives might themselves be socially differentiated based on power relations (Eichhorn *et al.*, 2020). Thus, our study contributes to sociological and psychological literature on the sharing economy.

The rest of the chapter is structured as follows. Section 2 reviews the literature on sharing economy motives and develops hypotheses about the relative importance of these motives under various circumstances. Section 3 discusses the data collection and analytical strategy. Section 4 presents the results. Section 5 concludes by discussing limitations of the study as well as implications for research on the sharing economy.

3.2 Literature Review

3.2.1 The Sharing Economy in Context

Regardless of the term used, the mutual focus when it comes to the sharing economy is on collaborative use of slack and poorly utilized assets and services, and how they can be used more efficiently (Stephany, 2015). In the sharing economy, ordinary people act as providers and offer services to consumers that used to be offered only by professional sellers (Narasimhan *et al.*, 2018; Sundararajan, 2016). Thus, the sharing economy is an economic system with emphasis on peer-to-peer exchange and sharing of slack and unutilized assets or services for free or for a fee. In this contribution, we follow Gerwe and Silva's (2020) definition of the sharing economy as *"a socioeconomic system that allows peers to grant temporary access to their underutilized physical and human assets through online platforms" (p. 71).*

(Belk, 2007, p. 126), in a frequently recalled definition, describes sharing as the "act and process of distributing what is ours to others for their use and/or the act or process of receiving or taking something from others for our use." Subsequent literature has differentiated the sharing of tangible or physical goods, such as cars, bicycles and apartments, and intangible goods, such as knowledge, emotions and ideas (Belk, 2010; Bucher *et al.*, 2016; Botsman and Rogers, 2010; Gansky, 2010; John, 2013). Sharing resources, whether they are tangible or intangible, is not a new phenomenon (Kemp and Olson, 2015), but rather something humankind has always been doing. The sharing economy in its present form is thus a technological transformation of an old phenomenon. More specifically, it is the result of a transformation of long existing concepts, such as flea markets, ride-sharing agencies, and neighbourly help, by information and communication technologies. ICT-enabled sharing allows strangers to share cars, homes, food, and tools with unknown individuals through online platforms, while previously sharing was mostly happening between known people. In this context, Belk (2014a) distinguishes 'sharing-in' and 'sharing-out'. Sharing within the family or between friends can be defined as 'sharing-in'. By contrast, when sharing involves strangers, it can be described as 'sharing-out'. The two types differ substantially in the degree of intimacy in the sharing process (Narasimhan *et al.*, 2018). Furthermore, ICT-enabled sharing economy is characterized by online platforms, hence two-way transactions turn into three-way transactions, where the platform acts as an intermediary between providers and consumers. Despite many benefits, which will be discussed in more depth below, sharing is tied to material and personal risks as it exposes one's possessions to the hazards of loss, damage and decreased utility (Bucher *et al.*, 2018; Lutz *et al.*, 2018). Sharing economy platforms attempt to address these risks leveraging ratings and reputation mechanisms (Frenken and Schor, 2017; Newlands *et al.*, 2019).

3.2.2 Motives for Sharing Economy Participation: Economic vs. Non-Economic

Considering its scale and growth, it is important to study the motives of participation in the sharing economy. The literature differentiates a plurality of motives, which depend on the kind of platform used for the exchange and on whether the exchange involves monetary compensation or not (Edbring *et al.*, 2016). Therefore, both non-economic and economic motives have been identified. Cost-savings and convenience (i.e., efficient access to goods and services) are classified as economic (Heo, 2016; Tussyadiah and Pesonen, 2016). The need for social interaction, the intrinsic and hedonic enjoyment of sharing, and intentions to help others and/or protect the environment are classified as non-economic. We will discuss economic and non-economic motives in turn.

Regarding economic motives, a major benefit for consumers in the sharing economy is the access to broader options and lower prices (Sundararajan, 2016). This is corroborated by substantive empirical research. A Eurobarometer study (2016) found that the benefits of sharing are largely monetary or related to convenience, and a Deloitte study (2015) on the sharing economy in Switzerland found that 65% of respondents considered lower costs as a key benefit of the sharing economy. Böcker and Meelen (2017) found that economic motives were particularly important for low-income users. Bardhi and Eckhardt (2012) showed how self-interest and utilitarianism (i.e., reducing expenses and increasing convenience) are frequent motives for access-based car sharing and that these motives were found to be more important than considerations about collective utility. Lamberton and Rose (2012) identified cost and utility factors, the perceived risk of product scarcity, and familiarity with sharing as key drivers. The studies by Bellotti *et al.* (2015), Möhlmann (2015), and Hawlitschek *et al.* (2018) also identify economic motives as the key drivers of sharing economy participation.

However, Botsman and Rogers (2011) argue that collaborative consumption is driven by motives that extend beyond economic considerations. Gansky (2010) suggests changing consumer attitudes towards consumption as a motivational factor that drives the sharing economy, as consumers are willing to try out new brands (Gansky, 2010) and are more open to new ways of accessing what they need (Botsman and Rogers, 2011; Bardhi and Eckhardt, 2012). Additionally, consumers are increasingly aware of the pressure that (over)consumption can pose to the environment. The idea of sharing excess capacity to reduce environmental concerns, the renewed belief in the importance of community, and cost-consciousness move consumers towards the practice of sharing, openness and collaboration (Gansky, 2010; Walsh, 2011). Botsman and Rogers (2011) suggest that social motives impact sharing economy participation as well. Sharing one's possessions with others is generally considered an inherently pro-social or even non-economic act, marked by feelings of solidarity and bonding (Belk, 2010; Benkler, 2004). Numerous studies refer in some ways to an alleged underlying anthropological or neuroscientific tendency for sharing (e.g., Schmidt and Sommerville, 2011; Tomasello and Warneken, 2008), showing the sharing economy's benefits for community building, social participation, and the creation of social capital (Belk, 2007, 2010; Botsman and Rogers, 2010; Hamari *et al.*, 2016). A study by Möhlmann (2015), for instance, on German users of Airbnb and the business-to-consumer service Car2Go, found that community belonging was a key driver for repeated use. In the context of accommodation sharing, Tussyadiah (2015) suggests that people engage in these activities because they want to interact with their local hosts. Benkler (2004) also stressed the importance of non-monetary factors such as social reputation, cooperation, and satisfaction. Applying qualitative research methods, Albinsson and Perera (2012) investigated drivers for participation in the sharing economy and identified a sense of community as both a driver and an outcome of participation. Furthermore, a variety of ideological and practical reasons was identified.

Previous research has also shown that motives to participate in the sharing economy can depend on the type of platform used and whether the exchange is commercial or non-commerical (Bucher *et al.*, 2016; Edbring *et al.*, 2016; Hawlitschek *et al.*, 2018). According to Edbring *et al.* (2016), non-profit platforms participants are driven by factors such as the desire to belong to a community, the need for reciprocity, and political and environmental ideals. Instead, in for-profit platforms, economic and convenience-related reasons together with the search for novelty and the desire for variation prevail over motives related to reciprocity and sustainability.

Taken together, the findings suggest the co-presence of economic and non-economic motives as drivers of participation in the sharing economy (Bellotti *et al.*, 2015; Shih *et al.*, 2015). The importance of each depends on the context (e.g., type of platform) and the characteristics of the participants (Davidson *et al.*, 2018). However, most previous research focuses on either consumers or providers but does not systematically contrast these groups. Moreover, aware non-users and their expected benefits are neglected in previous research. In the next sub-section, we will thus make the case that motives of consumers, providers and expected benefits of non-users should be differentiated. We will also introduce a rationale for studying the antecedents of motives.

3.2.3 Differentiating Providers, Consumers and Non-Users

Little research has differentiated user roles and compared providers and consumers as distinct groups. As an exception, Bellotti *et al.* (2015), through interviewing both users/consumers and providers of 46 different sharing economy systems, identified eight distinct motives for the use of sharing economy services: value/morality, social influence, status/power, empathy/altruism, social connection, intrinsic/autotelic reasons, safety, and instrumental motives. In their interviews of both consumers and providers, they found that while providers tend to stress idealistic motives, consumers are strongly driven by value and instrumental motives.

On the provider side, a frequently heard argument by sharing economy advocates is its expansion of micro-entrepreneurship opportunities. Sharing platforms can create new sources of employment and enable previously un-tapped sources of income (Ikkala and Lampinen, 2015; Lampinen and Cheshire, 2016). The relatively low entry-barrier is particularly beneficial for marginalized populations who may be traditionally excluded, such as those with criminal records or low education. Smith (2016), based on a representative survey in the United States, found that 80% of respondents identified job opportunities as a major benefit of ride-hailing services, whereas 85% of respondents considered a major benefit of home-sharing services to be a convenient source of income.

However, the public debate has been increasingly critical towards the greater proliferation of sharing platforms, with their legitimacy and practices frequently called into question (Newlands and Lutz, 2020). While the sharing economy has shown to open up new opportunities to make money, earnings on platforms are subject to significant diversity. In smaller scale initiatives, for instance, Fuster Morell *et al.* (2016) report that earnings are low and, in some cases, not even enough to cover basic needs. Critics have also argued that sharing services will undermine traditional employment relationships, leading to greater income inequality, poorer working conditions, labour uncertainty, and a tilt of power in favour of platforms in the creation of a 'new precariat' (Murillo *et al.*, 2017; Slee, 2013).

Economic motivation can be seen in people with lower involvement and commitment tied to their participation. This argument is supported by Shih *et al.* (2015) in the context of the less commercially-oriented sharing economy area of time-banking. The authors found that highly active time-bank users were more idealistic and participated because they believed in "equal time, equal value", whereas less active time bank users, who were mostly regular members, more frequently utilized time-banking in order to fulfil instrumental needs. Even in more commercially-oriented areas, such as peer-to-peer accommodation, the same pattern might hold. Dann *et al.* (2019), in a systematic overview of research on Airbnb, identified motives as a key theme. Out of 118 articles analysed in total (including topics other than motives), 31 look at motives from the guest (consumer) perspective and 16 from the host (provider) perspective. Among guests, "cost savings still remain the dominant motive" (p. 450) but for hosts, the motives seem to be somewhat more diverse, even though financial benefits play a key role. Extrapolating from these last elements, we question if consumers are exhibiting higher levels of economic motives compared to providers and providers to have higher levels of non-economic motives.

Beyond users, in the form of providers and consumers, non-users are also considered in studies on the sharing economy, even though rarely. However, an identification of their expected benefits should complement the analysis. Non-users constitute the largest group, as only a minority of the population uses sharing economy services. While the sharing economy has seen widespread growth and spans all socio-demographic categories in the European context, only 17% have used such services at least once (Eurobarometer, 2016). Thus, more than 80% are non-users. However, the majority (52% of the total population) of all EU citizens were aware of the services offered by the sharing economy, thus making aware non-users a key category. In our data, aware non-users and non-aware non-users are differentiated but we only include aware non-users in the analysis. Importantly, the term "motives" might not be appropriate for aware non-users since they have not experienced participation first-hand and could thus not give a substantiated account of motive-related questions. Therefore, we use the term expected benefits, rather than motives, when talking about aware non-users in the following.

While substantial research has looked into the question how motives affect sharing economy participation, less is known about the factors that affect participation motives themselves. In the next section, we present the research design and discuss our rationale for including demographics, three attitudinal antecedents – namely, trust, technological innovativeness, materialism, and a behavioural correlate, i.e. volunteering.

3.3 Methods: Data, Measures and Research Approach

3.3.1 Data

The analysis draws on a large survey conducted in 12 European countries on the state of the sharing economy (Andreotti *et al.*, 2017; Newlands *et al.*, 2018). A consortium of international researchers based in Norway, Germany, The Netherlands, Italy, Denmark, and Switzerland conducted the survey in summer 2017. The cross-national questionnaire was constructed to explore the prevalence, antecedents, and outcomes of participation, privacy, and power in the European sharing economy, and involved 6111 individuals across 12 countries (Denmark, France, Germany, Ireland, Italy, the Netherlands, Norway, Poland, Portugal, Spain, Switzerland, and the United Kingdom). This selection includes countries with both a higher and lower average income, as well as countries with a varied uptake of sharing economy services. The respondents were divided into users, who were further categorized into providers and consumers, and non-users, who were further categorized into aware and non-aware non-users. The research in this chapter is focused on the respondents who are either users (n = 1699) or aware non-users (n = 3983). Among the users, there were 1143 consumers and 556 providers.

3.3.2 Measures

The analysis considers demographics, three relevant attitudinal antecedents and one behavioural correlate as predictors of participation in the sharing economy.

We used the following demographics as independent control variables: age in years, grouped in five categories, gender, education based on the ISCED categories, and yearly gross household income in four categories (quartiles). These variables were selected because they represent the most common demographic indicators used in survey research on the sharing economy. For household income, originally between 13 and 17 relatively narrow categories in the respective local currencies in the survey were used, subsequently grouping the respondents based on the distribution and their distance from the mean in standard deviations for each country.

Trust, innovativeness and materialism were included as relevant attitudinal antecedents and volunteering as a behavioural one. Trust has been shown to be a key construct in the sharing economy (Ter Huurne *et al.*, 2017) and was measured based on the general disposition to trust, using the scale of McKnight *et al.* (2002). We expect trust to have a positive effect on motives or expected benefits, as it serves as a pre-condition for even being willing to participate in the sharing economy and develop motives. For technology innovativeness, which could indicate a higher propensity to try out sharing services, the scale by Agarwal and Prasad (1998) was adopted. Technology innovativeness should equally increase motives or expected benefits. As a key aspect of the sharing economy is platform mediation, those who exhibit higher levels of technological innovativeness should show stronger motives or expected benefits from participation. To measure materialism and volunteering, both attributes shown as important in the context of the sharing economy (Akbar *et al.*, 2016; Davidson *et al.*, 2018; Kornberger *et al.*, 2018; Lutz *et al.*, 2018), the scales from Bucher *et al.* (2016) were used. Materialism is particularly important for commercial sharing services and economic motives/expected benefits (Davidson *et al.*, 2018), while volunteering should play a key role for non-commercial sharing services and non-economic motives/benefits (Bucher *et al.*, 2016). Table 3.1 displays the individual items and measurement. All scales showed high loadings and good measurement properties (Cronbach's α between 0.74 and 0.90). The descriptive statistics (means, standard deviations) of the items are presented in Table 3.1.

The questionnaire used four items to assess motives or expected benefits of sharing economy participation: financial, meeting people/social interaction, fun, and social responsibility (Bellotti *et al.*, 2015; Bucher *et al.*, 2016; Möhlmann, 2015). Provider and consumers were asked about their motives for participation and non-participants about which benefits they would expect from using sharing services. The question prompt for providers and consumers was: *"How much did the following considerations affect your decision to use the sharing platform?"* The question prompt for aware non-users was: *"If you decided to use an online sharing platform, to what extent would you expect the following benefits?"* Respondents answered for each of the four items on a scale from 1 to 5 with 1-not at all, 2-to a small extent, 3-to a moderate extent, 4-to a large extent, 5-very much. For each item, some additional explanation was available in brackets: "Financial benefit (e.g., for additional income)", "Meeting people (e.g., to find company, to feel part of a community)", "Fun (e.g., adventure, distraction, entertainment)", "Social responsibility (e.g., contribution to healthy environment, helping others)". While the literature has stressed environmental aspects of sharing economy participation, the questionnaire unfortunately did not include a dedicated and separate item on environmental motives or expected benefits. Social responsibility carries a moral dimension and environmental considerations are mentioned in brackets for this item but overall, this item is

Table 3.1. Descriptive statistics of constructs and items.

Factor/Cronbach's α	Variables	Mean	St. Dev.	Factor Loadings
1. Innovativeness/0.90	Look for ways to experiment	3.24	1.14	0.908
	The first to try out	2.90	1.21	0.866
	Like to experiment	3.47	1.13	0.909
2. Trust/0.88	General trust in people	3.35	1.02	0.899
	General faith in humanity	3.34	1.01	0.875
	General reliability of people	3.30	1.00	0.907
3. Volunteering/0.83	Volunteering to help	2.59	1.32	0.830
	Getting involved in issues	2.70	1.22	0.845
	Working with a group to solve a problem	2.37	1.20	0.875
4. Materialism/0.74	Happier if I could afford more	3.33	1.19	0.747
	Like a lot of luxury	2.70	1.17	0.795
	Admire people with expensive things	2.48	1.19	0.831

Note: $N = 5682$.

more about the societal aspects rather than environmental ones. This is a limitation of the study.

3.3.3 Research Approach

We used descriptive analysis, one-way ANOVA and binary logistic regression to analyse the data. First, descriptive statistical analysis (mean and standard deviations for providers and consumers) was conducted. Then, the data was analysed to find whether there were statistically significant differences between providers and consumers (one-way ANOVA). Finally, two multivariate methods were used. Factor analysis was employed to reduce the number of variables and to determine the underlying structure of relevant self-reported attitudinal (trust, innovativeness, materialism) and behavioural (volunteering) constructs. This helped to find out whether the factors correspond to the pre-determined suggested structures. To test convergent and discriminant validity of the scales used to measure the independent variables, a principal component analysis with Varimax rotation was employed. The second multivariate method used was binary logistic regression. The binary logistic regression generates predicted probabilities of a case being in the category labelled (1) and is predicting the logit, that is, the natural log of the odds of having used sharing economy services.

3.4 Empirical Analysis and Results

The descriptive statistics for age, gender, household income and education by each category (providers and consumers, aware non-users) are shown in Table 3.2.

One-way ANOVA was used to detect if there are statistically significant differences between the groups in terms of demographics. Providers are younger and more likely to be male. Consumers have the highest level of household income and education. By contrast, aware non-users have the lowest level of household income and are less educated than providers and consumers.

The descriptive statistics of the four components by each group revealed statistically significant differences. Consumers are most innovative, they showed the most general trust in people, they are most materialistic, but they volunteer less frequently than providers (Table 3.3). By contrast, aware non-users are least trustful, innovative, materialistic and volunteer least frequently of all three groups.

Table 3.2. Descriptive statistics (demographics).

Provider, Consumer, Aware Non-user		**Age Band**	**Gender**	**Household Income**	**Education**
Provider*	Mean	2.54	1.59	2.33	4.73
	N	556	556	556	556
	Std. Deviation	1.194	0.492	0.983	1.135
Consumer*	Mean	2.76	1.48	2.41	4.78
	N	1143	1143	1143	1143
	Std. Deviation	1.297	0.500	1.013	1.026
Aware non-user*	Mean	3.36	1.50	2.24	4.31
	N	3983	3983	3983	3983
	Std. Deviation	1.295	0.500	1.020	1.062
Total	Mean	3.16	1.50	2.28	4.44
	N	5682	5682	5682	5682
	Std. Deviation	1.323	0.500	1.017	1.083

Note: * statistically significant at $p < 0.01$.

Table 3.3. Descriptive statistics of factors.

Provider, Consumer, Aware Non-user		Trust	Innovativeness	Materialism	Volunteering
Provider*	Mean	3.32	3.42	2.98	2.93
(n = 556)	Std. Deviation	0.985	1.023	0.953	1.045
Consumer*	Mean	3.43	3.46	2.99	2.78
(n = 1143)	Std. Deviation	0.913	1.004	0.955	1.022
Aware non-user*	Mean	3.31	3.11	2.78	2.43
(n = 3983)	Std. Deviation	0.932	1.065	0.956	1.070
Total	Mean	3.34	3.21	2.84	2.55
(N = 5682)	Std. Deviation	0.935	1.060	0.960	1.074

Note: * statistically significant at $p < 0.01$.

Descriptive statistics of the motives/expected benefits for each group are shown in Table 3.4. Despite the differences in how the questionnaire assessed motives among users (providers and consumers) and expected benefits among non-users (see 3.2), we think that the values are somewhat comparable, although we have to stress that motives were assessed in a past-directed way while expected benefits are future-directed. The ANOVA revealed statistically significant differences, with both providers and consumers mostly motivated by financial benefits. This could be caused by the pre-dominant platforms used. Most of the users (73%) declared that their most frequently used platform was Airbnb, Uber or BlaBlaCar, all of which are profit-oriented platforms.

Financial motives or expected benefits are apparent in all three groups. Even though (expected) financial benefits dominate in all groups, consumers showed more financial benefits as motives than providers. It seems that consumers are dominantly motivated by economic reasons and they declared more use of Airbnb and Uber. By contrast and in comparison to consumers, providers are more motivated by meeting people, fun and social responsibility. Interestingly, aware non-users expect more benefits from social responsibility and social interaction than consumers are motivated by these factors.

Table 3.4. Descriptive statistics on motives of users (providers and consumers) and expected benefits of aware non-users.

Provider, Consumer, Aware Non-user		Financial Benefit	Fun	Meeting People	Social Responsibility
Provider* (n = 556)	Mean	3.29	2.87	2.90	3.04
	Std. Deviation	1.19	1.12	1.16	1.18
Consumer* (n = 1143)	Mean	3.68	2.70	2.38	2.49
	Std. Deviation	1.07	1.12	1.13	1.11
Aware non-user* (n = 3983)	Mean	2.92	2.61	2.68	2.87
	Std. Deviation	1.12	1.07	1.08	1.09
Total (N = 5682)	Mean	3.11	2.65	2.64	2.81
	Std. Deviation	1.12	1.09	1.11	1.11

Note: * statistically significant at $p < 0.01$.

To further analyse the influence of demographics, the three attitudinal constructs and volunteering on motives/expected benefits, we used factor analysis to explore whether the motives can be reduced. The Kaiser-Meyer-Olkin (KMO) criterion and Bartlett's test were used to assess the goodness-of-fit of the solution. In our sample the KMO value was 0.770 and Bartlett's test was significant (p = 0.000), showing that the principal component analysis was appropriate. This analysis resulted in two components: the first one described economic aspects and consisted of financial motives with a loading of 0.971. The second component was named non-economic and included meeting people, fun and social responsibility. Convergent validity of the scales is supported by a Cronbach's α of 0.83 for the non-economic motives component. The factors are turned into binary variables by assigning a value of 1 if answers have a value of 3 or higher, and 0 for values below 3. Thus, the scale mid-point served as the split-point.

Logistic regression was then performed to test the predictive power of the demographic characteristics (gender, age, household income, education) as well as the three attitudinal constructs (trust, innovativeness, materialism) and volunteering as a behavioural correlate on economic and non-economic motives/expected benefits. This analysis was conducted separately for providers, consumers and aware non-users. Table 3.5 shows that providers with lower household income, who are more educated and innovative are more likely to be driven by economic motives. Moreover, providers who are younger, have higher trust and

Table 3.5. Results of logistic regression for providers.

	Economic				Non-economic			
	B	Wald	Sig.	Exp(B)	B	Wald	Sig.	Exp(B)
Age	0.10	1.25	0.26	1.10	−0.17	4.55	0.03	0.84
Gender	0.15	0.55	0.46	1.17	0.03	0.02	0.88	1.03
Income*	−0.21	3.80	0.05	0.81	−0.11	1.30	0.25	0.89
Education	0.26	7.67	0.01	1.30	0.11	1.76	0.18	1.12
Trust	0.21	3.53	0.06	1.24	0.37	12.58	0.00	1.45
Innovativeness	0.31	6.89	0.01	1.37	0.17	2.43	0.12	1.18
Materialism	0.08	0.47	0.49	1.09	0.18	2.50	0.11	1.19
Volunteering	0.03	0.07	0.79	0.97	0.47	21.08	0.00	1.60
Constant	−1.97	7.22	0.01	0.14	−3.20	21.30	0.00	0.04

Note: $N = 556$, * in the analysis we used household income.

Table 3.6. Results of logistic regression for consumers.

	Economic				Non-economic			
	B	Wald	Sig.	Exp(B)	B	Wald	Sig.	Exp(B)
Age	−0.08	1.42	0.23	0.92	−0.14	6.97	0.01	0.87
Gender	−0.16	0.77	0.38	0.85	0.03	0.06	0.81	1.03
Income*	−0.06	0.43	0.51	0.94	−0.06	0.95	0.33	0.94
Education	0.17	3.90	0.05	1.19	−0.15	5.36	0.02	0.86
Trust	0.20	4.60	0.03	1.23	0.19	6.31	0.01	1.20
Innovativeness	0.20	4.65	0.03	1.23	0.27	13.44	0.00	1.31
Materialism	0.09	0.81	0.37	1.09	0.30	16.41	0.00	1.34
Volunteering	0.03	0.12	0.73	1.03	0.48	48.47	0.00	1.62
Constant	−0.05	0.00	0.94	0.95	−3.00	31.92	0.00	0.05

Note: $N = 1143$, * in the analysis we used household income.

volunteer more frequently are more likely to be driven by non-economic motives.

Among consumers, economic motives are positively associated with education, trust and innovativeness. Thus, more educated, more trusting and more innovative consumers are motivated more strongly by economic benefits. By contrast, consumers who are younger, more educated, more innovative and volunteers are more likely to be driven by non-economic motives (Table 3.6).

In the group of potential users (in the survey recognized as aware non-users), income, gender and trust do not affect expected benefits that are economic, while

Table 3.7. Results of logistic regression: expected benefits among aware non-users.

	Economic				Non-economic			
	B	Wald	Sig.	Exp(B)	B	Wald	Sig.	Exp(B)
Age	−0.25	72.75	0.00	0.78	−0.16	37.93	0.00	0.85
Gender	−0.10	1.89	0.17	0.91	−0.20	8.95	0.00	0.82
Income*	0.02	0.23	0.63	1.02	−0.08	5.32	0.02	0.93
Education	0.10	8.85	0.00	1.11	−0.03	1.09	0.30	0.97
Trust	0.02	0.36	0.55	1.02	0.15	16.58	0.00	1.16
Innovativeness	0.16	20.57	0.00	1.17	0.20	37.31	0.00	1.23
Materialism	0.23	32.93	0.00	1.26	0.11	8.92	0.00	1.12
Volunteering	0.11	10.12	0.00	1.12	0.31	89.52	0.00	1.36
Constant	−0.12	0.20	0.66	0.89	−0.85	11.20	0.00	0.43

Note: $N = 3983$, * in the analysis we used household income.

only education does not impact the expected benefits in non-economic terms. (Table 3.7). Younger, more educated, more innovative, materialistic and volunteering aware non-users expect more economic benefits, while younger, female, with low household income, more trusting, innovative, materialistic and volunteering aware non-users expect more non-economic benefits.

3.5 Discussion and Conclusion

Based on an existing large survey, we studied demographics (age, gender, education, household income) as well as relevant attitudinal (trust, innovativeness, materialism) and behavioural (volunteering) antecedents of economic and non-economic motives or expected benefits in sharing economy participation. Using descriptive, univariate and multivariate statistics, we found that economic (expected) benefits outperform non-economic ones among providers, consumers and aware non-users. One-way ANOVA revealed statistically significant differences in demographic characteristics between providers, consumers and aware non-users. The analysis showed that the providers are younger and more likely to be male. Consumers have the highest household income and education level, while aware non-users are the oldest group and have the lowest household income and education levels. In terms of attitudinal and behavioural differences, we found that consumers are most innovative, they showed the most general trust in people, they are most materialistic, but they volunteer less frequently than providers. Thus, to a certain extent, the sharing economy seems to perpetuate existing inequalities and benefit those who are already privileged (Eichhorn *et al.*, 2020; Lutz, 2019; Schor *et al.*, 2016).

When it comes to the motives for participation, consumers were mostly driven by financial benefits. Economic motives were particularly prominent among more educated and trusting consumers, while younger, more educated, more innovative, materialistic and volunteering consumers were more likely to be driven by non-economic motives. This shows that economic and non-economic motives are not mutually exclusive and sharing economy participants can accrue multiple type of capital at the same time (Ladegaard, 2018). Users who are economically motivated can also be motivated by non-economic criteria and there can be a plurality of motives. We found that providers with lower household income, who are more educated and innovative are more motivated by economic benefits, while providers that are more trusting and that want to help voluntarily are more driven by non-economic motives.

Economic motives or expected benefits are obvious in all three groups of respondents. However, consumers had more pronounced economic motives, compared with providers. By contrast and compared with consumers, providers are more motivated by meeting people, fun and social responsibility. This is in line with Böcker and Meelen (2017), who found similar differences between providers and consumers in their study in the Netherlands. In our analysis, providers seem motivated by a broader set of motives, reflecting the results of Ladegaard (2018) from their interviews with Airbnb hosts. Interestingly, aware non-users expect more benefits from social responsibility and social interaction than consumers are motivated by these factors. Overall, economic motives are most prevalent among consumers, while non-economic motives tend to be more salient among providers and aware non-users.

Our findings have *implications* for theory and practice. In terms of theory, our study identifies important antecedents of motives, something which previous literature (Albinsson and Perera, 2012; Bardhi and Eckhardt, 2012; Bellotti *et al.*, 2015; Bucher *et al.*, 2016; Hawlitschek *et al.*, 2016a, 2018; Tussyadiah, 2015; Tussyadiah and Pesonen, 2016) has mostly overlooked, as it focused more on the types and outcomes of motives in different sharing economy domains and contexts. Analysing not only motives but also their antecedents enhances our knowledge of sharing economy participation and allows for a more holistic understanding of its social dynamics. Particularly, our study contributes to research that studies the sharing economy in terms of power dynamics and digital inequalities (Eichhorn *et al.*, 2020).

The importance of trust across the analyses, with significant effects for all three groups considered (providers, consumers, aware non-users), solidifies the crucial role of this construct in the sharing economy (Hawlitschek *et al.*, 2016b; Ter Huurne *et al.*, 2017). Particularly, the fact that trust mattered not only for aware non-users but also for users indicates that trust constitutes an important pre-condition for continued motivation to stay active in the sharing economy. Innovativeness proved similarly important, as it had a significant – and positive – effect on economic motives across all three groups, and only proved to be insignificant for non-economic motives among providers. Thus, the sharing economy seems to cater particularly well to technologically innovative groups and might leave behind those who lack the drive to try out new technologies, thus *potentially exacerbating social inequalities* between different social groups (Ladegaard, 2018; Lutz, 2019; Schor *et al.*, 2016). This conclusion is supported by the demographic profile of aware non-users, who are older and have lower levels of household income and education than providers and consumers.

Volunteering was the strongest predictor of non-economic motives/expected benefits across all three groups. Non-economic sharing economy motives or expected benefits are particularly prominent among individuals who engage civically by volunteering and helping others (Kornberger *et al.*, 2018). This indicates that *different segments of the sharing economy follow partly different logics*, with certain platforms – and service categories within platforms – catering to a more bottom-up, non-commercial and social experience, while others target a more convenience-oriented and materialistic audience (Guttentag *et al.*, 2018; Lutz and Newlands, 2018). This is to be taken into consideration in any design and/or policy intervention.

A further contribution of our research to the sharing economy literature is the differentiation and comparison of providers, consumers and aware non-users. Previous research on motives for sharing economy participation has either looked at providers or consumers (Dann *et al.*, 2019) but rarely contrasted these two groups systematically (see Böcker and Meelen, 2017, for an exception), let alone included aware non-users. Our results show that the expected benefits of aware non-users are similarly pronounced as the motives of users. This is somewhat surprising as we had expected lower values. Future research could follow up on this and compare providers and consumers for specific services as well as the transition from consumers to providers (Angelovska *et al.*, 2020) and what makes individuals transition from aware non-users to users (as either consumer or provider), and from non-aware non-user to aware non-user.

The findings have *practical implications* and relevance for platform managers, policymakers and users. Platform managers can foster desired motives by leveraging key antecedents identified in our analysis. For example, a platform such as Uber can tap into heightened economic motives among young, educated, innovative, materialistic, and volunteering groups (e.g., students) that do not yet use the platform. Uber could leverage the motives of such aware non-users by designing targeted promotions and campaigns specifically for that group, for example student discounts or recommendation rewards. Similarly, a bottom-up sharing platform that caters strongly to non-economic motives should create a climate of trust and volunteering, potentially encouraging and supporting such volunteering outside the platform to keep their providers motivated in the long run. For policymakers, our findings might prove useful to steer the growth of platforms through supporting conditions that tap into distinct motives. For example, if a city wants to promote non-economic motives and participation (and corresponding platforms), it can drive up such motives by creating a climate of trust, offering skill training and information to foster innovativeness, as well as lowering the threshold for volunteering. Finally, users themselves might find the results helpful to reflect on their own practices. Those who use sharing platforms as consumers might be confronted with a broader range of benefits, especially non-economic ones, that could be reaped if they started using the platform as a provider.

This study comes with several *limitations*. Namely, our survey is cross-sectional and does not allow for temporal and strong causal claims. Moreover, it lacked a strong comparative framing. Future research should use longitudinal data to study participants' and potential participants' demographics, trust, innovativeness, materialism, volunteering and motives over time. Such research could adapt a comparative scope to map the differences in the adoption of the sharing economy across different industries or countries.

Acknowledgement

The authors would like to acknowledge the contribution of – From Sharing to Caring: Examining Socio-Technical Aspects of the Collaborative Economy – (COST Action, CA16121://https://www.cost.eu/actions/CA16121/#tabs|Name:overview/). They would also like to thank the editors of the volume, Chiara Bassetti and Maurizio Teli, for their excellent organization of the review process, for giving very helpful feedback throughout the revision process and for making this volume possible in the first place through their commitment and hard work. Finally, they would like to thank the peer reviewers who critically assessed this chapter and made valuable suggestions that increased the quality of the chapter: Filip Majetic, Rodrigo Perez Vega, Eglantina Hysa, Sindiola Koka and Alba Demneri Kruja.

References

Agarwal, R. and Prasad, J. (1998). A conceptual and operational definition of personal innovativeness in the domain of information technology. *Information Systems Research, 9*(2), 204–215.

Akbar, P., Mai, R., and Hoffmann, S. (2016). When do materialistic consumers join commercial sharing systems. *Journal of Business Research, 69*(10), 4215–4224.

Albinsson, P.A. and Perera, B.Y. (2012). Alternative marketplaces in the 21st century: Building community through sharing events. *Journal of Consumer Behaviour, 11*(4), 303–315.

Andreotti, A., Anselmi, G., Eichhorn, T., Hoffmann, C. P., Jürss, S., and Micheli, M. (2017). European perspectives on participation in the sharing economy. *SSRN Electronic Journal,* 17 October 2017. https://papers.ssrn.com/sol3/papers.cfm?abstract_id=3046550

Andreotti, A., Anselmi, G., Eichhorn, T., Etter, M., Fieseler, C., ... and Vermeulen, I. (2017). Ps2Share – Participation, Privacy, and Power in the Sharing Economy (Version 1.0) [Data set]. Zenodo. http://doi.org/10.5281/zenodo.1122633

Angelovska, J., Čeh Časni, A., and Lutz, C. (2020). Turning consumers into providers in the sharing economy: Exploring the impact of demographics and motives. *Ekonomska Misao i Praksa, 29*(1), 79–100.

Balck, B. and Cracau, D. (2015). Empirical analysis of customer motives in the shareconomy. *Working Paper Series, University of Magdeburg.* https://ideas.repec.org/p/mag/wpaper/150002.html

Bardhi, F. and Eckhardt, G. (2012). Access-based consumption: The case of car sharing. *Journal of Consumer Research, 39*(4), 881–898.

Belk, R. (2007). Why not share rather than own? *The Annals of the American Academy of Political and Social Science, 611*(1), 126–140.

Belk, R. (2010). Sharing. *Journal of Consumer Research, 36*(5), 715–734.

Belk, R. (2014a). Sharing versus pseudo-sharing in web 2.0. *The Anthropologist, 4*(2) 7–23.

Belk, R. (2014b). You are what you can access: sharing and collaborative consumption online. *Journal of Business Research, 67*(8), 1595–1600.

Bellotti, V., Ambard, A., Turner, D., Gossmann, C., Demkova, K., and Carroll, J.M., (2015). A muddle of models of motivation for using peer-to-peer economy systems. In *Proceedings of the 33rd Annual ACM Conference on Human Factors in Computing Systems* (pp. 1085–1094). ACM.

Benkler, Y. (2004). Sharing nicely: On shareable goods and the emergence of sharing as a modality of economic production. *Yale Law Journal, 114*, 273–358.

Benkler, Y. (2006). *The Wealth of Networks: How Social Production Transforms Markets and Freedom.* Yale University Press.

Boar, A., Bastida, R., and Marimon, F. (2020). A systematic literature review. Relationships between the sharing economy, sustainability and sustainable development goals. *Sustainability, 12*(17), 6744.

Böcker, L. and Meelen, T. (2017). Sharing for people, planet or profit? Analysing motivations for intended sharing economy participation. *Environmental Innovation and Societal Transitions, 23*, 28–39.

Botsman, R. and Rogers, R. (2010). Beyond Zipcar: Collaborative consumption. *Harvard Business Review, 88*(10), 30.

Botsman, R. and Rogers, R. (2011). *What's Mine is Yours: How Collaborative Consumption is Changing the Way we Live.* Harper Collins.

Bucher, E., Fieseler, C., and Lutz, C. (2016). What's mine is yours (for a nominal fee) – Exploring the spectrum of utilitarian to non-economic motives for Internet-mediated sharing. *Computers in Human Behavior, 62*, 316–326.

Bucher, E., Fieseler, C., Fleck, M., and Lutz, C. (2018). Authenticity and the sharing economy. *Academy of Management Discoveries, 4*(3), 294–313.

Constantinides, E. and Fountain, S.J. (2008). Web 2.0: Conceptual foundations and marketing issues. *Journal of Direct, Data and Digital Marketing Practice, 9*(3), 231–244.

Dann, D., Teubner, T., and Weinhardt, C. (2019). Poster child and guinea pig–insights from a structured literature review on Airbnb. *International Journal of Contemporary Hospitality Management, 31*(1), 427–473.

Davidson, A., Habibi, M.R., and Laroche, M. (2018). Materialism and the sharing economy: A cross-cultural study of American and Indian consumers. *Journal of Business Research, 82*, 364–372.

Dervojeda, K., Verzijil, D., Nagtegaal, F., Lengton, M., and Rouwmatt, E. (2013). The sharing economy: Accessibility based business models for peer-to-peer markets. *European Commission. Business Innovation Observatory.* http://ec.europa.eu/growth/industry/innovation/business-innovation-observatory_en

Dillahunt, T.R. and Malone, A.R. (2015). The promise of the sharing economy among disadvantaged communities. In *Proceedings of the 33rd Annual ACM Conference on Human Factors in Computing Systems* (pp. 2285–2294). ACM.

Edbring, E.G., Lehner, M., and Mont, O. (2016). Exploring consumer attitudes to alternative models of consumption: motivations and barriers. *Journal of Cleaner Production, 123*, 5–15.

Eichhorn, T., Jürss, S., and Hoffmann, C.P. (2020). Dimensions of digital inequality in the sharing economy. *Information, Communication and Society*, online first. https://doi.org/10.1080/1369118X.2020.1791218

Eurobarometer. (2016). The use of collaborative platforms. *Flash Eurobarometer: Vol. 438*. Luxembourg: Publications Office. http://ec.europa.eu/COMMFrontOffice/publicopinion/index.cfm/Survey/getSurveyDetail/instruments/FLASH/surveyKy/2112

Frenken, K. and Schor, J. (2017). Putting the sharing economy into perspective. *Environmental Innovation and Societal Transitions, 23*, 3–10.

Fuster Morell, M., Salcedo, J. L., and Berlinguer, M. (2016). Debate about the concept of value in Commons-Based Peer Production. In *International Conference on Internet Science* (pp. 27–41). Springer.

Gansky, L. (2010). *The Mesh: Why the Future of Business is Sharing*. Penguin.

Gerwe, O. and Silva, R. (2020). Clarifying the sharing economy: Conceptualization, typology, antecedents, and effects. *Academy of Management Perspectives, 34*(1), 65–96.

Grassmuck, V.R. (2012). The sharing turn: Why we are generally nice and have a good chance to cooperate our way out of the mess we have gotten ourselves into. In Sützl, W., Stalder, F., Maier, R., and Hug, T. (eds.), *Cultures and Ethics of Sharing* (pp. 17–34). Innsbruck University Press.

Gurven, M. (2006). The evolution of contingent cooperation. *Current Anthropology, 47*(1), 185–192.

Guttentag, D., Smith, S., Potwarka, L., and Havitz, M. (2018). Why tourists choose Airbnb: A motivation-based segmentation study. *Journal of Travel Research, 57*(3), 342–359.

Hamari, J., Sjöklint, M., and Ukkonen, A. (2016). The sharing economy: Why people participate in collaborative consumption. *Journal of the Association for Information Science and Technology, 67*(9), 2047–2059.

Hawlitschek, F., Teubner, T., and Gimpel, H. (2016a). Understanding the sharing economy – Drivers and impediments for participation in peer-to-peer rental. In *2016 49th Hawaii International Conference on System Sciences (HICSS)* (pp. 4782–4791). IEEE.

Hawlitschek, F., Teubner, T., and Weinhardt, C. (2016b). Trust in the sharing economy. *Die Unternehmung, 70*(1), 26–44.

Hawlitschek, F., Teubner, T., and Gimpel, H. (2018). Consumer motives for peer-to-peer sharing. *Journal of Cleaner Production, 204*, 144–157.

Heo, Y. (2016). Sharing economy and prospects in tourism research. *Annals of Tourism Research, 58*, 166–170.

Holmström, J. and Stalder, F. (2001). Drifting technologies and multi-purpose networks: the case of the Swedish cashcard. *Information and Organization, 11*(3), 187–206.

Ikkala, T. and Lampinen, A. (2015). Monetizing network hospitality: Hospitality and sociability in the context of Airbnb. In *Proceedings of the 18th ACM Conference on Computer Supported Cooperative Work & Social Computing* (pp. 1033–1044). ACM.

John, N.A. (2013). The social logics of sharing. *The Communication Review, 16*(3), 113–131.

Kemp, J. and Olson, M. (2015). Sharing economy: An in-depth look at its evolution and trajectory across industries. http://collaborativeeconomy.com/research/sh

aring-economy-an-in-depth-look-at-its-evolution-and-trajectory-across-in dustries/

Kornberger, M., Leixnering, S., Meyer, R.E., and Höllerer, M.A. (2018). Rethinking the sharing economy: The nature and organization of sharing in the 2015 refugee crisis. *Academy of Management Discoveries, 4*(3), 314–335.

Ladegaard, I. (2018). Hosting the comfortably exotic: Cosmopolitan aspirations in the sharing economy. *The Sociological Review, 66*(2), 381–400.

Lamberton, C.P. and Rose, R.L. (2012). When is ours better than mine? A framework for understanding and altering participation in commercial sharing systems. *Journal of Marketing, 76*(4), 109–125.

Lampinen, A. and Cheshire, C. (2016). Hosting via Airbnb: Motivations and financial assurances in monetized network hospitality. In *Proceedings of the 2016 CHI Conference on Human Factors in Computing Systems* (pp. 1669–1680). ACM.

Luchs, M., Naylor, R.W., Rose, R.L., Catlin, J.R., Gau, R., Kapitan, S., ... and Subrahmanyan, S. (2011). Toward a Sustainable Marketplace: Expanding Options and Benefits for Consumers. *Journal of Research for Consumers, 19*, 1–12.

Lutz, C. (2019). Digital inequalities in the age of artificial intelligence and big data. *Human Behavior and Emerging Technologies, 1*(2), 141–148.

Lutz, C., Hoffmann, C.P., Bucher, E., and Fieseler, C. (2018). The role of privacy concerns in the sharing economy. *Information, Communication and Society, 21*(10), 1472–1492.

Lutz, C. and Newlands, G. (2018). Consumer segmentation within the sharing economy: The case of Airbnb. *Journal of Business Research, 88*, 187–196.

Lutz, C., Newlands, G., and Fieseler, C. (2018). Emotional labor in the sharing economy. In *Proceedings of the 51st Hawaii International Conference on System Sciences*. https://scholarspace.manoa.hawaii.edu/handle/10125/49968

McKnight, D.H., Choudhury, V., and Kacmar, C. (2002). Developing and validating trust measures for e-commerce: An integrative typology. *Information Systems Research, 13*(3), 334–359.

Möhlmann, M. (2015). Collaborative consumption: determinants of satisfaction and the likelihood of using a sharing economy option again. *Journal of Consumer Behaviour, 14*(3), 193–207.

Mont, O. (2002). Clarifying the concept of product–service system. *Journal of Cleaner Production, 10*(3), 237–245.

Murillo, D., Buckland, H., and Val, E. (2017). When the sharing economy becomes neoliberalism on steroids: Unravelling the controversies. *Technological Forecasting and Social Change, 125*, 66–76.

Narasimhan, C., Papatla, P., Jiang, B., Kopalle, P.K., Messinger, P.R., Moorthy, S., ... and Zhu, T. (2018). Sharing economy: Review of current research and future directions. *Customer Needs and Solutions, 5*(1–2), 93–106.

Newlands, G. and Lutz, C. (2020). Fairness, legitimacy and the regulation of home-sharing platforms. *International Journal of Contemporary Hospitality Management*, online first. https://doi.org/10.1108/IJCHM-08--2019-0733

Newlands, G., Lutz, C., and Fieseler, C. (2018). European perspectives on power in the sharing economy. *SSRN Electronic Journal*, 3 January 2018. https://papers.ssrn.com/sol3/papers.cfm?abstract_id=3046473

Newlands, G., Lutz, C., and Fieseler, C. (2019). The conditioning function of rating mechanisms for consumers in the sharing economy. *Internet Research, 29*(5), 1090–1108.

Novel, A.S. (2014). Is sharing more sustainable? The environmental promises of the sharing economy. In Grosclaude, J.-Y., Pachauri, R.K., and Tubiana, L. (eds.), *Innovation for Sustainable Development* (pp. 139–144). IDDR SciencesPo.

Schmidt, M.F. and Sommerville, J.A. (2011). Fairness expectations and altruistic sharing in 15-month-old human infants. *PloS One, 6*(10), e23223.

Schor, J.B., Fitzmaurice, C., Carfagna, L.B., Attwood-Charles, W., and Poteat, E.D. (2016). Paradoxes of openness and distinction in the sharing economy. *Poetics, 54*, 66–81.

Shih, P.C., Bellotti, V., Han, K., and Carroll, J.M. (2015, April). Unequal time for unequal value: Implications of differing motivations for participation in timebanking. In *Proceedings of the 33rd Annual ACM Conference on Human Factors in Computing Systems* (pp. 1075–1084). ACM.

Slee, T. (2013). Some obvious things about Internet reputation systems. *TomSlee.net*, 29 September 2013. http://tomslee.net/2013/09/some-obvious-things-about-internet-reputation-systems.html

Smith, A. (2016). Shared, collaborative and on demand: The new digital economy. *Pew Research Center: Internet & Technology*, 19 May 2016. http://www.pewinternet.org/2016/05/19/the-new-digital-economy/

Stephany, A. (2015). *The Business of Sharing: Making it in the New Sharing Economy*. Palgrave Macmillan.

Sundararajan, A. (2016). *The Sharing Economy: The End of Employment and the Rise of Crowd-Based Capitalism*. MIT Press.

Ter Huurne, M., Ronteltap, A., Corten, R., and Buskens, V. (2017). Antecedents of trust in the sharing economy: A systematic review. *Journal of Consumer Behaviour, 16*(6), 485–498.

Tomasello, M. and Warneken, F. (2008). Share and share alike. *Nature, 454*(7208), 1057–1058.

Tussyadiah, I.P. (2015). An exploratory study on drivers and deterrents of collaborative consumption in travel. In *Information and Communication Technologies in Tourism 2015* (pp. 817–830). Springer.

Tussyadiah, I.P. and Pesonen, J. (2016). Impacts of peer-to-peer accommodation use on travel patterns. *Journal of Travel Research, 55*(8), 1022–1040.

Walsh, B. (2011). Today's smart choice: Don't own. Share. *Time*, 17 March 2011. http://content.time.com/time/specials/packages/article/0,28804,2059521_2059717_2059710,00.html

DOI: 10.1561/9781680838411.ch4

Chapter 4

In the Scenario of Sustainable Mobility and Pandemic Emergency: Experiences of Bike- and E-Scooter-Sharing Schemes in Budapest, Lisbon, Rome and Vilnius

By Vera Diogo, Venere Stefania Sanna, Aniko Bernat and Egle Vaiciukynaite

4.1 Introduction

Within the collaborative economy, bike sharing and e-scooter sharing are relevant services that have been associated with increases in wellbeing, health (Woodcock *et al.*, 2014) and quality of life, as well as with the creation of (often temporary) employment (De Groen *et al.*, 2017),[1] It is therefore important to understand the ways in which these sharing practices transform economic, social and cultural values related to mobility, and how they foster, rather than disrupt, social relationships. The impact of these services on the quality of urban life and on a mobility shift towards sustainability, is the specific focus in this chapter.

1. As mentioned in Chapter 13 of the current publication.

The connections between sharing practices and urban systems are multidimensional and inter-influential. Sharing practices have grown considerably in Europe in the 2010's, particularly in urban areas (cf. also Salvia *et al.*, 2019), and many local authorities have increasingly defined plans to become "sharing cities":

> Sharing cities make use of (often smart) technologies to connect a larger number of users to idling assets, hence to be 'shared' by a wider population, rather than being individually owned. Within this trend, assets that are typically shared include vehicles and rides, bedrooms and accommodation, as well as tools and competences (Salvia *et al.*, 2019, p. 1).

This shift from ownership to access concurs to the implementation of mobility as a service, which is the aim of the European Green Deal (European Commission, 2019), voted for by the European Parliament on 15 January 2020. Light sustainable mobility, such as walking, cycling and e-scooter riding, can contribute quite significantly to The Green Deal target of reducing 90% of transport-related greenhouse gas emissions by 2050. Bike and e-scooter sharing can play a major role in this by improving multimodality, particularly in urban areas. Additionally, the digitalization of these services shows great "potential for collection of mobility data", which can be integrated to municipal or even national level, depending on political will and administrative capacity.[2] This connection between bike and e-scooter sharing, on the one hand, and urban planning, on the other, requires that when analysing the evolution of such sharing initiatives, the socioeconomic, cultural and urban contexts in which they are located are taken into consideration, along with the policies and legislative frameworks that regulate, support or limit these forms of transportation.

In this chapter we focus on four countries and their capital cities: Rome (Italy), Lisbon (Portugal), Budapest (Hungary) and Vilnius (Lithuania) to illustrate how shared light sustainable mobility spread in different political, social and economic contexts. The current status of sustainable mobility practices, policies and discourses in these countries is characterized by a series of common trends but also by elements of absolute divergence. The existence of light mobility sharing schemes in these capitals is an indicator of a potential to expand the rates of *active mobility*. Furthermore, it is also a relevant factor of digitalization and commodification of mobility as a service, in line with the European Green Deal, hence it concurs to simplify transportation systems and urban logistics, to free public space, while reducing environmental, social and economic costs. With an increased rate of 45% per year,

2. https://ecf.com/news-and-events/news/innovation-keeps-driving-bike-sharing-sector-forward

bike sharing has been the fastest growing mode of urban transportation since 2007 (Lopes, 2015). It is therefore interesting to discover the specificities that characterize the landscape and the policy framework of the analysed countries and capital cities, what limits are found in the use of bicycles and/or electric scooters (e-scooters) sharing schemes, and where to leverage to favour the transition towards more sustainable modes of transport.

The primary goal of the comparative research we propose, is to identify similarities and differences between these countries and cities, in order to understand how shared light sustainable mobility is developed in European capital cities of different characteristics and opportunities. For this purpose the analysis compares and contrasts cultural, societal, institutional and political traits related to light mobility, with a specific focus on the bike sharing and e-scooter initiatives, in order to understand and identify various evolution patterns as well as key institutional actors and measures. The following section defines sharing mobility and describes the reality of bike and e-scooter sharing in the four cities; Section 4.3 sheds light on the broader mobility cultural contexts and institutional actors; in Section 4.4 we analyse national policy frameworks; in Section 4.5 we discuss the impact of the pandemic, before presenting some final remarks.

4.2 Bike and E-Scooter Sharing: A Four-City Comparison

Shared mobility represents a subset of the larger sharing and collaborative economy. Building on the elaborations of other authors (Jin *et al.*, 2018[3]; Shaheen and Chan, 2016[4]), bike and e-scooter-sharing systems are a particular form of shared mobility in which what is facilitated is "the sharing of a vehicle". Figure 4.1 offers a classification of shared mobility options existing at the time of publishing. Although this chapter is focusing only on bike- and e-scooter-sharing services, the figure illustrates the wide range of possible shared mobility modes that, to some extent, might be rivals to attract their target population and might overlap among the various sharing modes.

The first bike-sharing projects were initiated by various local communities and organizations in the Netherlands, and the earliest well-known community bicycle programme was started in 1965 in Amsterdam. Since then, the growth of bike

3. https://www.researchgate.net/publication/322605779_Ridesourcing_the_sharing_economy_and_the_future_of_cities

4. https://www.researchgate.net/publication/311973901_Mobility_and_the_Sharing_Economy_Potential_to_Facilitate_the_First-_and_Last-Mile_Public_Transit_Connections

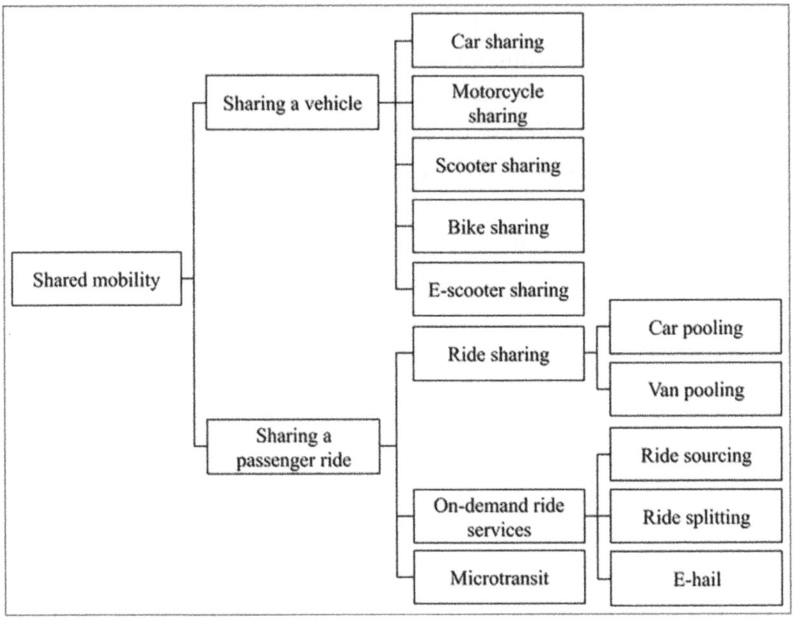

Figure 4.1. Categories of shared mobility.
Source: Elaboration of the authors on Jin S.T. *et al.* 2018.

sharing has been exceptional worldwide, and together with the recent explosion of e-scooter-sharing is changing how urban travellers access transportation, and how cities are planned and built (Cohen and Shaheen, 2016[5]). In fact, the wide range of sharing schemes, represents innovative transportation means that enable users to gain short-term access to transportation modes on an as-needed basis for either passenger trips or goods delivery.

Parallel to their success, controversies about light mobility sharing schemes potential impacts and externalities have arisen. There have been a number of studies discussing the social, economic, and environmental impact of the broader "sharing economy" (e.g.; Frenken and Schor, 2017; Martin, 2016; Schor, 2016). Nevertheless the specific area of light mobility still represents a niche of this growing body of research, particularly for e-scooter sharing schemes. It is worth mentioning that given the novelty of this phenomenon, there is far less knowledge built about it in comparison with bike-sharing. As for the latter, potential economic effects (Otero *et al.*, 2018; Qiu and He, 2018; Ricci, 2015), principal externalities on the environment (Qiu and He, 2018; Shaheen *et al.*, 2010; Zhang and Mi, 2018), people's health (Otero *et al.*, 2018; Qiu and He, 2018), urban efficiency (Ricci, 2015) and

5. https://escholarship.org/content/qt0dk3h89p/qt0dk3h89p.pdf

traffic congestion reduction (Fan and Zheng, 2020; Fiedler *et al.*, 2017; Fishman *et al.*, 2015) – also during the COVID-19 pandemic (Teixeira and Lopes, 2020) – have been documented. However, their social impacts (e.g. equity and inclusivity), have been largely overlooked and the debate is still open (Qiu and He, 2018; Ricci, 2015; Teixeira *et al.*, 2020).

If on the one hand, as resulted from a wide study on major bike sharing schemes in Europe, these "provide health and economic benefits" and their promotion "can significantly increase the health benefits", and "can be used as a tool for health promotion and prevention" (Otero *et al.*, 2018, p. 7), on the other hand, bike sharing benefits "are unequally distributed, since users are typically male, younger and in more advantaged socio-economic positions than average" (Ricci, 2015; p. 1). Moreover, even when in the process of planning bike-sharing systems the "spatial equity" was considered as a key factor for fostering social inclusiveness, "maximising accessibility or coverage alone, without considering equality", still produced discrimination between different groups (Caggiani *et al.*, 2020, p. 1; see also[6]).

4.2.1 Bike Sharing Services

Bike-sharing systems (BSS) represent not only a sustainable mobility tool but also a means of urban intermodal transport (Caggiani *et al.*, 2020). According to their evolution over time, these systems can be grouped into five categories, or generations:

1. Staffed stations (zero-generation system): bicycles can be rented or borrowed from a location and returned to that location.
2. White bikes (first-generation system): bikes are made available for free and are simply released into a city or given area for use by anyone.
3. Coin deposit stations (second-generation system): the bicycle can be borrowed free of charge and for an unlimited time. A coin deposit is needed but the coin can be retrieved by returning the bicycle to a station.
4. Automated stations (third-generation system): bicycles can be borrowed or rented from an automated station or "docking stations". These are bike racks that lock the bike, and only release it by computer control. The bike can be returned at any station belonging to the same system.
5. Dockless bikes (fourth-generation system): free floating bikes are available on demand using mobile phone apps and GPS technologies.

6. https://pdxscholar.library.pdx.edu/cgi/viewcontent.cgi?referer=&httpsredir=1&article=1138&context=trec_reports

Table 4.1. Population size, density and city areas in 2020.

European Cities	Budapest	Vilnius[7]	Rome	Lisbon[8]
City Population	1 696 128	617 000	2 782 858	508 368
Population density/km^2	3 366	1 446	2 166	5 081
City area (km$^{2)}$	525.2	401	1 285	100.05

Source: Elaborated by authors.

Table 4.2. Number of bike sharing providers and bicycles in the four cities.

Bikesharing Provider Type/ Platforms	Budapest	No. of Bikes	Lisbon	No. of Bikes	Rome	No. of Bikes	Vilnius	No. of Bikes
Local service	1	2,071	1	1,000	NA	NA	NA	NA
Global private company	1	200–300	2	1150	1	3,300	1[9]	300

Source: Elaborated by the authors.

According to recent data, bike-sharing services are offered in several areas, mostly in the biggest cities or specific areas of the four capitals, given their cultural traits or functional uses (recreation, tourism). Table 4.1 displays the area, population size and density of each city, to serve as reference for the comparison of bike and e-scooter sharing distribution. At the time of writing, while in **Budapest, Lisbon** and **Vilnius**, BSS providers are using automated stations (third-generation system), in **Rome** only free floating bikes (fourth-generation BSS) are available. In Vilnius and Budapest only mechanical bicycles are available, while in Lisbon both mechanical and electric bicycles are offered, and in Rome, only electric.

As shown in Table 4.2, Budapest and Lisbon count on both local public and private multinational bike sharing providers, while Vilnius has only schemes managed by global private initiative, and at the time of writing, Rome entirely relies on a private multinational operator. This difference is probably related with Budapest and Lisbon municipalities' strong pro-bike policies, given that public bike sharing services (BSS) provide governmental administrations with direct influence on how BSSs are promoted and managed. In particular, the Portuguese BSS run by the municipal company has received the capital investment of $16 million by the City

7. Vilnius data were only available for 2018.

8. Lisbon data were only available for 2019 (PORDATA).

9. https://www.jcdecaux.com/partners/supplying-self-service-bikes

of Lisbon (Christensen, 2019) and some of these city shared bicycles are assigned to the Municipal Police of Lisbon. The main Budapest bike sharing company is also run by the municipality of Budapest, incorporated into the Budapest public transportation company (BKK), supported by the largest Hungarian company, which is the leading integrated Central & East European oil and gas corporation and which is partly owned by the Hungarian State.

The number of bicycles available seems insufficient in all cities, more importantly in **Budapest** and **Rome**, given the larger size of their populations; however, the higher population density of **Lisbon** makes it crucial to increase the forms of shared light mobility, in order to free public space from traffic and parked vehicles. In Italy, between 2015 and 2019 the available fleet is more than tripled and (as reported by the Italian Observatory of Sharing Mobility according to a sample study of about 31 cities in Italy) 5,413 electric bikes were available for sharing in 2019. Nevertheless, **Rome** is the largest of the four cities, with the second lowest population density, and not all areas of the city are served by the sharing services.[10] With reference to the infrastructure needed at the city level to promote the BSS, **Budapest** is showing growing networks of dedicated paths and fleets of shared bicycles.

Despite their blooming, these systems are still lacking in terms of accessibility and equity; issues that could probably be addressed by a reinforced public-private dialogue and reinforcing people's participation in the co-design process behind these services. In terms of accessibility, a common feature of the four analysed cities (which also resemble the situation which characterizes most major urban areas in Europe and in the US), is the uneven geographical distribution of the services between "centre" and "periphery". These services are in fact offered predominantly in the most touristic areas and/or wealthier neighbourhoods and are completely absent in peripheral and/or less-connected areas. In terms of social inclusion and equity, docks, stations and free floating bikes are rarely placed in low-income areas, and when this happens, they do not fit within the overall urban transport system. Payment systems do not consider free memberships or special discounts for low-income or disadvantaged categories, etc. Moreover, by relying on smartphones and digital platforms (from registration, to access and service payment), digital divide, access to the Internet, smart-phones and credit cards determine an unequal distribution of accessibility among the population and represent some key factors that limit the use of BSS to the elderly, low-income and minority populations.

Table 4.3 presents a comparison of bike-sharing services regarding pricing, ticket modalities, discounts and penalties, published in the fall-winter period of 2020. It is

10. https://www.romeing.it/car-bike-scooter-sharing-in-rome/

Table 4.3. Price scheme comparison of bike-sharing services in the four European cities.

	Rome	Lisbon			Budapest		Vilnius	
Service ownership	Private	Public	Private	Private	Public	Private	Private	Private
Bicycle type	Electric	Both	Electric	Electric	Mechanical	Mechanical	Mechanical	Mechanical
Long-term tickets	none	25 €/ year for rides up to 45 min.; 15 €/ month for rides up to 45 min.			0.3 €/ month for rides up to 30 mins	14 €/ month OR 8,4 €/ month	19,90 €/ season	2.90 €/3 days; 3.90 €/30 days; 19,90 €/ season
Daily tickets	13,99 €	2 €				11,6 €	N/A	N/A
Price per minute/ extra time	0,20 € min +1 € to unlock the bike	from 45–90 min.:1 €; more than 90 min.: 2 €/ additional 45 min.	0,20 €/min	0,20 €/min	after 30 mins: 1.4 Euro/ 30 mins up to 3 hours, from 3 hours 3,6 €/30 mins	1,4 € up to 15 mins, 2 € for 15–30 mins, 2,9 € for 30–60 mins; 5,7 € for 1 h or more	trips from 30 to 1 h: 0.39 €; trips longer than 1 h: 0.50 €/ additional 30 min; Long-term users extra km: 0.39 €/ 30 min.	30 min/ 0.50 €
Penalties			If parked in restricted area: 20 €				If the bicycle disappears or is irreparably damaged: 159 €	
Discounts		First 45 min. free with yearly and monthly ticket; all trips up to 45 min. free with daily ticket			The current price is discounted, since March 2020 due to the pandemic	14 € per month – 12 hour free with every rental OR 8,4 € per month – 1 hour free with every rental	First 30 min free with membership; Special discount for seasonal ticket: 13.90 € for young people; 3 day ticket available	

highlighted that only one company, in Vilnius, gave a discount to a specific population group: young people between the ages of 14 and 26. This provider has over 6,000 long-term subscribers – mostly aged from 25 to 34 who use services as a connection in the city area (JCDecaux, 2020). With the exception of Rome, all cities have long term ticket options for bike-sharing, although in Lisbon only the local public service is providing that option, as a brief experience of a monthly pass by a private operator was recently interrupted, despite its high demand.[11] The specific modalities of long term ticket vary, along with the way it is marketed. For instance, in Vilnius, the two providers[12] define long term purchase as membership. Favouring this type of costumer-provider relationship, for each trip, the first 30 minutes are free for users with memberships. Similar privileges are also given by the public service in Lisbon and the private one in Budapest, to holders of long-term tickets. Finally, three of the providers offer a daily ticket, in Rome, Lisbon and Budapest.

The information regarding price per minute or extra time shows that four of the eight providers, distributed in these four cities, apply charges to longer trips, possibly promoting the use of bike-sharing as last mile option. However, combined with the evidence that these services are mostly available in the city centres, such practice underlines the limited socio-geographical inclusion of bike-sharing. Once the legal contracts implied in these purchases were not analysed, we assume there must be more situations where penalties can be charged than what the data collected can indicate. However, two of the providers display more concern for making users clearly understand they are responsible for damages to the bicycles, or to their adequate parking.

4.2.2 E-Scooter Sharing Services

Recently, micro mobility and in particular e-scooters, have become very popular across all Europe,[13] as bike lanes, common use zones and wider pavements[14] can also be used for this means of transportation. It was only in 2018 that the first European e-scooter sharing scheme appeared, in **Lisbon**,[15] where this type of modality has

11. https://www.publico.pt/2020/01/21/local/noticia/jump-termina-passes-mensais-bicicletas-1901140

12. Cyclocity (2020). Mūsų narysčių pasiūlymai [Offers for members]. *Cyclocity*. https://www.cyclocity.lt/en/offers/groups

13. https://www.eu-startups.com/2020/09/battle-of-the-european-e-scooter-startups-dott-tier-voi-wind/

14. Even though most e-scooter service delivers advice against it, once pavements are for pedestrians. However, the poor regulation of e-scooters in most countries has generated a grey area in this regard.

15. https://www.dn.pt/lusa/trotinetes-eletricas-lime-chegam-a-lisboa-como-meio-alternativo-de-transporte-9916788.html; https://insider.dn.pt/em-rede/lime-trotinetes-eletricas-lisboa-como-usar/6527/

Table 4.4. Number of e-scooter sharing providers and e-scooters in Budapest, Lisbon, Rome and Vilnius in Winter 2020–2021.

	Budapest		Lisbon		Rome		Vilnius	
E-scooter Sharing	**Local Service**	**Global Private Company**	**Local Service**	**Global Private Company**	**Local Service**	**Global Private Company**	**Local Service**	**Global Private Company**
No. of services/ companies	NA	2	NA	4	NA	6	0	2
No. of e-scooters	NA	330	NA	12.000	NA	8,000	0	1,100

been increasingly promoted ever since by several companies.[16] In Vilnius, e-scooter services were opened in the Spring of 2019,[17] as well as in Hungary, in the centre of the capital city. In Budapest the infrastructure is also adequate, with an extensive network of bike lanes and paths, while the target group is not only the 1,7 million inhabitants but thousands of tourists, too, but the number of scooters is low, around 300. Between July 2019 and January 2020, e-scooter users of one of the schemes available in **Budapest** have travelled a total of 1.3 million kilometres.[18] **Rome** is following this trend only since Spring 2020, during the pandemic. As recently stated by the mayor, e-scooters represent a small revolution for the city in terms of sustainable mobility.[19]

The widest offer of e-scooter sharing service providers located in **Rome** and **Lisbon**; a smaller number of e-scooter sharing operators is active in **Budapest** and **Vilnius** (see Table 4.4). These numbers refer to the reality between December 2020 and January 2021. As companies and users try to adapt to the evolution of the pandemic restrictions, the demand and offer of e-scooters-sharing has changed very rapidly. For instance, before March 2020, both Rome and Lisbon had local start-ups providing e-scooter-sharing; at the outset of the pandemic, these services were interrupted. Therefore, data reported in Table 4.4 and following description just provide an overview at a particular moment in time.

Comparing this distribution with geographical and demographical aspects (Table 4.1), **Budapest** has the most inadequate number of e-scooters available considering the size of its population, while **Lisbon** is the city with the widest offer. However the poor regulation and control regarding the parking of these vehicles,

16. https://www.eltis.org/sl/node/49528

17. https://www.themayor.eu/ro/scooter-sharing-system-to-be-launched-in-vilnius

18. https://www.themayor.eu/ro/lime-reports-record-figures-in-budapest

19. https://www.thelocal.it/20200622/a-small-revolution-for-our-city-electric-scooters-come-to-rome; https://www.intelligenttransport.com/transport-news/100411/lime-rolls-out-1000-e-scooters-in-rome/

combined with the lack of responsibility demonstrated by some users, led to tensions in the public space, with no clear evidence so far if these services contributed to free public space, particularly from cars.

In Table 4.5, we can see that the prices for using shared e-scooters on a "pay as you go" regime are very similar between these four capitals, with an expected while Vilnius and Budapest have the lowest prices of the analyzed capital cities. However, these differences are not that significant, considering that the minimum wage in Portugal and Lithuania are similar, while in Rome this is about 200 euros higher, and in Hungary 200 euros lower than in Portugal. Rome and Budapest are the only cities where long term tickets are available. Most companies do not apply penalties, nor discounts for specific population groups, with only one service in Lisbon promoting inter-modality more actively by attributing a discount to holders of the city intermodal public transportation card "Viva". The penalties active in Lisbon and Vilnius are similar to those applied in the bike-sharing services.

Overall, following the same trend reported above for BSS, e-scooter sharing schemes are mostly used in the historical centres of the European cities, and this uneven distribution of the services – together with the unequal distribution of accessibility among the population due, for example, to the costs and technologies required to use these services – is producing discrimination between different geographical areas and social groups (Caggiani *et al.*, 2020).

Official statistical information about these light shared mobility services is scarce, given the lack of regulation and even recognition of these forms of transportation by the national legislations. Their expansion must be perceived in the broader cultural mobility context of these countries and cities, which is exposed in Section 4.3. The continuity of these initiatives depends extensively on the implementation of sustainable mobility frameworks, with particular focus on active and micro mobility, which we discuss in Section 4.4.

4.3 How Far Must We Come From?

Overall, the four European capital cities show a general positive attitude towards light and sustainable urban mobility. However, this growing interest in sustainable micro-mobility shown by national and local governments, mostly in bicycles and e-scooters,[20] including sharing schemes, must overcome the cultural car dominance, solidified since the mid-XX century. Italy, Portugal, Hungary and Lithuania face a

20. It should be noted that neither of these means of transportation is new: the bicycle was invented in the XIX century and the e-scooter in the beginning of the XX century.

Table 4.5. Price scheme comparison of e-scooter-sharing services in Rome, Lisbon, Budapest and Vilnius.

	Rome					Lisbon					Budapest	Vilnius	
Long term tickets	13,99 € for 24 hours	29,99 € per month (up to 20 trips per day, each trip up to 30 minutes)	10% discount for packages of trips bought in advance			6,99 € Daily pass, unlimited unlocks					Monthly pass: 5.5 € for unlocking; Daily pass: 13.9 €*		
Pay as you go	0.25 €/ min. + 1 € to unlock	0.20 €/min. + 1 € to unlock	0.25 €/min. + 1 € to unlock	0.19 €/min. + 1 € to unlock	0.25 €/ min. + 1 € to unlock	0.19 €/min. + 1 € to unlock	0.15 €/min. + 1 € to unlock	0.15 €/min +1 € to unlock	0.15 €/min. + 0.5 € to unlock	0.20/min + 1 €	0.14 €/min +0,7 € to unlock	0.10 €/min. + 0.50 € to unlock	0.15 €/ min. + 0.60 € to unlock
Penalties/ Discounts							20 min free for users of Lisbon intermodal card					30 € for leaving scooter outside the dedicated area	100 € for leaving scooter outside the dedicated area

Table 4.6. Comparison of motorization and road safety indicators.

Indicators/year	Italy	Portugal	Hungary	Lithuania
Passenger car ownership/1,000 inhabitants (2018)	646	514	373	512
Hours spent in traffic/year (2017)	37,7	29	26,4	21,5
Road fatalities/million inhabitants (2017)	56	58	64	67
Number of cyclist fatalities/million inhabitants (2016)	275	73	33	22

number of challenges to various extents, due to a combination of a less developed cycling culture that is hindered by the scarcity of a proper infrastructure, exacerbated by the influence of decades of car-oriented policies.

Table 4.6 displays evidence of what limits the spread of sustainable light mobility: the high dependency on the use of the car; the time spent in road congestion; road fatalities and cyclist fatalities. Despite the developments in urban cycling and the recent (and still unquantifiable) boom in e-scooter usage, in general, passenger cars remain the dominant mode of transport by far. Peculiarly, in our limited sample, **Italy** and **Hungary** are positioned respectively at the two extremes of the European statistics of car ownership (Table 4.6). Among the EU-27 Member States, in fact, Luxembourg (with 676 passenger cars/1,000 inhabitants in 2018) is the country with the highest motorization rates, followed by Italy (646), while Romania (332) and **Hungary** (with 373, just over one car per five inhabitants) show the lowest rates. Regarding the time spent in road congestion, **Italy** has the third highest in EU-27, while the lowest was registered in Sweden.

It is probably road safety figures that most indicate the need for urgent change. In this scope, although road fatalities have decreased[21] in the four countries, in 2017 their numbers were still above the EU-27 average (49 per million inhabitants). The number of cyclist fatalities by country however, shows different trends in the countries under investigation (while information about e-scooter fatalities is still not available). Considering the period 2007–2016, **Italy** (352 people were killed per million inhabitants in 2007 versus 275 in 2016) and **Hungary** (158 to 73) demonstrate a substantial decrease in fatalities, **Portugal** remain almost stable

21. European Commission (2018, September). Reduction in Road Deaths 2010–2017. European Commission. Retrieved from https://ec.europa.eu/transport/road_safety/sites/roadsafety/files/mapcare_chng2010_2017.pdf

(34 to 33), while **Lithuania** (for which data are not available for all years) shows a relative increase passing from 18 fatalities in 2013 to 22 in 2015.[22]

All these factors contributed to urban safety and sustainability becoming a primary objective, and more action is being taken at different government levels to implement measures that facilitate greater awareness and changes in mobility practices. This may be indicated either by the development of urban bike and e-scooter sharing systems, or by the development of the infrastructure.

4.3.1 Cycling as Part of a Sustainable Mobility Culture

The use of the e-scooter is a new phenomenon and there are (still) no associations or social movements that primarily promote this new means of transportation, the voice of stakeholders in the field of cycling is more consolidated. Regarding the particular role of urban cycling, the three major European cycling associations (European Cyclists' Federation – ECF, Cycling Industries Europe – CIE, and Confederation of the European Bicycle Industry – CONEBI) who participated in the public consultation on the roadmap for the European Strategy for Sustainable and Smart Mobility, jointly agreed that investments in policy framework for the promotion of cycling and infrastructure are crucial to success, and must play a pivotal role in achieving the ambitious objectives set by the EU Green Deal. On May 6th 2020, these organizations together with other three cycling industry, logistics and users associations sent a letter to the European Commission, advising on measures to promote cycling, simultaneously as a response to the urgency of the multidimensional crisis that the COVID-19 pandemic has triggered and a means to accelerate the path to accomplishing the goals of the European Green Deal (CIE, ECF, CONEBI, ECLF, EBMA, IMBA-Europe, 2020). At the time of writing, mid-pandemic, this is a more timely issue than ever before, so it is important to understand how and if the share of urban cycling and e-scooter usage is improving.

Bike-sharing practices, with a strong influence in promoting urban cycling (Teixeira *et al.*, 2020) are getting to be the cornerstone of sustainable urban mobility across Europe: "recent cycling innovations are transforming the cityscape and contributing to the broader acceptance of cycling in society. Bicycle sharing schemes offer a valid alternative cycling mobility in urban areas and can be combined with public transport for longer distances" (EPRS, 2014, p. 4). While bike and e-scooters' sharing schemes are gaining ground in many cities around Europe due to the private initiative of sharing economy organizations, political decisions on

22. *Source*: https://ec.europa.eu/transport/road_safety/sites/roadsafety/files/pdf/statistics/dacota/bfs20xx_cyclists.pdf

issues around sustainable transport and mobility are still lagging behind in some EU states.

Cycling is by far the least utilized means of transport in the four capital cities considered, which presented low modal shares: Rome and Lisbon show only 1% cycling[23] in 2020, Vilnius 1.5% in 2018,[24] while Budapest showed 2% in 2010 and 4% in 2020 (Bucsky, 2020). Cities where the cycling modal share is below 10% and with limited expertise on developing strategies to include cyclists as road users in their urban planning and to consider bicycles as transportations in their intermodal network are considered beginner cities (BYPAD, 2008). According to available data, the four capitals analysed can be considered as beginners. However, particularly in the case of Budapest, the modal share seems to have increased considerably, as a recent study[25] revealed that 16% was the national modal share, while 71% of adults "are used to cycling – especially in Budapest" and the pandemic has increased this tendency. The other three capitals are possibly on the first step to an evolution towards including cycling within their regular mobility choices.

The improvement of policies on cycling promotion are a necessary stepping stone to further boost this evolution, and the provision of BSS is a very relevant point, proved in previous research to be a motivator for people who were not used to cycling to consider changing their modal choice (Felix, 2019; Pucher and Buehler, 2008, 2012). According to Felix (2019, p. 15), between 2000 and 2018, Lisbon's path to active cycling mobility has been uphill. The Lisbon Municipality has been leading a political shift regarding sustainable mobility, not only with investment in cycling infrastructure but also with the promotion of sharing mobility, including bike and e-scooter sharing services, and with several campaigns to promote active mobility, within school communities by creating school mobility plans. The current cycling patterns are embedded into different contexts in the four countries.

4.3.2 The Voice of Stakeholders and Social Movements

As well as their counterparts at the European level, social movements and their particular actors in the four analysed countries had and continue to have a major role in this cultural shift to sustainable mobility.

In **Italy**, cycling activists are increasingly demanding for an improvement to the country's road infrastructure in order to prevent a rise in casualty figures. In the major cities, environmental campaigns to boost bicycle usage take place every year,

23. https://www2.deloitte.com/content/dam/insights/us/articles/4331_Deloitte-City-Mobility-Index/Lisbon_GlobalCityMobility_WEB.pdf

24. https://www.cities-multimodal.eu/sites/cmm/files/cmm_fact_sheet_vilnius_nov_2018.pdf

25. https://kerekparosklub.hu/kerekparoskutatas_2020

and the first Critical Mass[26] in Rome took place in 2002. From then on, associations such as the "Salvaiciclisti" (Save the Cyclists) organize regular cycling protests[27] in order to warn the government about the high number of cyclist fatalities, while demanding shared space on the roads and a "highway code" that caters for all road users, and not just car drivers. The city counts on a number of important cyclist non-profit organizations, including the Federazione Italiana Ambiente e Bicicletta – FIAB (Italian Federation for the Environment and Bicycle) or grassroots initiatives like Ciclofficine Popolari (which stands for Community Bike Workshops) and Associazione Ciclonauti, which make an important contribution by mobilizing people participation and organizing advocacy activities and political pressure.

As well as in other **Portuguese** cities, Critical Mass events[28] began to be held in Lisbon around 2003. Since this movement started "there was a growing trend in the Lisbon cycling community", which has diminished since 2012. In the same period, "the formal bicycle (…) organizations also increased their memberships, and played an important role in advocating for cycling infrastructure, cyclists' rights in the road code legislation, educational programs, and other bicycle promotion initiatives" (Felix, 2019, p. 54[29]). The Federação Portuguesa de Cicloturismo e Utilizadores de Bicicleta (FPCUB),[30] Associação pela Mobilidade Urbana em Bicicleta (MUBi)[31] and Federação Portuguesa de Ciclismo (FPC)[32] are the three strongest national organizations (Felix, 2019). The founding cores of these initiatives are located in Lisbon, which concurs to this city's highlight, within the country, in cycling promotion movements and general civic activities related with mobility. Probably as a consequence of the municipality's efforts and the presence of the cycling movement, the cycling modal share has increased from "0.2% in 2011 (INE, 2011)", "far below the EU average of 8% (European Commission, 2014)", to an estimated rate "of 0.6% of daily trips made by bicycle" in 2017, according to the INE (National Statistics Institute) (Felix, 2019, p. 52).

In **Hungary**, civilian actors actively contributed to the development of urban cycling during the past decade with consulting or elaborating a National Cycling

---

26. A traffic jam on bikes', protest by cyclists reclaiming the streets originated in 1992 in San Francisco.

27. Called 'Bicifestazione'.

28. Massa Crítica, also known as Bicicletada.

29. Felix, R. (2019). Barriers and motivators to bicycle in low cycling maturity cities: Lisbon case study. Phd Thesis, INSTITUTO SUPERIOR TÉCNICO, UNIVERSIDADE DE LISBOA, Lisboa.

30. Member of European Cyclist Federation (ECF).

31. Member of European Cyclist Federation (ECF).

32. Member Union Cycliste Internationale (UCI).

Concept for 2014–2020, or organizing various programmes to promote urban cycling. The largest bike promotion civilian actor is the Hungarian Cyclists' Club (Magyar Kerékpáros Klub – HCC) which influences cycling policies and the implementation of new, mainly infrastructural developments in strategic and professional partnerships with cities and companies all over Hungary, but mainly in Budapest. Besides its ongoing promotional campaigns (Bike to work, Bike to school etc.), it also organized the Critical Mass in Hungary from 2004, but the initiative ended in 2013 as, according to the HCC, the critical mass of urban bikers has realized and thus the movement has achieved its goals.

In **Lithuania**, social movements related to fighting for the protection of rights of pedestrians and cyclists are growing in membership numbers and activists are using social media to voice their concerns and anger at city planning that ignores their needs at the expense of motor vehicle road users (EU Country Profile, 2016, p. 5).

If, on the one hand, the structure of the Lithuanian government incorporates neighbourhood governance through local 'elderships' (*Seniunijos*), enabling local community-based organizations to raise their issues and act directly for the promotion of a wide range of social, economic, political and environmental improvements and rights, on the other hand their participation to the development and implementation of the 2015 Sustainable Urban Mobility Plan (SUMP) is still weak.

While cultural patterns, major stakeholders and the field's dynamics are fundamental muscle, the sustainable mobility policy frameworks are a backbone to the shape of bike- and e-scooter-sharing services.

4.4 Policy Framework and Legislation on Sustainable Mobility

Sustainable mobility is a pressing issue, framed under European Union policies, such as the imminent Strategy for Sustainable and Smart Mobility announced in the European Green Deal (European Commission, 2020), therefore it is not surprising that it is present in the legislation of the member states under analysis. Cycling is a significant axis of sustainable mobility, included in "the European agenda for sustainable urban and regional mobility" in line with the desired shift to "sustainable consumer choices and zero and low emission practices" (European Commission, 2020; p. 3). Bike and scooter sharing, in particular, concur to the European Green Deal's goal of creating smart solutions of "mobility as a service" (European Commission, 2019, p. 1). Therefore in order to understand these sharing economy practices, we need to consider the mobility policies and legislation in which they are framed. Despite their common framework, in the four countries, the use of scooters and e-scooters is almost invisible in the policy reports, less regulated than cycling, and existing regulations are less known and enforced.[33] Specifically the organization and control of the parking of these vehicles has been poor, causing several conflicts.

Sustainable mobility was already considered in the late 90's in **Italy**, while in **Hungary**, **Lithuania** and **Portugal** it only began to be taken more seriously in the 2000's. The national frameworks vary given this historical discrepancy that implies a deeper level of institutionalization of the matter in the Italian political and administrative system. The Ministerial Decree of 27 March 1998[34] represents the main regulatory instrument in favour of sustainable mobility in Italy, while in Portugal the National Active Cycling Mobility (ENMAC 2020–2030[35]), was approved by decree only on August 2nd 2019. In Hungary, the central piece of legislation on sustainable mobility is embedded into tourism strategies: the Active Hungary programme (2019) and The National Tourism Development Strategy 2030[36] (2017), both focusing on a wide variety of tourism development measures, and the improvement of bicycle use (either as mobility or a leisure activity), mainly by the development of rural bike lanes.

In **Italy** these interventions concerned, among others, the introduction – at the national level – of eco-incentives, with the aim of supporting the use of low environmental impact vehicles and to discourage the use of the most polluting means of transportation, as well as at local level, financing sustainable mobility projects. In the latter case, measures to limit car use in certain areas ("blue lines" parking lots, Limited Traffic Zones (ZTL), eco-pass, pedestrian areas), on the one hand, and, on the other, sustainable mobility tools have been promoted through the creation of cycle paths, safe home-school "foot-bus" routes, preferential lanes, as well as through the enhancement of public transport and the implementation of mobility management, ride-sharing, bike-sharing and more recently scooter-sharing.

In this regard the situation in **Portugal** is different. Although the Active Mobility Strategy was published in 2019, during 2020 it was not scheduled nor initiated, which has mobilized cycling promotion associations and the Parliament in a recommendation. Only between April and May 2021, regional online sessions were organized for municipalities and other stakeholders to discuss cycling

33. https://www.eltis.org/resources/case-studies/overview-policy-relating-e-scooters-european-countries

34. https://mtu.gov.hu/documents/prod/mtu_strategia_2030-english.pdf

35. Diário da República, 1ª Série, nº 147, Resolução do Conselho de Ministros n.º 131/2019 de 2 de Agosto de 2019. Available at: https://dre.pt/web/guest/pesquisa/-/search/123666113/details/normal?q=mobilidade+ativa

36. Active Hungary Program: https://aktivmagyarorszag.hu/; National Tourism Development Strategy 2030 https://mtu.gov.hu/documents/prod/mtu_strategia_2030-english.pdf

promotion initiatives.[37] Nevertheless, at municipal and intermunicipal levels, a cycling infrastructure has been included in Municipal Master Plans, some deactivated railways have been restored as cycling roads, and paths within natural landscapes have been renewed, complementing the recent efforts of several cities that promote sustainable mobility by introducing car-free days in certain areas, areas of 30 km speed limit and the opening of bike- and scooter-sharing services.

Lithuania's political support on issues around sustainable mobility is considered as leading among European countries, as structural funds are used in a way that helps to support a consistent national approach to EU Sustainable Mobility Plans (SUMP) (Eltis, 2019). The capital Vilnius, and a number of minor locations introduced restrictions to entering the city with large vehicles and bans on coaches without proper emissions certification, and endorsed innovative public transport vehicles to meet EU emission targets in urban centres, by promoting and subsidising the adoption of low emission modes of transportation such as electric vehicles (EVs) and bike-sharing systems. Indeed, Vilnius has set the goal to increase the cycling modal share up to 7.5% (Judu, 2020).

In **Hungary** the landscape of sustainable mobility is ambiguous. During the past decade urban cycling, including bike sharing schemes and more recently e-scooters, gained popularity predominantly in Budapest. Even though urban micro mobility became the topic of heated political debates in the past years, a number of strategies, policies and practices have been introduced by various political actors, both at national and city level. At national level, the commissioner for cycling and active recreation is developing mainly rural bike paths, while subsidising e-bikes to enhance sustainable mobility. In Budapest, a new green pro-biking mayor and administration have been elected in 2019, that have further boosted the infrastructural developments by opening more bike lanes and adopting pro-cycling policies. These core policies have been further enhanced by the prolonged pandemic. The growing demand for sustainable mobility generated public and political debates, as it was seen as "a threat" for the traditionally car-dominated urban mobility regime. Nevertheless, cycling is a traditional means of transportation in rural Hungary, particularly in smaller settlements and in appropriate (mostly flat) topographical conditions, but urban cycling, especially in the past decades in Budapest, is on the rise.

The considered countries have all made major, albeit often initial steps forward in the promotion of sustainable mobility, both at national and city level in the past

37. Instituto da Mobilidade e dos Transportes, IP (2021). Estratégia Nacional para a Mobilidade Ativa Ciclável. Encontros Regionais. Available at: http://www.imt-ip.pt/sites/IMTT/Portugues/Noticias/Paginas/EncontrosRegionaisENMAC-22042021.aspx

decade, and thus arrived at the pandemic in 2020 with already existing strategic frameworks, which could be further improved to address the challenges of interpersonal distancing and mobility safety in pandemic times. The promotion of bicycle and scooter use as part of micro-mobility is a relevant part of all the strategy plans, and often combines infrastructural (the improvement of cycling networks) and fiscal (subsidies for e-bicycles) measures. It is relevant to notice that such fiscal benefits are given to promote ownership and not the sharing of light sustainable vehicles, which displays the legislators disregard for sharing mobility as a service. At national level, fiscal incentives were highlighted more in **Italy** since years, while they were missing from the policy instruments in **Hungary** up until September 2020, when the subsidy for e-bicycle purchase was triggered by the pandemic in order to enhance cycling. On the contrary, Italy initially placed less emphasis on developing bicycle infrastructure, while Hungary focused mostly on the improvement of bike lanes, mainly in rural areas. **Portugal** seems to have applied the most comprehensive approach by covering both fiscal and infrastructural aspects with various measures, although the latter aspect has been less expanded at national level.

There are also some peculiarities in the governmental scale of the conception, public consultation and execution of such regulations, given these countries' diverse administrative structure. In **Italy**, the main sustainable mobility interventions are implemented at the local level, with the possibility at state and regional level to draw a picture of reference, in which to design the legislation of local authorities. In **Portugal**, although the implementation of measures and the specific regulations, such as plans for bicycle networks, is an attribute of the local authorities, the general legislation is defined by the central government, not often with representative participation of all municipalities in its discussion and definition of coordination mechanisms. As an added complication, the larger cities of Lisbon and Porto as well as Italian cities are framed within metropolitan areas that have their own jurisdiction, although they have no power over the decisions of elected municipal assemblies. **Hungary** also applies a mixed approach, but in a different way: cities are usually limited to improving local cycling networks and introducing bike-sharing schemes, but have less influence on the surrounding developments (except for the capital city), while state level agencies are in charge of improving bike lanes among settlements, and thus municipalities are conditioned by state bodies in the development of the sustainable mobility modes outside the municipality. These differences also have impact on the decision-making regarding the activation of European policies, such as the SUMP, which in Portugal were assumed on national scale but, so far, the decision to implement and regulate them was left to the municipalities, while in **Lithuania** these were mandated at national scale.[38]

Despite the general positive attitude, in these four countries, we found more production of general plans or wide strategies that define high goals and significant interventions, at the national scale, than effective legislation that implements, regulates and schedules such changes. This tendency is stronger in **Hungary**, **Lithuania** and **Portugal**, where the more concrete measures that the legislation has defined were fiscal benefits to purchasing bicycles and other sustainable vehicles and the promotion of the construction of cycling networks. The latter are, as mentioned above, under the arm of local authorities within the borders of the municipality and there is limited information on what kind of support is given by the central government. In contrast, in **Italy**, the central government established, in the Law of 19 October 1998, n. 366 "Rules for the financing of cycling mobility", a structural funding for interventions by local authorities and associations of municipalities, both of infrastructural type and aimed at spreading the culture of cycling as an alternative to motorized means of transportation. This way, local authorities have contributed to the construction of the regional cycle network, as part of the Territorial Provincial Coordination Plans (P.T.C.P.) and General Urban Plans (P.U.G.).[39] More recently, various laws on sustainable mobility were adopted and special funding has been dedicated to the 2016, 2017 and 2018 budget laws. Specific attention was paid to cycling, through the allocation of resources to the national system of tourist cycling routes, and with the approval of law no. 2/2018 which promoted the use of the bicycle as a means of transportation. The **Italian** government is showing readiness for progress.

4.5 Impacts of the Pandemic: Favouring or Discouraging Bike/E-Scooter Sharing?

Just as it has affected all societal dimensions, the pandemic has had a wide impact in transportation systems and mobility patterns, highlighting the need for

38. On a similar note, car-free days are celebrated in Lithuanian cities during the European Mobility Week, while in Italy similar events were defined by a national decree published in 2000 by the Environment Minister, opening the first of four successive Ecological Sundays, to take place on the first Sunday of the month. In Portugal car-free zones were defined during parts of the weekend, but only by local authorities' decisions and programs. On the contrary, Hungary applies these measures only occasionally, albeit European Mobility Week is also celebrated.

39. Additional funding for cycling has been provided for by the Law 27 December 2006, n. 296 (so-called 2007 Finance Law) which reserved less than 5 percent of the Fund for sustainable mobility for the interventions referred to in the Law 19 October 1998, n. 366.

connectivity, intermodality and public-private partnerships,[40] for which bike and e-scooter sharing can be pivotal. In Europe, six cycling organizations have prompted the EU Commission to acknowledge that the new COVID-lanes" combined with support for e-bikes can relieve pressures on public transport and stimulate green growth in line with the EU Green Deal".[41] Scientists of several fields also called on governments to promote conditions for safe walking and cycling in order to promote public health.[42] As the ECF (2020) points out, the experience of "new" traffic-free "soundscapes" during the lockdown periods has presented us "a great opportunity to make people aware of the real impact of noise on our lives",[43] in addition to all the other factors, mentioned above, that have proved the need for a change. Indeed, one of the positive outcomes of the pandemic is the resurgence of cycling (ECF, 2020).[44] In fact, the lockdown motivated cycling as it facilitates social distancing and contributes to maintaining health. This increased interest in cycling and forms of locomotion that allow interpersonal distancing can concur with a higher demand for bike and e-scooter sharing services.

So, let us examine the cycling patterns and the policy interventions in the four cases, considering both the historical background and the COVID-19 pandemic, in which we are still immersed. For that purpose, despite their different degrees of restriction, it is relevant to declare which were the confinement periods in the four countries. On the occasion of the first wave of the pandemic, Italy established a confinement period between 9th March and 19th May, while in Portugal it was from 22nd March and ended on 2nd May.[45] In Hungary it lasted from 16th March to 4th of May (in Budapest until 25th May) and in Lithuania, from 16th March 2020, until 31st May 2020.[46]

In **Italy**, bicycle use has been traditionally popular in the flat northern cities (e.g. Parma, Bologna) but is now also becoming more frequent in cities further

40. https://www.weforum.org/agenda/2020/03/a-covid-19-transportation-adapt-lessons-learned/

41. https://cyclingindustries.com/news/details/cycling-is-a-fast-track-from-the-eu-covid-recovery-package-to-the-eu-green-deal

https://ecf.com/news-and-events/news/cycling-against-covid-19

42. https://www.iass-potsdam.de/en/blog/2020/04/covid-19-pandemic-researchers-and-scientists-call-government-enable-safe-walking-and

43. https://ecf.com/news-and-events/news/coronavirus-lockdown-mutes-traffic-noise-and-new-soundscapes-rise

44. https://ecf.com/dashboard [03 October 2020].

45. In two subsequent declarations of Emergency State.

46. https://lrv.lt/en/news/lockdown-restrictions-continue-to-relax.

south. **Rome** has an unexpressed potential to tap into walking and cycling. Unfortunately, in many cases, cycle paths fail to protect cyclists, because they have been sometimes poorly planned, tapering off into the oncoming traffic or dead-ends; cars and motorcycles often fail to respect bike lanes. With the current mayor, in charge since 2016, the cycling policy is changing. The length and quality of the infrastructure has increasing, but cultural barriers remain. Despite good weather, tracks are not used as they could be, given the need to overcome cultural attitude and generalized beliefs, and the fact that safety conditions still represent a big issue for riders.

However, the COVID-19 pandemic crisis determined a series of behaviour changes, a significant growth of cycling and a massive surge in bicycle sales.

Some 540,000 bikes have been sold nationwide since shops across the country reopened in early May 2020, according to sector lobby Ancma, a 60% increase in the first month compared to the same period in 2019. To keep people off public transportation and avoid road congestion, the government has offered to contribute up to 500 euros for city-dwellers who buy traditional or 'pedal-assisted' electric bicycles. The subsidy, which kicked in on May 4 and runs to the end of the year, has accelerated a trend in place even in small centres where it is not available.[47]

During the pandemic, e-scooter services popped up in Rome. However, since their blooming, one has already been interrupted, and one the BSS was also closed. In this period, there have been no public measures regarding the promotion of bike and e-scooter sharing. Nevertheless, it is too early to understand whether this is just a temporary effect or a more radical shift.

In **Portugal**, in addition to the bicycle promotion movements, the increase in urban cycling only became expressive after the beginning of the twenty-first century, when at both local and national scale governments also started to consider this practice, although in most municipalities bike lanes were firstly built in leisure areas, either by the shore or near natural landscapes of interest. However, in May 2020, the Ministry of Environment opened a call for municipalities to propose measures, within this strategy, to promote the use of bicycles as a response to the pandemic situation, maintaining distance and physical activity. The current mayor of **Lisbon**, in charge since 2015, is an advocate of cycling and walking in the city, and has been documented as a bicycle commuter, particularly in COVID times. The pandemic also increased the sales of bicycles, mechanical and electric exponentially,[48] as well as the applications for fiscal benefits on their purchase, under

47. *Source*: https://www.reuters.com/article/us-health-coronavirus-italy-bikes-idUSKBN23U1UF

48. https://www.publico.pt/2020/05/14/economia/noticia/mobilidade-suave-trazida-pandemia-veio-ficar-1916632

central and more recently, local measures.[49] In Lisbon, the public BSS was also a means for the municipality to promote cycling as a strategy to fight the pandemic, firstly by attributing bikes to the delivery services, secondly by providing them for free to health workers and other first line responders, and finally, since July 2020, being free to use by all residents, workers and students.[50]

Hungary is a peculiar case where cycling is ambiguous. On the one hand, cycling is traditionally part of life in rural Hungary, further boosted by recent policies and a governmental commissionaire that are enhancing developments in cycling, both as mobility and a leisure activity at national level. On the other hand, cycling turns to a battlefield when it comes to **Budapest**, even though all political actors are in favour – to various extents – of urban cycling in the capital city, but car-dominated urban mobility is still considered as the default context by conservative politicians and actors. Although the previous right-wing conservative municipality developed the cycling infrastructure, introduced the first bike-sharing scheme in Budapest and also elaborated strategic plans to enhance urban cycling and inter-modality, it also prioritized cars over bikes constantly in urban development programmes during an almost decade-long leadership. On the contrary, the newly elected oppositional municipality (green-left-liberal) holds a coherent vision on sustainable mobility with a strong focus on micro-mobility, including urban cycling and with less dominance of cars. Especially during the COVID-19 pandemic, the new municipality reacted by opening new bike lanes on major car routes and made the use of the municipal bike-sharing system free of charge for all people. At a national level, a subsidy for e-bicycle purchase was launched in September 2020. According to a recent survey, seven out of ten Hungarians are cycling with varying frequency, and their share has increased especially in Budapest after the first wave of the pandemic by mid-2020.[51] The government has also introduced the first ever bicycle subsidy for e-bikes from late 2020, thus further enhancing cycling.

---

https://www.jn.pt/nacional/boom-na-venda-de-bicicletas-gera-escassez-em-todo-o-mundo-12337002.html

49. Central Government Fiscal benefits, firstly introduced in 2019, and improved in 2020: https://www.fundoambiental.pt/avisos-2020/mitigacao-das-alteracoes-climaticas/incentivo-pela-introducao-no-consumo-de-veiculos-de-baixas-emissoes-2020.aspx; Lisbon Municipality launched fiscal benefits in 2020: https://www.lisboa.pt/programa-de-apoio-aquisicao-de-bicicletas;

50. https://www.publico.pt/2020/07/09/local/noticia/pcp-propoe-estrategia-alternativa-apoio-bicicleta-lisboa-1923773; However there is no further information about this experience, other than the news about the decision.

51. *Source*: *The same number of pro-government and opposition cyclists, many cycling because of the pandemic – national research 2020* (Ugyanannyi kormánypárti és ellenzéki kerékpározik, sokan bicikliznek a járvány miatt – országos kutatás 2020), Hungarian Cyclists' Club, 2020, in Hungarian. https://kerekparosklub.hu/kerekparoskutatas_2020

The coronavirus pandemic has also triggered some improvements regarding investment in cycling in **Lithuania**. Over the past four years, Vilnius has renovated over 50 km^2 cycle lanes. Thus, it is expected that by 2023, the reconstruction of the main cycle lanes will be complete (Judu, 2020). Indeed, the city seeks to increase the cycling share up to 7,5%.[52] In July 2020, the **Vilnius** municipality, taking advantage of the opportunities offered by confinement, closed four streets and its sections for pedestrians at the heart of the capital. Moreover, the decision entailed a diverse type of traffic organization that organized the traffic in loops in order to make temporary walking and cycling lanes. Notably, based on data collected by the city, 40% of the traffic was diverted during the peak hours from the centre of Vilnius, and thus, all streets and their sections were free-up for residents and Vilnius's guests. The mayor of Vilnius highlights that the main ideas came from the experiences of many Western European cities, but the confinement due to COVID19 brought the process further and has allowed the city to implement the project a year earlier (Vilnius, 2020). Neither bike- nor e-scooter sharing were expanded or made available for discounted prices, nor any other measure to promote its use was taken.

Common measures introduced during or after the lockdown periods related initially to the infrastructure, by extending the bike lane network partly by converting roads previously used only by cars to mixed modes, where cyclists can ride on separate safe lanes. This was feasible due to the drastically decreased car traffic during the lockdown. Therefore, the challenge is how to keep or maintain these new improvements after the confinement, when car traffic returns to the same level. This issue has generated a heated political and public 'cars vs bicycles' debate in the case of Budapest, where, finally, most newly created bike lanes remained after the lockdown. In Lisbon, reports of situations of conflict between drivers and cyclists have increased considerably, in the last few months, along with some outbreaks against new cycle paths created during the lockdown (MUBi Forum, 2020).[53]

In every index that compares cities' conditions for cycling, **infrastructure** is a major factor taken into consideration,[54] as it is proven that the existence of a structured network of bicycle lanes is behind the increase in cycling as a regular means of transportation (Marques *et al.*, 2015). The length and functionality of cycling infrastructure is a criterion to define a city's cycling maturity, to label them as starter

52. It's Official: Vilnius Introduced Its Plan for Combating After-Effects of the Pandemi. Vilnius. Retrieved from https://vilnius.lt/en/2020/05/05/its-official-vilnius-introduced-its-plan-for-combating-after-effects-of-the-pandemic/

53. Fórum da MUBi – Associação pela Mobilidade Urbana em Bicicleta. Available at (subjected to membership): https://forum.mubi.pt/

54. https://copenhagenizeindex.eu/about/the-index; https://ecf.com/resources/cycling-facts-and-figures/ecf-cycling-barometer

Table 4.7. Comparison of cycling networks and their improvements during the COVID-19 pandemic.[55]

European Cities	City Population	Population Density/km^2	City Area/km^2	Cycling Paths (km)	New Cycling Paths (km)
Budapest	1 696 128	3.366	525.2	325	20
Vilnius	617 000	1.446	401	93	N/A
Rome	2 782 858	2 166	1.285	225	150
Lisbon	508 368	5.081	100.05	105	90.7

Source: Elaborated by the authors.

or champion cities, and to evaluate their cycling potential (e.g., the possibilities of a starter city to increase its cycling modal share) (Felix, 2019; Silva *et al.*, 2018, 2019). The infrastructure can cover four major components: network links, intersections and crossings, parking, and public transport (Dufour, *et al.*, 2010). Here, we focus on the first one. Many European cities have announced some infrastructure changes in the city centre to promote walking and cycling during the COVID-19 period. According to the latest data provided by the ECF (2020), the largest number of new cycling kilometres was announced and implemented in Rome (150), followed by Lisbon (90.7) and Budapest (16.83) (see Table 4.7).

As shown in Table 4.7, **Budapest** (325 km) and **Rome** (225 km) have the largest network of cycle paths in comparison with **Lisbon** (105 km) and **Vilnius** (93 km). However, all the considered municipalities have decided to extend their cycle paths, taking advantage of the pandemic situation to implement this earlier. Lisbon is expected to expand up to 90,7 kilometres of cycle paths,[56] and the main cycle lanes should be completed in Vilnius by 2023 (Judu, 2020). Regardless of city areas, which would justify a wider cycling network in Rome, and based on population density, which is higher in Budapest and Lisbon, it would be expected for these municipalities in particular to densify their cycling networks, in order to reduce motor traffic. According to the ECF (2020), additional cycling funding in euros per person has risen in many European countries during COVID-19 period (Vandy, 2020). For instance, Finland spends the most significant amount of money – 7.76 EUR in comparison with other European countries: Italy – 5.04 EUR, Lithuania – 2.61 EUR. The smallest amount of money was spent in Portugal (0.29 EUR). It is

55. Data source: https://ecf.com/dashboard [03 October 2020].

56. https://jornaleconomico.sapo.pt/en/news/camara-de-lisboa-will-create-95-kilometers-of-bike-paths-by-2021-596789

worth noting that cycling infrastructure is useful also for e-scooters and other light sustainable transportation modes.

The changes in mobility habits during the pandemic are not limited to cycling – the use of e-scooter-sharing, which is still more common than using private e-scooters, is expanding. According to recent data from a private provider, users switch to using e-scooter services from leisure trips to work trips, especially during working days and hours in Vilnius. Thus, the number of users of e-scooter-sharing services since the COVID-19 pandemic has increased.[57] Consistent with this trend, in March 2020, a new international player in this sector arrived in Vilnius, offering 100 e-scooters.[58] In a similar vein, another private company active in Budapest, expanded its services to rural cities.[59] In summary, the increasing number of e-scooter-sharing service companies indicate a trend to shift mobility habits, especially prompted by COVID-19. However, during this period some services in Rome and Lisbon have also been disrupted, some of them were then restarted, other two did not. Therefore, the impacts of the pandemic both in e-scooter and bike-sharing practices can not be interpreted unidimensionally, and any indicator of a trend in these sharing practices must be considered within a highly uncertain and ever rapidly changing scenario.

4.6 Conclusion

In the four European capitals considered in this chapter – Rome, Lisbon, Budapest and Vilnius – bike- and e-scooter-sharing services have recently become a reality, a possibility for mobility as a service, favouring access over property, and active travel over motorized sedentary mobility. The cultural context of mobility is similar among the four cities, with private car use still being the norm. At the policy level, it is clear that mobility as a service is still not a priority in the analysed countries, as bike and e-scooter sharing services are merely local measures, that have not been subsidised or included in any national sustainable mobility plan or promotion initiative; in contrast, fiscal benefits have been given for purchase of bicycles. Therefore, the rise in bike- and e-scooter-sharing has caused intense debate – further

57. blog.Bolt.eu (2020a April 06). Micro-mobility as a safer method of essential urban travel. In *Bolt*. Retrieved from https://blog.bolt.eu/en/micro-mobility-as-a-safer-method-of-essential-urban-travel/ [2020 09 21]

58. https://www.delfi.lt/verslas/transportas/vilniuje-ankstinama-elektriniu-paspirtuku-sezono-pradzia.d?id=83792393 https://www.lrt.lt/naujienos/verslas/4/1201360/kelione-i-darba-paspirtuku-pigiau-nuomotis-ar-tureti-savo; https://www.vz.lt/transportas-logistika/2020/03/17/vilniuje-veikla-pradeda-trecioji-paspirtuku-nuomos-kompanija-scoot911;

59. https://www.themayor.eu/cs/lime-reports-record-figures-in-budapest

fuelled by the poor regulation and fiscalization of these services – about the impacts on public space and traffic. Their potential for breaking cultural ground and changing mobility patterns needs to be further explored. Even though there is evidence of the expansion of their use, in the analysed countries, except for Italy, there are no official statistical data about bike- or e-scooter-sharing.

Within bike-sharing, two of the providers are public – Lisbon and Budapest – and they have been the only cities to use bike-sharing as a measure to promote active mobility for interpersonal distancing and health during the pandemic. Scooter-sharing systems are all run by private enterprises, but despite the lack of information on any discount or promotion measure of these services, during the pandemic there is some evidence of its expansion. The lack of public regulation and support for expanding the inclusivity of light sharing services, even in a pandemic context, calls for action on **enlarging the accessibility of the public BSS and improving public-private partnerships** within mobility systems.[60] At the same time, it leads us to question **how much caring is in these sharing services,** if their private promoters do not take on their social responsibility. The potential for social inclusion of light mobility sharing is strong, as these are transportation means that most people can use and which can be adapted to people with disabilities. However, business models, pricing policies, geographical distribution, the level of technology included and its user-friendly traits must all be taken under consideration.[61] Further research is needed in order to understand, not only the ethos and the concerns for social inclusivity of these sharing mobility services, but also the relations between municipalities, private providers and stakeholders of social movements within the cycling and active mobility scenarios of each city and country.

Acknowledgement

We thank Ph.D. Candidate Jonas Damidavicius, Eimantas Balta, Karolina Tomkeviciute for the helpful material, and suggestions.

References

Bucsky, P. (2020). Modal share changes due to COVID-19: The case of Budapest. Transportation Research Interdisciplinary Perspectives 8 (2020) 100141. Available at: http://dx.doi.org/10.1016/j.trip.2020.100141

60. https://www.weforum.org/agenda/2020/03/a-covid-19-transportation-adapt-lessons-learned/

61. https://ecf.com/news-and-events/news/velo-city-day-2-bike-sharing-tool-social-inclusion

BYPAD (2008). Cycling, the European approach: Total quality management in cycling policy. Results and lessons of the BYPAD-project. Available at: https://ec.europa.eu/energy/intelligent/projects/sites/iee-projects/files/projects/documents/bypad_platform_bypad_cycling_the_european_approach_en.pdf

Caggiani, L., Colovic, A., and Ottomanelli, M. (2020). *An Equality-based Model for Bike-sharing Stations Location in Bicycle-public Transport Multimodal Mobility.* Transportation Research Part A Policy and Practice. https://doi.org/10.1016/j.tra.2020.08.015 https://www.researchgate.net/publication/344207704_An_equality-based_model_for_bike-sharing_stations_location_in_bicycle-public_transport_multimodal_mobility

Christensen, C. (2019). An Inquiry Into the Potential and Sustainability Benefits Derived From the Expansion of Bike Sharing System in Porto, Portugal. Master's thesis, Harvard Extension School. http://nrs.harvard.edu/urn-3:HUL.InstRepos:42004143

CIE, ECF, CONEBI, ECLF, EBMA, IMBA-Europe (2020). Investing in cycling to fast track the EU green recovery and EU Green Deal. Available at: https://cyclingindustries.com/news/details/cycling-is-a-fast-track-from-the-eu-covid-recovery-package-to-the-eu-green-deal

De Groen, W.P., Kilhoffer, Z., Lenaerts, K., and Salez, N. (2017). The Impact of the Platform Economy on Job Creation, in Regulating the Platform Economy: How to Protect Workers While Promoting Innovation, *52*(6), pp. 345–351.

Eltis (2019, May). Practitioner briefing. national Support frameworks for sustainable urban mobility planning. National SUMP Supporting programmes. Eltis. Retrieved from: https://www.eltis.org/sites/default/files/national_support_frameworks_for_sustainable_urban_mobility_planning.pdf [p. 18] and National support frameworks for sustainable urban mobility planning. National SUMP Supporting Programmes (May, 2019).

EPRS, European Parliamentary Research Service (2014). Moving cycling forward. A coordinated approach to cycling for local and regional authorities in the EU. Available online at: www.greencycle.si/wp-content/uploads/2018/01/Moving-cycling-forward.pdf

European Commission (2016). Identifying barriers, hooks and opportunities. Country profile report, Lithuania. (Report completed and finalized in July 2015/Revised in late 2016). Retrieved from https://ec.europa.eu/energy/intelligent/projects/sites/iee-projects/files/projects/documents/evidence_-_10_wp4_lithuania_country_profile_report_final.docx

European Commission (2018, September). Reduction in Road Deaths 2010–2017. European Commission. Retrieved from https://ec.europa.eu/transport/road_safety/sites/roadsafety/files/mapcare_chng2010_2017.pdf

European Commission (2019). Sustainable Mobility. The European Green Deal. *European Commission*. Retrieved from https://ec.europa.eu/commission/presscorner/detail/en/fs_19_6726

European Commission (2020). *Communication from the Commission on the EU Strategy for a Sustainable and Smart Mobility*. Available at: https://ec.europa.eu/info/law/better-regulation/have-your-say/initiatives/12438-Sustainable-and-Smart-Mobility-Strategy

Fan, Y. and Zheng, S. (2020). Dockless bike sharing alleviates road congestion by complementing subway travel: Evidence from Beijing, Cities, Volume 107.

Felix, R. (2019). Barriers and motivators to bicycle in low cycling maturity cities: Lisbon case study. Phd Thesis, INSTITUTO SUPERIOR TÉCNICO, UNIVERSIDADE DE LISBOA, Lisboa.

Fiedler, D., Cáp, M., and Certický, M. (2017). Impact of mobility-on-demand on traffic congestion: Simulation-based study, 2017 IEEE 20th International Conference on Intelligent Transportation Systems (ITSC), Yokohama, pp. 1–6.

Fishman, E., Wasington, S., and Haworth, S. (2015). Bikeshare's impact on active travel: Evidence from the United States, Great Britain, and Australia, Journal of Transport & Health, *2*(2), pp. 135–142.

Frenken, K. and Schor, J. (2017). Putting the sharing economy into perspective. *Environmental Innovation and Societal Transitions*, (23): 3–10.

JCDecaux (2020). Cyclocity Vilnius. In *JCDecaux*. Retrieved from: https://www.jcdecaux.lt/klientams/cyclocity-vilnius

Judu (2020). Vilnius pristato vieningà judėjimo paslaugų vardą – JUDU [Vilnius presents a unified name of movement services – JUDU] *Judu*. Retrieved from: https://beta.vilniustransport.lt/lt/naujienos/vilnius-pristato-vieninga-judejimo-paslaugu-varda-judu

Lopes, D. (2015). An intelligent bike-sharing rebalancing system. https://estudogeral.uc.pt/bitstream/10316/35509/1/An%20Intelligent%20Bike-Sharing%20Rebalancing%20System.pdf

Marques, R. *et al.* (2015). How infrastructure can promote cycling in cities: Lessons from Seville. Research in Transportation Economics. Volume 53, November 2015, Pages 31–44. Available at: https://doi.org/10.1016/j.retrec.2015.10.017

Martin, C.J. (2016). The sharing economy: A pathway to sustainability or a nightmarish form of neoliberal capitalism? *Ecological Economics*, 121: 149–59.

Otero, I., Nieuwenhuijsen, M.J., and Rojas-Rueda, D. (2018). Health impacts of bike sharing systems in Europe. *Environment International, 115*, 387–394. doi: 10.1016/j.envint.2018.04.014

Pucher, J. and Buehler, R. (2008). Making Cycling Irresistible: Lessons from The Netherlands, Denmark and Germany. *Transport Reviews, 28*(4), 495–528. https://doi.org/10.1080/01441640701806612

Pucher, J. and Buehler, R. (2012). Promoting Cycling for Daily Travel: Conclusions and lessons from across the globe. In City Cycling (pp. 347–363). Cambridge, MA: MIT Press.

Qiu, L.Y. and He, L.Y. (2018). Bike Sharing and the Economy, the Environment, and Health-Related Externalities. *Sustainability, 10*, 1145.

Ricci, M. (2015). Bike sharing: A review of evidence on impacts and processes of implementation and operation. *Research in Transportation Business & Management, 15*, 28–38.

Salvia, G., Morello, E., and Arcidiacono, A. (2019). Sharing Cities Shaping Cities. *Urban Science, 3*(23), 1–5. doi: 10.3390/urbansci3010023; www.mdpi.com/journal/urbansci

Schor, J. (2016). Debating the sharing economy. *Journal of Self-Government Management Economy, 4*, 7–22.

Shaheen, S.A., Guzman, S., and Zhang, H. (2010). Bikesharing in Europe, the Americas, and Asia: Past, Present, and Future, Transportation Research Record. *Journal of the Transportation Research Board, 2143*, 159–167.

Silva, C. (2019). New applied research project on Starter Cycling Cities in Portugal. https://www.eltis.org/discover/news/new-applied-research-project-starter-cycling-cities-portugal

Silva, C. *et al.* (2018). O Potencial Ciclável Em Cidades Principiantes. 8º Congresso Luso-brasileiro Para O Planeamento Urbano, Regional, Integrado E Sustentável (Pluris 2018) Cidades E Territórios – Desenvolvimento, Atratividade E Novos Desafios, Coimbra – Portugal, 24, 25 E 26 De Outubro De 2018.

Teixeira J.F. and Lopes, M. (2020). The link between bike sharing and subway use during the COVID-19 pandemic: The case-study of New York's Citi Bike, Transportation Research Interdisciplinary Perspectives, Volume 6.

Teixeira, J.F, Silva, C., and Moura e Sá, F. (2020). Empirical evidence on the impacts of bikesharing: a literature review. *Transport Reviews*, doi: 10.1080/01441647.2020.1841328

Vandy, K. (2020). Coronavirus: How pandemic sparked European cycling revolution. In *BBC*. Retrieved from: https://www.bbc.com/news/world-europe-54353914

Vilnius (2020). Vilnius Municipality. *Vilnius*. Retrieved from https://vilnius.lt/

Woodcock, J., Tainio, M., Cheshire, J., O'Brien, O., and Goodman, A. (2014). Health effects of the London bicycle sharing system: health impact modelling study. *BMJ, 348*:425. https://www.bmj.com/content/348/bmj.g425

Zhang, Y. and Mi, Z. (2018). Environmental benefits of bike sharing: A big data-based analysis. *Applied Energy, 220*, 296–301.

DOI: 10.1561/9781680838411.ch5

Chapter 5

Why Are We Still Using Facebook? The Platform Paradox in Collaborative Community Initiatives

By Chiara Rossitto, Airi Lampinen, Ann Light, Vera Diogo, Aniko Bernat and Penny Travlou

5.1 Introduction

Over the past fourteen years, Facebook has grown from a social network site for elite college students in the United States, to its current position as a dominant global hub for online sociality; a platform for an ever-increasing range of daily activities, and a staple in efforts to arrange and coordinate local civic initiatives (Bennett and Segerberg, 2011; Bennett and Segerberg, 2012; Crivellaro *et al.*, 2014; Berns *et al.*, 2019). Civic and people-led initiatives sometimes express opposition to the values that Facebook is perceived to represent, or explicitly seek out alternatives to the platform. Yet, the lack of dedicated budgets, together with initiatives' concern to gain visibility, often result in the adoption of Facebook – and arguably other social media platforms – to mobilize resources, plan collective actions and coordinate them, or to manage both internal and external communication (Tayebi, 2013; Costanza-Chock, 2020).

Facebook arouses mixed emotions. The company is known to engage in unscrupulous advertising practices and shows both a reluctance to distance itself from distasteful political causes (Gillespie, 2018) and ambivalence about data use; in common with other social media, it is criticized for destroying democracies and communities (e.g. Zuboff, 2019). However, alongside the squeamishness about using a global neoliberal platform that resists all attempts to rein in its uglier aspects, Facebook has also become a major platform for community solidarity and care work across Europe. While it might be ironic that one of the prime agents credited with undermining democracy is also a major provider of tools for accomplishing resilience and support, this is just one of many tensions in how Facebook is received, which belies its overall contribution.

In this chapter, we examine how five different, community, bottom-up initiatives across Europe use Facebook, the reasons for choosing this platform, and what kind of challenges arise from adopting it. Rejecting polarizing narratives – of social media as the sole instrument of social change, on the one hand, and pessimistic views of surveillance and mistrust, on the other – we draw attention to how these narratives do, or do not, play into use in practice. In doing so, we focus on the tensions that stem from using Facebook as a platform for community initiatives, not at the theoretical level of media studies, but by analysing the situated use of the platform on the ground.

We do this through five empirical case studies: a network for self-organizing co-working days in homes in Sweden *(Hoffice)*, migrant solidarity grassroots groups *(Migration Aid)* in Hungary, short-let accommodation for foreign volunteers in Greece *(Athens Volunteers' Accommodation and Ride-sharing)*, a neighborhood-centered community group in England *(Egg Club)*, and a cycling promotion group in Portugal *(Cicloficina do Porto)*. With a strong emphasis on practices of care among those involved – both for oneself and others – the cases provide alternative visions to what have become mainstream examples of platforms, and platform use, in the collaborative economy. Rather than adopting bespoke digital technologies to advance their causes, all five cases rely primarily on Facebook.

Through a meta-analysis of previous, qualitative investigations that were independently carried out, we explore empirical examples that illustrate what a European caring economy might (and already does) look like, particularly in terms of how community initiatives self-organize using a global platform as a central, digital infrastructure. Our analysis unpacks what we regard as the *platform paradox*. While all five initiatives rely on Facebook to organize and infrastructure their actions, the values of these networks are hardly aligned with a global corporation like Facebook.

As the chapter illustrates, all the cases are participatory in nature and feature an ethos of solidarity, rather than focusing on enabling exchange or matching people together for the purposes of transacting. This, we argue, leads to a set of tensions that are not easily resolved through design. Illustrating both the ups and downs resulting from the initiatives' choice to use Facebook to support their activities, we explore the nature of this compromise that can be seen as 'dining with the devil'.

5.2 Methodology and Case Studies

While different in nature and scope, the five cases are all instances of local initiatives that use a global platform like Facebook to promote community practices that embody caring. Despite their different domains and approaches, each initiative aims at fostering the non-monetary co-creation of more meaningful and sustainable lives, from collectively managing work arrangements to responding to refugee crises, from nurturing small scale business to promoting bike use in urban areas.

Below, we briefly introduce the five initiatives that the authors of this chapter have previously engaged with and developed as individual case studies (Bryman, 2012). The cases have all been qualitative in their methodological approaches and featured, in varying constellations, participant observations both on- and off-line, interviews, focus groups and co-design workshops. We invite readers of this chapter to turn to our previous publications (Rossitto and Lampinen, 2018; Lampinen *et al.*, 2019; Diogo and Rosa, 2018; Light and Briggs, 2017) for the methodological details of each case study.

Specifically to address the themes of this chapter, we held online meetings from January to June 2020 to inductively analyze each of the collaborative initiatives' use of Facebook and the specific relationships to the social networking site. This has entailed, for instance, comparing examples of situated practices, along with the specific pros and cons that each of the collaborative initiatives experiences in using the platform. Driven by reflections on the many commonalities, the final meta-analysis outlines the reasons why these initiatives still use Facebook, despite the concerns they have with the proprietary digital platform.

5.2.1 *Hoffice*: Self-organizing Co-working Events in Stockholm, Sweden

A merger between the words home and office, Hoffice is a self-organizing, volunteer-driven network that brings together people who wish to co-create temporary workplaces (Rossitto and Lampinen, 2018; Lampinen *et al.*, 2019). The network was founded in Stockholm, Sweden, at the beginning of 2014, with the main intention to facilitate the collective use of private homes as shared offices. Core to the Hoffice network is the idea that resources, like private homes, can be collectively used as coworking spaces open to friends, acquaintances, or even strangers. The Hoffice concept entails a co-working methodology, and a set of practices inherent in opening up one's home as a temporary, shared workplace. The Hoffice network in Stockholm uses a Facebook group[1] as a hub for advertising and organizing coworking events. This is done by using the Facebook event function, which any member of the group is allowed to create. Here information is usually provided about the date and address, the number of attendees allowed and any infrastructure available at the hosting home – from wi-fi to lunch facilities. Hoffice is an interesting example of a local, collaborative economy initiative that aspires to co-create an alternative social model – encouraging trust, self-actualization, and openness – by relying on a commonly available digital platform to coordinate efforts to manage flexible work arrangements. As we write this chapter, the current Facebook group is still in use, although the intensity of activities varies over time.

5.2.2 *Migration Aid*: Refugee Solidarity Grassroots Groups in Hungary

Refugee/migrant solidarity grassroots groups are volunteer-run collectives that emerged in some major Hungarian cities, during the summer of 2015, as a response to the refugee crises (Bernát *et al.*, 2016). While official care providers, such as public institutions, were reluctant to respond to the unmet needs of migrants and refugees, local civilians self-organized to express solidarity, to provide aid and immediate relief and to contest the government anti-immigration policy. The aid groups all used Facebook as the central platform for sharing information, developing contacts, organizing activities, collecting and distributing donations during the entire crisis.

Migration Aid (MA), the largest and most influential of the refugee solidarity groups, organized and promoted its work via a hierarchical structure with a core open Facebook page[2] (44k followers) for raising awareness about the migration crisis and activities contending with it. Other Facebook groups were also created; a closed group,[3] with a national scope, targeting potential volunteers and providing them with specific information on how to become involved; a number of local closed groups, restricted to active volunteers, for the daily operation of aid activities.

1. https://www.facebook.com/groups/240395772788705

2. Migration Aid open page: https://www.facebook.com/migrationaid.org

3. Migration Aid national closed group: https://www.facebook.com/groups/1602563053360018

Besides some smaller groups in other cities, in Budapest three groups operated their own Facebook groups linked to the travel hubs of the refugees: Keleti,[4] MA Nyugati,[5] MA Déli/Déli Csillagszálló.[6] While the level of activities within each group has significantly decreased, due to the lower flux of migrants, the core Facebook page and the group pages have remained active and concerned with migration and politics on a broader scale.

5.2.3 *Athens Volunteers Accommodation & Ride-sharing*: Short-let Accommodation in Athens, Greece

As a response to the refugee crisis going on in Greece since 2015, a solidarity network of non-profit organisations and self-organised groups has emerged in Athens to support refugees and migrants, and the large number of volunteers relocating to Athens to manage the emergency (see Travlou, 2020). In response to the lack of appropriate accommodation for volunteers, a number of groups appeared on Facebook to provide necessary information and guidance to find accommodations and provide flat/rooms-to-let listings. These groups can be seen as alternatives to local newspapers and websites, oftentimes written in Greek only and generally used to find long-term accommodation, and platforms – e.g., Airbnb or Booking.com – more commonly associated with tourism and short-term rental. The case study focuses on a Facebook group called "Athens Volunteers' Accommodation & Ride-sharing[7]" that was created in April 2016 to coordinate accommodation and transport for, mostly, foreign volunteers in Athens. The group is still being used by landlords and tenants, and by foreign volunteers looking for accommodation, or for people to share a flat with. The group page administrators are foreign volunteers who are familiar with the rental market in the city.

5.2.4 *The Egg Club*: Collaborative Buying in Brighton, UK

The Egg Club grew out of an initiative (The Roundhill Community Noticeboard[8]) that used the Facebook "community" function for groups to set up pages and which became an active hyperlocal site for part of Brighton in southern England.

Roundhill is a compact area bounded by major roads. It has approximately 700 properties, a society that conducts matters of common interest like planning

4. Migration Aid Keleti: https://www.facebook.com/groups/835984696454826
5. MA Nyugati: https://www.facebook.com/groups/490046001145489
6. MA Déli/Déli Csillagszálló https://www.facebook.com/groups/1612866438993255
7. https://www.facebook.com/groups/236125173408995
8. https://www.facebook.com/groups/1278271078868009

permission disputes, and a public noticeboard that is barely used. About three years ago, one of the residents added a community group on Facebook, which now has more than 450 members. This has led to a number of initiatives that supply the neighbourhood and have a material existence independent of the community group page but which could not have existed without it – for instance, a micro-baker and a sustainability-focused dress agency have been launched. In this chapter, we detail how an egg run to fetch organic eggs from a local farm, developed into the Egg Club in response to conversations on the community site and how it became tied in with general area improvement.

5.2.5 *Promoting Bicycle Self-repair*: Cicloficina do Porto, Portugal

Cicloficina do Porto[9] is a bike kitchen, that is an informal cycling group, related to DIY cultures, that promotes the development of skills to self-repair cycles and encourages the use of this vehicle in urban contexts. Concretely, such groups gather together to repair cycles, share tools, space and knowledge. *Cicloficina do Porto* was founded in 2006, and bike repair activities take place in several parts of the city – for free or at a symbolic price – in properties occupied, or owned, by other collectives. This organization has been shaped by a fluctuating group of volunteers, with some of the founding members still active. Their organization's first website was created in 2008, but in 2012 it changed to a blog and then to a Facebook page, which is now the central online platform. The Facebook page is used for public communication: for instance, to disseminate Cicloficina's activities, share photos of the repairs that are done on different open days, provide some information about bicycle mechanics, and also to promote activist campaigns and related cycling journeys. Facebook Messenger is also used for queries about bike repairs; Cicloficina's typical answer is an invitation for people to attend in person: "Come by on the open day and we'll see". For interaction among volunteers other forms of communication are used, such as a mailing list and telephone contacts.

5.3 Findings

Facebook's size and popularity, and the underlying logic of network effects that is central to platform capitalism (Srnicek, 2017) makes it a pragmatic choice for community initiatives. As noted elsewhere (Costanza-Chock, 2020), although an entire ecosystem of dedicated activist Constituent Relationship Management

9. https://www.facebook.com/CicloficinaDoPorto

systems (CRMs) exists, such platforms remain niche services. Despite having built-in features, interface elements, and capabilities that match the core processes that community organizers and political campaigners seek, adopting them tends to cost money, while using them oftentimes requires a significant investment of time and energy to learn their functions and convince people to use unfamiliar systems. Instead, like in the cases we focus on in this chapter, it is common to turn to the most popular corporate social networking site, which many people are already using, and find ways to work within its affordances and limitations.

While Facebook provides support for the initial organizing – and arguably success – of local sharing initiatives, relying on it can require compromises and result in practical struggles. All of the five initiatives considered in this chapter use Facebook to organize and infrastructure their actions. Notably, however, the values of these collectives – each aiming at fostering reciprocally caring relationships in their own way – are not aligned with a global corporation like Facebook. As previously noted, we refer to this overarching tension as the *Platform Paradox,* which we articulate with the help of our findings. In the analysis that follows, we first discuss the various reasons the different initiatives have for using Facebook to support their activities. We then turn to consider both the design and the broader political tensions that stem from these initiatives' use of the platform.

5.3.1 Reasons for Using Facebook

Across our five cases, Facebook comes up as a medium that allows community initiatives to come together, attract new participants, experiment with ideas for community practices, mobilize widespread collaboration, organize events and collective action. This utility matters for different reasons in each of our cases, and different aspects of it can be manifested in different stages of each initiative's evolution. In the following sections, we discuss gaining visibility and organizing action, even amid emergencies, as two core concerns that make Facebook a pragmatic resource to turn to.

5.3.1.1 Gaining visibility

The five collectives discussed can be regarded as *"go online to act offline"* communities: the respective Facebook pages and groups are mainly hubs for announcing events, possibly making people interested in participating or even committed to attend. The low entry threshold to using Facebook in general, and the open groups and pages more specifically, makes it easy for grassroots initiatives to gain visibility and easily attract audiences/participants: having access to a Facebook account might be the only requirement for participating in such initiatives, and even this aspect becomes less important if people get involved in offline activities first. The different cases illustrate various ways in which visibility comes to matter and is enabled by the platform.

In the **Hoffice** case, Facebook played an important role in the initial success of the network by allowing early Hoffice enthusiasts to coordinate their activities and promote the network – and the Hoffice concept – broadly and across various groups of people. Starting on Facebook was a pragmatic success for Hoffice in that it allowed the network to get started quickly and without any specific funds to support it. Despite the shared sentiment that the platform constitutes a challenge to the continuance of the community's activities (this point is further addressed in the section "Design tensions"), key participants are hesitant to migrate to another platform to organize co-working days. As they see it, the wide-spread use of the platform makes it easy for newcomers to join the network, organize and participate in events. As expressed in the words of an active participant: *"Facebook is where everyone is"*.

In the case of the **Egg Club**, the initiative owes its existence to Facebook, if obliquely. The platform for the Egg Club is the community noticeboard. The noticeboard enables encounters between attention-paying, socio-ecologically-minded, hyper-local members who use the member-only page for sharing news, organising local support, seeking lost cats, redistributing unwanted household goods and so on. The Egg Club was born when someone shared a Facebook post that brought attention to the fact that a local, organic egg farm had lost its contract with a major supermarket and faced closure and the rehoming of hundreds of chickens (June 2019[10]). Jane,[11] a retired social worker who had already established, and administered, the group as the community noticeboard, then suggested that she might offer a regular fortnightly service bringing eggs to Roundhill and making them available for collection. She tested this out, the initiative was popular and, at time of writing, she has made it work for a year and a half, including right through the British lockdown in summer 2020. She brings about 400 eggs into Roundhill each time and people collect their order from the side of her house during the same afternoon (Figure 5.1a). She provides fresh eggs at near bulk price, taking a small contribution for petrol and another levy for community improvements (such as the flowers to be seen in the planter in Figure 5.1b). In this way, not only does the neighbourhood support enlightened farming practices (the farm has been able to continue, based on this and other local supply lines), but participants benefit from the provision of cheaper healthy food, while the small fee is put into other forms of local wellbeing.

Despite its birth on Facebook, much of the day-to-day running is now organized through email. Jane has a list of regulars and checks with a message every fortnight

10. https://www.facebook.com/themacsfarmsussex/

11. Jane has given consent to use her real name.

Figure 5.1. a/b: The eggs for collection, and flowers in a communal planter in Roundhill, bought with money from the Egg Club.

if anyone wants to change their order, announcing which hours she will be available on that occasion for collection. Very few people have been found unreliable, but the club does not seek new participants: Jane believes her initiative is ill-equipped for unlimited growth (Rossitto *et al.*, 2020) and that it could not be scaled much more without causing her additional inconvenience.

The **Hungarian refugee solidarity** groups show how using Facebook can make visible alternatives to mainstream, governmental politics. These collectives were born and operated in a strong political headwind determined by comprehensive anti-immigration state policy. While using Facebook was instrumental to infrastructure these groups' activities, it also contributed to making visible a number of solidarity initiatives, driven by humanitarian concerns, that were sometimes considered at the very edge of legality by formal political groups. As noted elsewhere (Bernát, 2019; Berns *et al.*, 2019), during the crisis, the Hungarian government amended laws to discourage (and even criminalise) civilian support of asylum seekers. Activities concerned with providing information for onward travels or refugee rights were regarded as means to traffick and support illegal flows of people. In this hostile context, Facebook was perceived as a platform independent of national, structural politics, which enabled the organization of activities contesting the governmental approach to the refugee crisis.

5.3.1.2 Organizing action

For all five initiatives, online activity organized via Facebook has seemed the most effective way not only to promote, but also to plan and coordinate offline activities. And, while dealing with unexpected circumstances is not the only reason to adopt an existing digital platform, both the Hungarian and Greek cases illustrate that using Facebook was central to generating and infrastructuring quick responses to socio-political emergencies.

For the refugee solidarity groups in **Hungary**, the Facebook groups served as centers for the volunteers working with refugees offline to organize core aid activities, such as the collection and distribution of in-kind donation (e.g. food, clothes, medicines), or providing refugees with legal and practical information. Closed groups for the volunteers were established for the management and the most active members to organize daily operations at a street level and to coordinate with other grassroots all over the country. Additionally, open groups for wider activist communities were established to allocate donations and enroll volunteers.

As noted above, for the **Athens Volunteers Accommodation & Ride-sharing group** Facebook provided a space to share information about available flats, property rental regulations, advice on how to avoid bogus landlords, tips on where to purchase cheap furniture and offers for transport/ride-sharing. Transactions between prospective tenants and landlords were generally easy, straightforward and fast. In most cases, there were posts about available accommodation with photos of rooms and/or the whole flat, the type of tenancy (i.e. single or multiple occupancy), and information about amenities, neighbourhoods, and vacancy duration. Prospective tenants could respond directly via comments and then arrange to discuss the vacancy with property owners (or subletters) via Messenger. There were also posts from volunteers looking for accommodation in Athens prior to their arrival, describing their budget, preferred areas and duration of their stay in the city.

Facebook facilitates the organization of **Hoffice** days for any member who has joined the group. The platform makes it easy to reach potential members and provides the basic tools for coordinating the organization of co-working days. In particular, members use Facebook events to issue invites to Hoffice days, coordinate participation, and communicate details necessary for coming together face-to-face. However, the continued use of Facebook for the same purposes, as the network has grown rapidly, has caused challenges (explored below as design tensions). Moreover, the use of the platform raises issues of peripheral versus more active forms of participation. While a click is all it takes to join the online network, more "active" participation, such as organizing or attending physical events, requires a level of connection to the community that membership in the initiative on the platform alone does not provide.

Both in the Egg Club and Cicloficina cases, Facebook was not the only platform adopted and its use is combined with other digital channels. In the case of the **Egg Club**, the Facebook community pages provided the means to identify local needs and organize the initial response to the chicken farm crisis. However, after the community was formed, action was often coordinated through other means.

In contrast with the previous initiatives, in **Cicloficina**, Facebook appears around six years after the group's constitution. Since then it has became central to Cicloficina for public communication, but the collective is not fully reliant on it, and other channels – e.g. mailing lists, Messenger and telephone contacts – are used by active participants for internal communication, to plan and schedule meetings and events, or discuss and make decisions about materials and tools needed. The later development of the Facebook page indicates that this group had a previous existing practice of organization and collaboration.

5.3.2 Design Tensions

Despite the many reasons for turning to Facebook, there are also a number of pragmatic design tensions that complicate the initiatives' relationships to it. A key issue is that, in relying on Facebook, community organizers experience a practical powerlessness: they cannot adapt the platform to fit the needs of their initiative and they have no guarantees that a feature they depend upon will remain available and continue to function in the same way. Rather, they are left to adapt to unexpected changes that may either help or hinder their activities. This makes sustained community governance and everyday organizing more difficult. In this section, we consider three common design tensions: difficulties in managing growth, clashes between platform culture and community norms, and challenges in broadcasting that stem from algorithmic filtering.

5.3.2.1 Managing growth

As noted above, one of the aspects that makes Facebook attractive for local initiatives, at least initially, is that the platform offers an opportunity to foster broad participation and broadcast messages to a quickly growing network. When community organizers are just getting started, being able to get the word out and bring people together quickly is valuable. However, a different effort may be needed when the goal is to establish deeper community ties and develop continuous and systematic collective action, not only online but also offline. The story of Hoffice illustrates this point well. While starting on Facebook was a pragmatic success for the network, relying on the platform has had a role in stalling the activities of the community in Stockholm. While rapid growth in the number of participants would constitute a success for many initiatives on Facebook, and can be thought to align well with Facebook's incentives, for an initiative like Hoffice it has significant downsides. As the network's main purpose is to facilitate in-person events, a rapid influx of new members makes it hard to coordinate participation and establish sufficient continuity for network members to get to know, and trust one another, to the degree that they desire. As one of the founders of the network put it in a meeting: *"Facebook is killing Hoffice."* This statement sums up how, over time, the tensions between what the community aims for and what the platform is geared to facilitate have become more apparent.

Similarly, the migrant solidarity grassroots groups in Hungary experienced a discrepancy between the growth of online membership and the offline activities of the community: while Facebook features were good enough to mobilize resources, including volunteers, they did not support the articulation work needed for the workings of these groups. This was an unexpected challenge for both the leaders and ordinary members of these groups as it made the coordination of offline activities more difficult, especially when it came to allocating tasks. In other words, the low entry threshold to joining the groups often resulted in an oversupply of volunteers.

As another example of challenges in managing growth, the Athens' Volunteers Accommodation & Ride-sharing group struggled with unwarranted members' expectations about what the group could do and what Facebook is suited to support. In particular, there were hopes that the group could facilitate renting properties by supporting negotiations between owners and prospective tenants. The problems stemmed from using a Facebook group to organize a two-sided marketplace, bringing into contact landlords and tenants who have not met before and who are left to deal with possible problems on their own. The Facebook group administrator(s) neither had the means to facilitate these interactions in a structured way, nor did they have the resources to help resolve eventual conflicts. There was little they could do to vet the participants to ensure good intentions. As a work-around, the administrators drafted a series of guidelines for prospective tenants to make them aware of possible pitfalls and disagreements with property owners: "*be aware that there can be unscrupulous people out there ready to take advantage, therefore please make sure to stay safe and check things out before agreeing to rent/stay somewhere. if something doesn't feel right, then it probably isn't.*" Moreover, as rent price was often not provided, the administrators started to request that the monthly rent price should be included on the post, otherwise they would be deleted.

5.3.2.2 Reconciling platform culture and community norms

There is a tension between Facebook platform culture and many of the community norms in play with these five initiatives. This can be seen in how Facebook templates participation in ways that often does not match the culture and needs of local initiatives. Again, it is worth emphasizing that community initiatives have no power over how the platform defines page and group outlines and structure contents over time. For instance Facebook's style of RSVP'ing to event invitations, can clash with community norms. While "Maybe attending" and agreeing to attend but failing to show up are relatively regular instances on the platform, they may hurt a local initiative's efforts. The mismatch of online and offline community norms leads to tensions in some cases. In the Hungarian solidarity groups, participants sometimes committed – for instance, volunteering for a daily shift – by answering to Facebook posts, but did not follow up. This posed challenges to the running of activities on the ground, or was a source of frustration for those who would have liked to apply for the shift, but could not as the list was already full.

For Hoffice, no-shows led both to frustration for those organizing co-working days, as they were left with a smaller than expected number of guests (which could undermine the event), and to disappointments for those who would have liked to participate but could not find a free slot. Overall, these types of trouble can push newcomers and active members alike to disengage from the network, deeming efforts to participate as "not worth it" and seeking other means to find the sociality that made Hoffice attractive in the first place – e.g. by advertising events to close contacts rather than making them public on Facebook. In the case of Cicloficina, situations when people RSVP that they will attend an event, but did not show up was never considered as problematic: such incidents do not disrupt activities because the members who did show up always had something to fix on their own bikes and also enjoyed each other's company.

As another example of how platform culture can expose participants to adverse outcomes is that while Facebook's design is geared to encourage information sharing, this can be very risky. For organizers of the migrant solidarity grassroots groups in Hungary, it was clear that Facebook should not be used for sharing sensitive data, including personal and contact information of either the fleeing asylum seekers and helping volunteers, or monetary information regarding either the collected or spent donations. They considered this to be the case even in closed groups. This lack of a trusted, private communication channel made effective operation more difficult and there was constantly the risk that someone might share sensitive information without having thought through the implications of posting it on Facebook. It also triggered the use of alternative communication tools (secured channels or email, chat, phone) in order to manage cases that involved any sensitive information.

5.3.2.3 Maintaining awareness amidst algorithmic filtering

A third concern is that maintaining awareness of network activities can become difficult when done via a platform that filters contents algorithmically as a strategy to personalize what is shown to individual users, cater for their interests and maximize their engagement. Depending on how Facebook is configured to work at any particular point in time, both community members and other audiences may be more or less likely to see information about upcoming events when they browse the platform. The algorithmic filtering to produce the personalized news feed that Facebook is famous for makes it hard to know who sees what and when, thus adding a level of troublesome ambiguity to all community communications. This tension came up particularly in the case of Ciclofícina. Here Facebook's structure for broadcasting – i.e. sharing photos, information or feedback on events – was constraining, particularly in giving visibility to present-moment posts. It worked better for more stable, general information and documentation about the initiative, for which the fixed layout was a more comfortable match. The practical troubles with broadcasting are part of the reason why the Ciclofícina organizers have kept using a blog, which is preferred among the more active members, given the higher level of autonomy that it allows.

Clearly, winding through these design tensions are also the politics of control. If the mechanisms were set up and maintained at local level, as Light and Miskelly argue in their analysis of supporting neighbourhood caring and sharing (2019), it would be possible for organisers to configure these problems away. Militating against this is the challenge of staying technically competent and keeping the site functional, overheads that have caused small organisations to reduce the degree to which their networks and services rely on digital components (see Light, 2019).

5.3.3 Political Tensions

Beyond the design tensions covered above, there are further political tensions that may be less discernible in the day-to-day functioning of local initiatives, but that are more fundamentally ironic. Drawing on different groups' principles, values, ideological orientations, and their experiences of using Facebook, we now reflect on political and ethical reasons for resistance to and/or discomfort with using the platform. We argue that Facebook has become a form of institutionalization for collaborative initiatives channeling, and sometimes fueling, the conflictuality that "is no longer contained" by the institutions "or happens in areas that can not be easily institutionalized" (Fernandes, 1993, p. 796). Political tensions come to light in these groups' dynamics while they use Facebook and in how they use it; such conflictualities are sometimes generated by pre-existing local tensions that are reflected in online interactions, and other times motivated by the collectives' vision of the platform and its policies.

Talking about trust and the collaborative economy, Light (2019) augments Hawlitschek *et al.* (2016) proposal of the 3 Ps (peer, product and platform) as sites of trust, with recognition that you can trust all three of these (i.e. be sure that the service will operate as intended), while still distrusting the company that provides the mechanism. The political tensions we discuss here reflect this distinction. Complementing the design tensions addressed above, they do not refer to design issues as such, but rather to people's perceptions of employing the *product* of a company that is not generally regarded as a good digital citizen – e.g. enabling fake news, polarising opinions. The tensions discussed manifest in the diffuse manner of late capitalism's conflictuality, not sufficient to force a change in the community's practices – i.e. to bring about the decision to stop using Facebook – yet nagging at users and influencing some of the relations that extend from it. As the cases chosen outline, this often happens once this platform is institutionalized as a means to an end, therefore it does not define the collective identities that use it. Given the tridimensional positioning/placement of trust – peer, product and platform (Hawlitschek *et al.*, 2016) – to use Facebook does not require trust in the company as an entity or in the way it runs its services (Light, 2019). As highlighted by the previous analysis, trust, and thus use, can simply be based on its utility to enlarge, maintain a network or make it visible to wide publics. These collectives use Facebook once they know their peers, their users, their members and followers are using it, and *trust them* to keep the initiatives alive.

In the Egg Club case, this is simply captured in Jane's comment that she does not like what Facebook gets up to, but it is convenient for a community group (and for supporting the egg run) because it is what a lot of people are already using. She is well aware of the political tensions and also that some of the neighbourhood will not use it because of what it represents. Nevertheless, Jane talks about what it enables: *"I want to live in a nice community. I like walking down a street and knowing that so and so lives here and so and so lives there. I like getting to know people. That can lead on to other things ... This strengthens community."*

Relatedly, in the Cicloficina case, a broad number of participants see Facebook as a company that has concentrated too much wealth and power within the social media sector. As a member put it, the fact that Facebook bought Whatsapp and Instagram further limits the possibilities of using social media with diverse ownership. The group is also critical of Facebook's lack of protection of users' rights in favour of its economic interests; here Facebook is considered the *"major instrument of personal data manipulation driven by financial goals"*. Given the group's anti-capitalist orientation, the platform's lack of concern for policies to handle personal data creates individual and collective tension. This conflictuality reflects mistrust in the economic system of which Facebook has become a telling example (Zuboff, 2019). The uneasiness that Facebook contradicts the group's core values has been discussed within the community. Suggestions were made to create profiles in other social media accounts (i.e. Twitter and Instagram), to be less dependent on Facebook. Nevertheless, while open source alternatives to Gmail and Doodle have been adopted, Facebook has been more difficult to replace, especially given the challenge of reaching out to wide audiences. Even if alternatives to the platform were to be found, it would still require effort to advertise the change in the technological setup and migrate to another platform. In sum, despite the group's rejection of the political and economic values materialised in the platform, Facebook, as a tool, remains in use because it requires "minimum effort" to be managed while easily providing visibility to the collective.

5.4 Discussion and Conclusion

Facebook has worked to support communities over the years, introducing features to support both closed and open groups and community action. Meanwhile, the chapter has illustrated that Facebook's pervasiveness and broad use often makes it a convenient choice for the organizing of volunteer-run collaborative initiatives. Despite issues such as the lack of control over core features of the platform, and the tension that might emerge from mistrusting the corporation behind the platform, the cases illustrate how performing community through Facebook is still attractive because of other qualities. *This is the platform paradox,* with both core volunteers and more peripheral members being aware of trade-offs between the utility of the platform and the compromises stemming from using it.

Given the inseparability of social and economic activities in community initiatives, it is perhaps not surprising that a platform initially designed for social networking gets taken up as a medium for grassroots/community-centred initiatives. Demonstrating the persistence in *"dining with the devil",* the findings have illustrated the subtleties of how this occurs from addressing social crises to sourcing food or co-organizing supportive work contexts. All these examples manifest care for community, for the environment and local neighborhoods and for more vulnerable groups. In all of the initiatives discussed, informal structures of support and transfer of resources unfold offline, through face-to-face interactions rather than being transactions of the visible parts of the platform-driven collaborative economy.

Despite the very specific design features of a social networking site, the possibility to tailor the platform to different practices, contexts and situations, makes its use inherently open. While, as noted, customization and control over central design features are rather limited – and a source of problems for collectives like Hoffice, for instance – the platform lends itself to diverse contexts and the flexibility to infrastructure action. The openness and lack of control over what Facebook can be used for (paradoxically, including the spreading of fake news or other anti-democratic values) is a reason for its success. Facebook, both as a platform and a corporation, does not seek to moderate local tensions. In fact, its 'neutrality' as a medium for, rather than a publisher of, information, means that it may look virtuous in the face of false viral messaging that undermines democracy. This can be observed, for instance, in the refugee solidarity groups in Hungary. As the initiatives grew and operated against the political headwind and the hostile public climate, Facebook closed groups provided political alternatives to the domestic, mainstream, public arena. In this context, Facebook's lack of ethical concerns (e.g. Gillespie, 2018; Zuboff, 2019) was not regarded as negatively as the governmental policy and the actions of some extreme right anti-immigration groups. Furthermore, it is worth noting that the Athens volunteer accommodation Facebook group enabled transactions that can be perceived as being at the edge of legality – e.g. no formal lease means tax evasion – but not a single post ever questioned these transactions. Possible explanations for this may be the urgency to find accommodation, the volunteers' lack of knowledge of the Greek tenancy regulations and property law, or the underlying support for a humanitarian aid cause. FInally, as we wrote this chapter, we learnt about a food-sharing community in Europe (the name and country are left anonymous to protect the identity of the people involved) that has decided to stop advertising events on Facebook after forming a partnership with a charity organization helping illegal refugees. Here, the lack of dedicated support for sensitive settings aspiring to more just futures (see Costanza-Chock, 2020) simply means not using the platform.

Although diverse in scope, the collectives described are embedded in new social movements that are said to be, on one hand, generated by the "contradictions of current capitalism" (Fernandes, 1993, p. 797), and on the other, based on moral, political and cultural values that do not necessarily question the institutionalized political system and the "market economy categories" (Fernandes, 1993, p. 811). However, in some cases, there were clashes between groups' ideological messages and online interactions through Facebook. As a new institutionalized arena, the platform provides new fuel to local conflictualities that are intertwined with both the perceived problems with the platform – given its policies regarding human rights and political liberties – and with their clash with the mainstream economic and political systems.

The cases illustrate that, despite differences in motivation and structure, the members and volunteers of these organisations value the sociotechnical mechanisms enabled by the platform, but do not necessarily trust the company behind it. Facebook provides templates for actions (organizing events, sharing posts, creating community and groups pages) that are widely recognised by heterogeneous audiences. Again, these mechanisms are adopted and integrated, even if there is ambivalence about control and flexibility. This makes it challenging for volunteers and more active members to envision alternative socio-technical setups or imagine how initiatives might migrate to them.

The collaborative initiatives discussed in this chapter show the ways much solidarity work takes place without bespoke technology. Moreover, while platforms like Facebook can provide basic support for collaboration and organizing action on the ground, supportive connections develop and unfold offline. Relatedly, previous work has shown that, in vulnerable settings, defining fixed roles and formally structuring matching mechanisms between "givers" and "those in need" might have unintended consequences, such as stigma and shame (Vyas *et al.*, 2015).

This questions mainstream design narratives that promote the platform model of sharing and the adoption of dedicated platforms for community initiatives. Given the constraints of limited budgets and dealing with emergencies, but also collectives' concerns to make visible inequalities and more sustainable lives, using what is available and popular has its virtues. Our argument to trouble the idea of technological innovation is not technology-agnostic. What we regard as a main design challenge for the settings discussed is the creative exploration of socio-technical practices that illustrate the use of alternative – yet existing – digital technologies, and how their use can be combined both with Facebook, and other means (see, for instance, Bødker *et al.*, 2016; Rossitto *et al.*, 2014), to support the different aspects and moments of organizing collaborative community initiatives.

References

Bennett, W.L. and Segerberg, A. (2011). Digital Media and the Personalization of Collective Action. *Information, Communication & Society*, DOI: 10.1080/1369118X.2011.579141.

Bennett, W.L. and Segerberg, A. (2012). The Logic of Connective Action. *Information, Communication & Society*, *15*(5), 739–768.

Bernát, A., Kertész, A., and Tóth, M.F. (2016). Solidarity reloaded: Volunteer and civilian organizations during the migration crisis in Hungary. *Review of Sociology of the Hungarian Sociological Association*, *26*(4).

Bernát, A. (2019). Solidarity powered via social media: Migrant solidarity grassroots groups in Hungary. In Travlou, P. and Ciolfi, L. (eds.), *Ethnographies of Collaborative Economies Conference Proceedings*. University of Edinburgh. ISBN 978-1-912669-11-0

Bernát, A., Fekete, Z., Sik, E., and Tóth, J. (2019). 'Borders and the Mobility of Migrants in Hungary'. CEASEVAL Research on the Common European Asylum System, *29*. Technische Universitat Chemnitz.

Berns, K. and Rossitto, C. (2019). From Commodities to Gifts: Redistributing Surplus Food Locally. In Travlou, P. and Ciolfi, L. (Eds.). *Ethnographies of Collaborative Economies Conference Proceedings.* University of Edinburgh, 25 October 2019. ISBN 978-1-912669-11-0

Bødker, S., Korsgaard, H., and Saad-Sulonen, J. (2016). A Farmer, a Place and at least 20 Members?: The Development of Artifact Ecologies in Volunteer-based Communities. In *Proceedings of the 19th ACM Conference on Computer-Supported Cooperative Work & Social Computing (CSCW '16).* Association for Computing Machinery, San Francisco, California, USA, 1142–1156. https://doi.org/10.1145/2818048.2820029

Bryman, A. (2012). Social Research Methods. Oxford University Press.

Crivellaro, C., Comber, R., Bowers, J., Wright, P., and Olivier, P. (2014). A pool of dreams: facebook, politics and the emergence of a social movement. In *Proceedings of the SIGCHI Conference on Human Factors in Computing Systems (CHI' 14).* ACM, New York, NY, USA, 3573–3582. DOI: https://doi.org/10.1145/2556288.2557100

Costanza-Chock, S. (2020). *Design justice: Community-led practices to build the worlds we need.* MIT Press.

Diogo, V. and Rosa, A. (2018). "Alternative Lifestyles in urban environment: the gender factor regarding the social usages of bicycles". Congress Woman XXI, Eventequalia, 20–21 September 2018, Porto. Available at: https://woman.eventqualia.net/pt/2018/inicio/publicacoes/

Fernandes, A.T. (1993). Conflitualidade e Movimentos Sociais. *Análise Social, Vol. xxviii* (123–124), 1993 (4.°–5.°), 787–828.

Gillespie, T. (2018). *Custodians of the Internet.* Platforms, Content Moderation, and the Hidden Decisions That Shape Social Media. Yale University Press.

Ganglbauer, E., Fitzpatrick, G., Subasi, Ö., and Güldenpfennig, F. (2014). Think globally, act locally: a case study of a free food sharing community and social networking. In *Proceedings of the 17th ACM conference on Computer supported cooperative work & social computing (CSCW '14).* Association for Computing Machinery, New York, NY, USA, 911–921. DOI: https://doi.org/10.1145/2531602.2531664

Hawlitschek, F., Teubner, T., and Weinhardt, C. (2016). Trust in the Sharing Economy, *Die Unternehmung 70*(1), 26–44.

Kallius, A. (2017). The East-South Axis: Legitimizing the "Hungarian Solution to Migration". Dire la violence des frontières. Mises en mots de la migration vers l'Europe. Dossier thématique. vol. *33* – n°2 et 3.

Lampinen, A., Rossitto, C., and Gradin Franzén, C. (2019). Scaling Out, Scaling Down: Reconsidering growth in grassroots initiatives. In: Travlou, P. and Ciolfi, L. (Eds.). *Ethnographies of Collaborative Economies Conference*

Proceedings. University of Edinburgh, 25 October 2019. ISBN 978-1-912669-11-0. Paper No. 2.

Light, A. (2019). Talking about Collaborative Economies: Platforms, Trust and Ethnographic Methods – Keynote Talk. In Travlou, P. and Ciolfi, L. (Eds.). *Ethnographies of Collaborative Economies Conference Proceedings*. University of Edinburgh, 25 October 2019. ISBN 978-1-912669-11-0

Light, A. and Briggs, J. (2017). Crowdfunding Platforms and the Design of Paying Publics. In *Proceedings of the 2017 CHI Conference on Human Factors in Computing Systems (CHI '17)*. Association for Computing Machinery, New York, NY, USA, 797–809. DOI: https://doi.org/10.1145/3025453.3025979

Rossitto, C., Bogdan, C., and Severison-Eklund, K. (2014). Understanding the Use of Constellations of Technologies in a Nomadic Setting. In JCSCW, the Journal of Collaborative, Computing *23*(2). Special issue on Work Practices, Nomadicity and the Mediational Role of Technology. Springer: The Netherlands, pp. 137–161. Online first in 2013.

Rossitto, C. and Lampinen, A. (2018). "Co-Creating the Workplace: Participatory Efforts to Enable Individual Work at the Hoffice." *Computer Supported Cooperative Work (CSCW)*27, no. 3 (December 1, 2018): 947–82. https://doi.org/10.1007/s10606-018-9319-z.

Rossitto, C., Lampinen, A., Bødker, S., Light, A., Berns, K., and Hui, J. (2020). Reconsidering Scale and Scaling in CSCW Research. In *Conference Companion Publication of the 2020 on Computer Supported Cooperative Work and Social Computing (CSCW '20 Companion)*. Association for Computing Machinery, New York, NY, USA, 493–501. DOI: https://doi.org/10.1145/3406865.3419409

Srnicek, N. (2017). Platform Capitalism. Polity Press.

Tayebi, A. (2013). Planning activism: using social media to claim marginalized citizens' right to the city. *Cities 32*, 88–93.

Travlou, P. (2020). From Cooking to Commoning: The Making of Intangible Cultural Heritage in *OneLoveKitchen, Athens*. In S. Lekakis (Ed.) *Cultural heritage in the realm of the commons*. Berkeley, CA: Ubiquity Press. https://www.ubiquitypress.com/site/chapters/e/10.5334/bcj.j/ (Open Access)

Vyas, D., Snow, S., and Mallett, M. (2015). More than just Food: Field Visits to an Emergency Relief Centre. In *Proceedings of the Annual Meeting of the Australian Special Interest Group for Computer Human Interaction (OzCHI '15)*. Association for Computing Machinery, New York, NY, USA, 662–666. DOI: https://doi.org/10.1145/2838739.2838787

Zuboff, S. (2019). The Age of Surveillance Capitalism. Profile Books Ltd. London.

DOI: 10.1561/9781680838411.ch6

Chapter 6

Designing for Motivations: Building on Two Local Communities Cases

By Veronica Cruciani and Myriam Lewkowicz

6.1 Introduction

In this chapter, we discuss the nuances of the terms sharing and sharing economy by looking at practical activities falling under the concept of "sharing", and the technologies related to those activities. Specifically, we are interested in how technologies can support groups of people that call themselves "communities", and reversely how communities shape the technology when adopting a triple-bottom line instead of a single profit line (Elkington, 1997) – getting profit, improving society, and respecting the environment, without emphasizing one motivation over the others.

Our interest for local communities comes from our reading of literature about food resilience and projects by communities of farmers, village's civilians, or little entrepreneurs (Light and Miskelly, 2015; Norström *et al.*, 2020; Ostrom, 1990; Teli *et al.*, 2017). We investigated two local communities. The first is an engineering students' association active in sustainability and social aid fields. The second one is a village community halfway between a volunteering-based association and a little enterprise exclusively selling local products. When looking at the missions and goals of the two associations, we found out that any form of capitalism or platform economy does not fully frame their activities. Members of both communities neither feel represented by the existing social media and tools available on Google, nor by the sharing economy platforms' philosophy, or generally any existing profit-based platform.

Bringing some practical examples, these associations are particularly concerned by the protection of members and customers' data from being sold; both associations are based exclusively on selling local or 0 km products, but they never ascribed this choice inside some doctrine or ideology: they just co-decided that this was the best solution to fully realize themselves as associate members and citizens. Further examples are the attempts of the students' association to find alternatives to social media and advertisements to sponsor, recruit, and engage students, and their choice of using Discord instead of Whatsapp. In fact, Discord is a proprietary freeware that offers instant messaging and VoIP features. First used by gamers, it reached 100 million active users in 2020. According to student association's members, the advantages compared to Whatsapp are privacy, as it is not *"owned by Facebook"*, and the possibility to create a thematic channel inside a group, with participants voluntarily joining if they are part of the group.

In the remainder of this chapter, we will first discuss the sharing economy for local communities, before presenting the two communities that we have observed, delineating their vision and mission, their structure, and the technological artifacts they are using. We then analyze the motivations of the members of these communities to participate, identifying four areas of motivation: sustainability, sociability, politics, and economy. These findings lead us to identify implications for designing for local communities in the forms of four personas.

6.2 Sharing in Local Communities

The sense of community, which is the sine qua non of every community, can be seen as made up of four elements (McMillen *et al.*, 2020): membership, influence, fulfillment of needs, and shared emotional connection. This sense of community is really contrasting with the platform economy models that are often related to "sharing economy": A platform, or a form of data usage that is capitalized, hardly promotes a sense of community and inclusiveness. In fact, capitalized platforms foresee transcending individual needs to recognize a common line of action (Keyes *et al.*, 2019).

This contrast between the sense of a community and the "sharing economy" may come from the fact that the term sharing is extremely broad and ambiguous. It could embrace different forms of sharing (Encyclopedia Britannica): reciprocity, demand sharing (Woodburn *et al.*, 1998), prosocial behaviours in general, which embraces mutualism (Feigin *et al.*, 2014) and sharing nicely.

Sharing nicely is the phenomenon we want to explore to offer suggestions on how to design for local communities. Sharing nicely has been used for the first time by the anthropologist Woodburn *et al.* (1998) in contraposition with the demand sharing. It might also be called altruism; it is innate for more developed species if not just the human (Lay and Hoppmann, 2015), and partly interiorized. Sharing nicely is aligned with the work by Mauss (2002) on gift giving. A gift is a connector between the giver and the receiver, acquiring among all, a symbolic meaning, going beyond the object itself, contributing to the strength of the community and building empathy among strangers. The mechanism to hold this is made of implicit giving and giving back. As Mauss writes: "In all this there is a succession of rights and duties to consume and reciprocate, corresponding to rights and duties to offer and accept. Yet this intricate mingling of symmetrical and contrary rights and duties ceases to appear contradictory if, above all, one grasps that mixture of spiritual ties between things that to some degree appertain to the soul, and individuals, and groups that to some extent treat one another as things" (Mauss, 2002, p. 17). This form of prosocial behavior might be considered as an elaborated form of reciprocity, with symbolic meanings and a more complex organization than the simple "returning the service" characterizing reciprocity. While psychologists still debate about the existence of pure altruism (Feigin *et al.*, 2014), the gift exchange might be considered one of the prosocial behaviors.

The relationships between different technologies and social practices in a community have been largely discussed, focusing on either their description, their conceptualization, or the design of technologies to better support social practices. Close to our interest in the critique of the capitalistic model of platforms for the sharing economy, Carroll and Beck (2019) have shown how to create a complementary service to platform capitalism, that they called platform collectivism. Platform collectivism is characterized by ownership and shared data (in contraposition of data possession), transparency and openness (in contraposition of buying and selling data), flat hierarchy, and peer-to-peer interaction (instead of pyramidal hierarchy and profit driven interactions). The central point in platform collectivism is to increase the value of data by giving priority to stakeholders' values instead of inserting or proposing new values for the platform. When data reflects stakeholders' values and needs, they are shared not just by the ones directly interested to them, but even by the community at large, increasing their visibility and access, as well as their educative value for local residents (Avram *et al.*, 2019; Carroll and Beck, 2019). However, before co-designing a platform conceived in a collective way, it is necessary to assess if the community, the artifacts, and the relationships between the two are fertile for a collective use of the platform instead of a capitalistic one (Bødker *et al.*, 2016).

Two interesting conceptualizations have been offered to describe the relationships between a community and its artifacts. Rossitto *et al.* (2014) suggest the concept of a "constellation of technologies", which is the entanglement of a group's private and shared artifacts, interactions, knowledge and skills, which take place to realize a performance. Bødker *et al.* (2016) define a "community artifact ecology" as "the particular constellation of artifacts that a community owns, has access to and uses in its activities. It is characterized by a high degree of shared understanding of the core activities and the role of the artifacts within the ecology" (Bødker *et al.*, 2016, p. 1144). The concept of community artifact ecology, then, includes the idea of a common level of knowledge about the artifacts-in-use. For example, we found that all students from the association that we have observed were highly skilled in informatics and at the same time knew a lot about sustainability and data protection, whereas the members of the village association were more prone to face-to-face relationships and low technology deployment. Thus, even if both associations have almost overlapping aims, they deploy very different artifacts ecologies, engaging with them in different ways, and using different sets of skills and knowledge.

In order to investigate these ecologies or constellations, a deep understanding of member's practices is required, especially the embodiment, domestication, routinization, overuse or, reversely the abandonment of artifacts. In the best cases, an appropriate community artifact ecology should transform a behavior into a social practice (Kuutti *et al.*, 2014). The introduction and appropriation of technology is particularly interesting to study for "low-tech" communities, in particular the way this appropriation could empower a community. For instance, Jayathilake *et al.* (2017) have studied the introduction of technology in subsistence agriculture, outlining how some forms of technology deployment give more autonomy and tools and knowledge to make decisions to a part of the community – like the Sri Lanka's farming one – that is usually not taken into account.

6.3 Two Local Associations Dealing with Food

The two associations we met are very different in their composition and their administration, but they share the same interest in community food and economic resilience. Both of them act for the reinforcement of local communities, work towards a socio-environmental transition, wish to gather multiple and diverse actors who have the same aim and ideals, and share an interest for participation and local democracy.

6.3.1 A Student Association

The student association, Ulisse, literally stands for "local union for social, solidary and environmental engineering" (Union Locale d'Ingénierie Sociale, Solidaire et Environnementale, in French). Ulisse was created in 2019 and is formed of 90 active members (and 137 participating in online group chats). It is the result of the merger of five student associations related to the promotion of a sustainable world, united world, and local ecological initiatives. The document presenting the merger (displayed on the facebook pages of the different associations) indicates that these five associations decided to meet because they all have similar values and projects, and had identified that this could lead to competition problems for subsidies, and waste of time and energy in administrative tasks. Ulisse conciliates the environmentalist and the locality causes in particular by managing projects aimed at supporting local producers and farmers. Most of the time, these farmers come from the local associations of farmers mostly linked to permaculture and organic agriculture, firmly opposing modern methods to cultivate and extensive agriculture.

6.3.2 A Village Association

The village association, Le Cerf à Trois Pattes (shortened into Cerf in the rest of the document), was created in 2018 after the failure of the local bakery. The reason for failure was very simple: in a 550 souls' village between two big cities – Reims and Tours – a bakery has to hire at least 4 people and be available from 7.00 to 18.00 to guarantee the service. As a result, the costs were higher than the gains.

From this experience, some people questioned themselves on how to make the village survive ("*we would like that people here do not feel forced to emigrate to town because there are no services and opportunities or facilities*"), on one side avoiding that young people abandon it, offering jobs, and on the other, keeping the elderly engaged with public life. After several meetings, the outcome was a hybrid activity that has put together volunteering and regular employment, with firstly social and then economic goals.

6.4 Data Collection and Method

Participant observation (DeWalt *et al.*, 1998) was adopted to study Ulisse: the first author, who came to Troyes as an exchange student, became a member of

Ulisse, actively participating in its initiatives, in order to gain their trust and have a deeper insight of both their culture and their community artefact ecology. We conducted six semi-structured interviews, and we participated in a general assembly. The COVID-19 pandemic made it impossible to keep performing active participation, so we pursued a virtual ethnography of both their social media and their Discord conversations during six months (approximately 1600 Discord messages), going on interacting with them while the university was closed and most of the activities suspended (Table 6.1).

During the lockdown, the second author met one member of Cerf (during a meeting of another association), who was telling how prolific Cerf was during the pandemic since it was taking care of each customer's order sending it in their own house. The second author then asked if it would be possible for the first author to come to the village and meet the members of Cerf. Indeed, since both associations presented very similar base ideas but very different ways of organising, we took the opportunity to follow Cerf to complement our findings coming from our work with Ulisse. We conducted five semi-directive interviews, observed how the activities were performed (we joined a local touristic activity and observed one members' assembly) when we visited the village during four days (Table 6.2).

We transcribed the interviews and our notes taken during field observations, and conducted a thematic analysis (Braun and Clarke, 2006), looking at the motivations of the associates, and the use of personal and community artifacts to realize community aims.

Indeed, since we outlined how sharing nicely through platforms is not an opportunism/altruism phenomena, but rather a multifaceted culture-dependent integration of both (Belk, 2014), and a complex assimilation of social motives and identities (Ryan and Deci, 2000), it is necessary to point out personal motives and how they are connected to the use of artifacts inside the community.

6.5 Findings

6.5.1 Organization and Activities of Ulisse

Ulisse has a president, two vice presidents, one for the projects, one for the contacts and the partnerships, one treasurer, one vice treasurer, one secretary, and two persons in charge of communication. Ulisse members are recruited and attracted through different means; first, during the general assembly, where at the start of each semester they introduce themselves and their projects to all the students.

Table 6.1. Synthesis of data collection for Ulisse.

Role	Technique	Duration	Place	Focus
Vice president	Face-to-face interview + notes	30 minutes	University room	General data and facts about Ulisse. His aims in the association
Vice president	Online interview	1 hour and 12 minutes	Online (Discord video -call)	Ulisse's use of technology and apps, Ulisse's changes after and during COVID 19
Univ. creator of a ridesharing initiative	Face to face + notes double check after the interview	30 minutes	Ulisse's room at University	General facts about Ulisse and the university regarding the ride sharing initiative: problems and ideation phase
2 vice presidents, 1 treasurer and three other members	Presence during the assembly + informal conversation after the assembly	45 minutes	Univ. hall	Knowing new projects, 2020 plan and getting in touch with members self introduction
Ulisse Discord members	Written question on Discord group + screenshot of each reply	//	Online	Knowing motivations, and self-perception as an engineer and Univ. member
Ulisse members	Informal interviews to 5 members after the 1st meeting for a particular project	Around 30 min	In person, group conversation in University room	Break the ice, opinions about Ulisse and the association with the university

Table 6.2. Synthesis of data collection for Le Cerf à 3 pattes.

Role	Technique	Duration	Place	Focus
President	Interview	72 min	Online (video call Whatsapp)	General information about association and artifacts the association uses
President	Interview while visiting the association's building + notes (mobile)	1 hour approx.	Guided visit to the association's venue	Visiting the venues, getting in touch with the members, understanding vision and mission
Members	Interviews	Several short interviews	House of the member and e-shop of the association	Field observation, technology use from elderly associates
Activity owner	Interviews	40 min approx.	WhatsApp video call	Presentation of his activity, work deontology and why joining Cerf
Website creator	Face to face + recorded interview	37 minutes	Web design company venue	Understanding web design, economical and marketing aspects of Cerf

Once students decide to join, they sign-in with their e-mail address and telephone number during or after the meeting. Students can also contact the associates, generally the leader or one of the vice presidents, by e-mail or Facebook, and they give their email address and phone number in order to be contacted again and, especially, added to the Discord channels. Indeed, even if Whatsapp and Facebook Messenger are used among some members for personal or informal communications, the official communication tool inside the association is Discord.

Decision-making happens face to face for most important decisions, and via Discord, where any member can submit a new idea to the vote and the majority wins. For important decisions (funding a project, leadership vacancy ...), they use the dot vote and the three winning proposals are re-evaluated, possibly merged and then (maybe) re-voted. The vice-president calls this process "the one that is really democratic".

Three kinds of meetings are organized:

- Project meetings, in which they discuss how to realize a project after it has been approved on Discord: people interested just need to join the meeting and the project channel.
- Head's meeting gathering the presidents, vice presidents and the accountant for monitoring the projects and their costs, or to check their progress. Most of the time, these meetings take place on Discord, but when an issue is raised for which no consensus can easily be found, an in-person meeting is organized.
- General meetings: official general assemblies, and thematic presentations.

Ulisse has different activities: projects, debates and lectures, and events.

Projects can be of different types (engineer without border, influencer – for instance supporting a hitchhiker and his YouTube channel, local food consumption, ride sharing). Project work is mainly accomplished using Discord, in which each project has a specific channel. Any student can decide to join a channel dedicated to a project. The proposer of a project will assume the role of coordinator of the Discord channel. When a project is concluded, the channel is dismissed. Ulisses members also create thematic Facebook pages for their projects. An example is the "Potager" (vegetable garden) group, in which members share information about University allotment, how to cultivate and how to compost, or the group "Panier" (basket), in which each Wednesday people put orders for seasonal and local food. The payment is done once they pick up the basket: in some exceptional cases the basket is left to the students common room (which is run by another association) and the sum is handed to the person in the entrance.

Debates and lectures are an important part of their activity and well reflect Ulisse's ideology – which they want to diffuse – and activism. They are held by an expert and are attended by the members who are passionate on the many topics that are discussed. Sometimes, associates intervene in external lectures to tell their experience or what they do for the community (ex: promote an "eco-responsible" food consumption).

The last type of activity are the events. Usually they do it "Just to have some fun" or to raise money for their projects. It might consist of any kind of initiative: selling crepes, mask parties, hitchhiking marathons, etc.

6.5.2 Ulisse's Ecology of Artifacts

When it comes about the artifacts they use, Ulisse has an Instagram and a Facebook account accessible to all the members. Although they would rather not use social media, especially "If owned by Zuckerberg" (Ulisse vice president), they use social media to:

- Advertise their events (debates, lectures, public speeches, parties charity fundraising and projects' fundraising) and projects.
- Invite to "presentation meetings" and recruit new members.
- Communicate about social engagement, environment and sustainability, especially through posters, ads and leaflets created with a shared Canva account.
- Show the weekly basket of available vegetables to order online and buy at the University every Wednesday.
- Present stories on Instagram, and follow other people.

When it comes to official communications among members, they use Discord since it is possible to create thematic channels and privacy is respected. There, they propose projects, meetings and thematic conversations: each project has a dedicated channel and ends with a face-to-face meeting. Every member of Ulisse has access to the Discord group, and sometimes, even non-members can join some channels if they have the possibility to contribute somehow to the association. They collectively participate in the conversation, and they make "group calls" when they cannot do it in person. Only some of the newcomers are familiar with Discord (usually gamers), so others take time to show the newcomers how to use it. Some others do not get familiar and they rather keep themselves updated through friends or Messenger conversations.

Lastly, as one goal of Ulisse is to spread as much knowledge as possible, each week a different member records a fifteen minutes podcast in a studio made available by the university (examples of topics are: collapsology, permaculture, ocean pollution, freedom of speech in France). The podcast is publicized on Facebook and the university radio. It is accessible free for all Ulisse's members.

6.5.3 Organization and Activities of Cerf

Le Cerf à 3 pattes owns a shop, where they sell food, drinks and handmade products from 30 local producers. If a consumer wants to substitute the mainstream products from big chains (toothpaste, soaps, Coca Cola etc.) they can find alternatives there. The shop is temporary and has an outdoor part with flowers and plants, and an indoor part. They want to base their selling only on local and well-selected products which production respects nature; they carefully choose their partners and visit their places. They even pay attention to the way their partners present themselves to sellers and engage clients. In the shops, there is the highest attention to the way packaging is used (at the minimum level and often of recycled materials), every product has the origin, and the main features reported: a casual customer would easily understand the boutique identity. Another aspect of their economy is that they want to promote local and sustainable tourism, especially out of the usual eno-gastronomic tourism, since their area is famous for Champagne production. They would like to prove that there is a lot more to be discovered like the mountain, panoramas, countryside tours, outdoor sports etc. Children and families are their main targets and the activities they offer foresee contact with nature, discoveries and learning by doing (and having fun).

Out of the shop itself, the associates organize a "*socio-cultural or touristic entertainment, proposed at least once per month: convivial dinner, theater, concert, night walk, philosophic café, producers' open air market ... we're among 60 to 500 participants, depends on the event*".

They also have a meeting room "recycled" from the local school's space that is used for decision-making processes.

When it comes to ordering, an offline solution is still used for elderly and people who usually swing by the shop. They use a paper-based list with all the seasonal items they might order: the customer makes the order and handles it to the shop assistant in the boutique.

6.5.4 Cerf's Ecology of Artifacts

In terms of artifacts, Cerf has a Facebook page and a website to give them visibility, and a newsletter sent every two weeks to keep the most affectionate customers informed.

Facebook is mainly used to sponsor the events that are organized and the touristic or educational activities. Some special announcements are made for last minute selling and special offers presented as a full-page picture with few words to explain the post. The ones who have access to the shop's laptop run the page and by the closest associates; it gives a sense of belonging and creates interest and curiosity among association's sympathizers but even enthusiasts of local products and sustainable tourism. Paraphrasing what the associates said in their meeting, they do not share the page management but they share the involvement, the values and the knowledge (especially related to local economy and environment) among whoever is interested in.

A Google folder is used for listing the available products; the association's administrators in charge of selling the products are the ones who manage this folder.

Customers who see these announcements, online documents, or receive the newsletter and want to buy some products, can put an order by sending an email and paying by transferring money to the bank account of the association.

The website has involved a web design agency and is now run by the President and the most motivated and expert associates. The agency followed the directions given by the President and tried to create a website that is easy to run (in their words: "very front office based"). The website presents the association in order to attract people, and gives the possibility to get in touch with Cerf's staff, to ask questions and know the events hosted by the association (Facebook posts about the events are depicted). But the aim of the website is mostly to sell the products from the little autonomous local enterprises that cooperate with the association.

6.6 Analysis - Motivations for Sharing Nicely

What has been clear since the very first interviews was that the economic activity of the association and the members' aims were perceived as "*for good*" and ethical: there was no perception of discrepancy between opportunistic goals (meet new people, make money and learn new skills), and altruistic ones (environment, society and volunteering).

An emblematic example is the way Cerf chooses local businesses to create a partnership with their shop: '*We visit all the enterprises, which must work in a specific way … Before any contract, we take our time, we investigate and then we base our relationship even on trust and dialogue*'. Another example is the will to find other people who share the same ideals: "*I am in a lot of associations here and I try to do some good around me […] Moreover I had a band and I've been playing for 6 years now: if there are some eco-social-musicians we can try some riffs during the weekend!*"

By analyzing the interviews, we found four main motivations for Ulisse and Cerf members: sustainability, sociality, technology and economy.

6.6.1 Sustainability

Sustainability can be a consequence like in Cerf's case, or a cause, like in Ulisse. In the first case, sustainability is not always what associates seek, but it comes out when actors follow the precise and ethical steps required by a local and volunteer-oriented association. Sustainability as a cause implies that it is the cause of members' actions. In this case, sustainability is learnt and studied because, as many students affirm, they look for a "*change of paradigm*".

The president of Cerf often mentioned a sustainability shift as the main drive of their customers: "*our customers are willing to pay even a little bit more (not too much) to have a reliable and good quality product, linked to the environment and the territory*". Sustainability is a very rooted drive for Cerf: for example, they are building an electric car park to encourage carpooling and green mobility as the president said *"This car park has been thought specifically for environmental and sustainability purposes. Germaine's citizens must think about the environment. This is perfectly in line with our beliefs and at the very base of community's survival".*

These pieces of conversations correspond to the "planet" motivation in (Böcker and Meelen, 2017). As a customer of Ulisse said: *"I try to live a life as consistent as possible with my ideas…I think this is the strongest revolution we can propose […] I could go to the local market but the Ulisse knows better than me, they select the sellers, I wouldn't be able to choose the right seller at the local market since I would like to respect millions of criteria: independent farmer, no pesticide, km0 etc. I refuse to buy from big chains and eat fast food, even if sometimes I am extremely tempted by 50 cents tomatoes from the supermarket wrapped in 1 kg of plastic".*

Words appear to be followed by practice since in the field observation we had the opportunity to observe how vegetables are grown in the University's yard with their own compost, and also how the vegetables used to be picked from the farmer's house on the way back home by one of the members: these two practices aim to cut the carbon emissions of selling local products. Discord is always used to agree on who is "driving around there" (the farmer's house). Another example is the absence of plastic in Cerf's shops and in all Ulisse's initiatives.

The idea of a local community is always related with the issue of sustainability: to start something sustainable or to oppose lobbyism and consumerism – which are both not conform to the sustainability principles – it is necessary to start from a local community or from a group of stakeholders which have an expertise on the ongoing situation. The concept of sustainability is fundamental, not just because of the recent challenges and emergencies, but even as an idea that should always be included and embedded in every social or socio-technical research (Volpato *et al.*, 2019; Poderi and Dittrich, 2018; Dourish, 2010; Cinderby *et al.*, 2014).

Whether as a cause or as a consequence, having members motivated in making the world more sustainable, lead both associations to have behaviours labelled as "sustainable", which is aligned with the kind of products they sell.

6.6.2 Sociality

"*We're not just an enterprise*". Social motivation is slightly different for the two associations. While Ulisse's members want to find new friends with a similar vision or with which being engaged in projects, Cerf members are oriented towards socio-economic mutual aid, with the aim of reinforcing the local community through a sense of community and relationships among members. An example from Ulisse is The Discord channel group for organising a Secret Santa *"all in line with our ideals and food km0"* and the requests to *"have a beer with someone downtown because tonight I am bored".*

While a member of Cerf said "*We're partners with two Foundations that deal with social and solidarity economy [...] We're mostly made up of volunteers and each one of them has a different position. There is the café group that organises all related to the bistro; there is the shop group [...], there is the tourism group [...]. We take care first of our citizens and we want to take care of youngsters' needs. We want to promote links and connection among people. Not by chance meetings are always in physical presence, we talk face-to-face, especially because we can take advantage of our spatial proximity. Human links are important*".

Another very important part related to Ulisse sociality is linked to the way they use Discord conversations. In the general channel or sometimes in thematic ones, they ask for recommendations related to environment or eco-friendly behaviours, like some permaculture techniques, for instance: *"has someone bought a BeeHotel? Do you know if it is better to orientate them to the north or to the south?"* or *"You should pass from Bank X to bank Y, which have very interesting projects related to wildlife conservation, even if unfortunately they are very small [...] everything is better than banks A and B"*. This means that Discord and informal conversations ended up being a huge and very important knowledge repository.

The attention of social and relational aspects led members to create very close ties whether friendship-like (especially in Ulisse) or more supportive like in Cerf, enhancing the sense of community among members.

6.6.3 Technology

"*Close to reality*". This sentence has been said by the website designer when she explained the kind of technology Cerf needed. We think that a human-centered vision of technology is what Ulisse and Cerf have in common. For Ulisse, the ecology of artifacts is technology-centric, tries to be independent, and extremely creative and open source thanks to every contribution: "we're a little bit nerds" as one member said. For Cerf the ecology of artifacts is low tech, both as an ideological choice – "we want to rely on people" – and a forced choice, since a lot of members and customers are not familiar with digital technology.

Technology is even seen as a challenge: some members of Ulisse are intrigued by the perspective of matching their passion for coding, design or technology with the possibility to be useful. Members would like to be independent, both technologically and ideologically: *"We wouldn't like to use Facebook or Whatsapp to communicate: we do not see ourselves on the same page [...] But it's impossible so far not to use them, especially because we are small and we want to be reached by everyone"* said the organizer of the ride-sharing project. Similarly, a piece of Discord conversation shows: *"Oh, if you know how we could share documents without Google, you're welcome!"*. But even on the choices people make for their everyday life, both for themselves and inside the association, like "*I don't have Whatsapp, actually I don't even have internet on my mobile*" (showing a very old mobile). The members of Ulisse also agree on "*... using Discord. First of all your data is protected there. And you can create textual or voice channels for different topics. You can find people with your interests faster. You can even make video conferences. The problem is that it is not free and not open source*".

All the members agree to use their own device for "public benefit" and to share the meeting rooms they manage to book at the university with others so that they can have spaces for meetings or debates.

Their initiatives as an association are moved by their values: social benefit, sustainability, and human centered technology: "*We believe in an Engineering that can change how things are now, in engineers' engagement*". Said the president of the students association.

Similar to technological independence is the right claimed by the president of Cerf of being disconnected and not using technology. More than technological independence here is a wish for an independence *from* technology, but these are two faces of the same medal. Indeed, online interactions can be seen as a complement and a support of offline ones, and often, what is classified online such as social coordination, invitation, offer, request, or collective action, encourage offline action (López and Farzan, 2015). As assessed by Lòpez and Farzan, the local activities are mostly off-sites ("*we privilege face-to-face relationships*", confirms the president of Cerf) and community goals are achieved by an entanglement of on-site and off-site, rising at any rate, users' social capital. This is even synthetized by the President's words: "*our advantages are mostly based on people's energy, their competences, especially from our partners and the volunteers [...] The importance is harmonizing logistic and conviviality*".

We observed that Facebook was used as an update tool (like a newsletter) for off-site actions, and the low response level on Cerf's Facebook page did not correspond to an equal low participation in events and activities taking place offline: the President himself stated that their events ranged from 50 to 120 participants and their touristic events are as well very popular.

Moreover, as noticed by (Satchell and Dourish, 2009), non-use is a form of use and can have a nuanced meaning, instead of being "labelled" or "pathologized". In Cerf, some forms of non-use happened because of lagging, especially with elderly members: "*I have Whatsapp just to communicate with my daughter in Scotland, and some younger expert friends help me with computer and smartphone problems*". Some of them might even be discouraged by some form of physical impairment like illness or low sight, falling partially in the category of the disenfranchisement (Satchell and Dourish, 2009). The website creator for Cerf said that she kept in mind to create a website that leaves the possibility to have a paper substitution to internet orders.

The second non-use category happened with the displacement, very familiar to rural HCI and technological stewardship (Jayathilake *et al.*, 2017; McMillen *et al.*, 2020). In our case, people decided to rely on others, namely sons/daughters or neighbours when it came about using the technology, especially buying online. Similarly, in Cerf the "heavy" use of technology (design, website, creation of advertisements ...) is mostly on the hands of the professionals, namely the website creators.

A more complex example of non-use comes from Ulisse: despite they are all engineers' students, they try to limit the technological part of their community ecology of artifacts because of their environmental convictions. Even more interesting is their collaboration with a group of university professors and philosophers which are part of the low-tech movement. Taking an example of a Discord conversation about a new Facebook post on reading recommendations for the summer, a member said: "*I recommend Philippe Bihouix: Low tech age and Happiness was for yesterday. We must promote our adhesion to the low–tech movement. And then a little bit of Collapsology won't harm anyone: read some books from Pablo Servigne*".

The very wide range of technology related behaviours make it difficult to predict and label a typique user's relationship with his/her constellation of technologies: each time is necessary to "go to the field", interview, and avoid oversimplifications of this very complex issue.

6.6.4 Economy

Cerf is supporting forms of sharing that do not fall into two opposite poles – capitalized sharing economy or pure altruism. The president of Cerf presents the association as aiming to revive the local *economy*; he says that they have created the association/community to help people and their territory, and even that they are not just economy-based, implying that the economic and generally opportunistic component is present. As shown even by Bødker *et al.* (2016), to speed up or make possible some necessary processes in order to make a volunteering association work properly, some members had to be externally motivated: altruism (volunteering in this case) and ideals was not enough. Specifically they report the example of the web designer, which had to be paid in order to have the right mix of intrinsic (beliefs and ideas) and extrinsic (reward) motivation to pursue his job which had become too demanding and time consuming to be done in due time just in the name of ideals (which anyway, were present).

Cerf and Ulisse's sellers have chosen such partners and they do not sell to other more lucrative lobbies, already embed sustainability and local-loyalty values. The president of Cerf is clear on this purpose: their partners are strictly selected and very highly motivated to pursue a certain idea of production and marketing.

A stable and self-run infrastructure would allow members of local communities to be in line with their motivations, and, reversely, it would allow members to shape their artifacts ecology. Our aim is then to go from this analysis of motivations to implications for designing for a community ecology of artifacts.

6.7 Motivation-centered Design

The first way to support local communities in designing their ecology of artifacts may be to help them in identifying the different motivations that drive their members to act in the community. For so doing, we suggest four personas (Nielsen, 2018) that represent the four kinds of motivations that we have identified. Personas should help the communities to combine motivations, the artifacts in use and socio-economic background, in order to co-design meaningful products or services, aligned with their common mission, and to define a common future.

Motivations are what represent the stakeholders' push to participate in the community, leaving very little space for assumptions. In addition, clearly defined motivations can be tested and refined in successive design phases, during workshops for instance.

Concretely, we think that the four personas could be used in role-play workshops[1] (Seland, 2006) with some of the community members to allow them both to identify themselves with the various personas and to reflect on what they want as a community at large. This kind of workshop could even take place without the presence of a professional designer. The advantage of role-playing is that participants have to adopt different points of view: It would be a precious resource for associations with members having very different aims, and to foster discussion among stakeholders. This could help different stakeholders to understand each other, to be able to frame a common problem (Steyaert *et al.*, 2007), and to be able to create a story usable in other situations, for instance in marketing and social media management of the association.

The four personas we have identified from our analysis of motivations to participate in local communities are: user (Figure 6.1), volunteer (Figure 6.2), producer (Figure 6.3), and worker (Figure 6.4). We now explain the link between motivations and personas, and describe each persona.

The sociality motivation consists of a desire for affiliation, a wish for networking, for feeling part of a community, a group or something more close-knitted than society-at-large. We then created a persona called *volunteer* since volunteering is

1. https://www.designkit.org/methods/36?ref=publicdesignvault

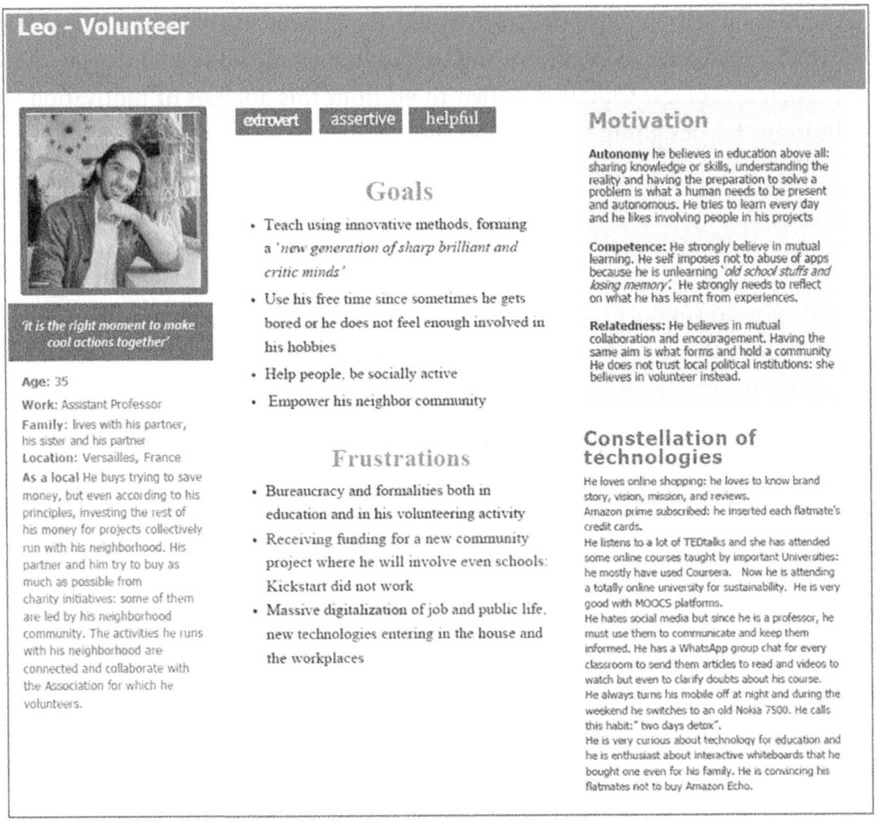

Figure 6.1. Persona volunteer.

the best way to fulfill all the listed aims consistently with one own values (in Ulisse and Cerf cases, environmental-related values).

The sustainability motivation led us to suggest a persona named the user, in other words the consumers of the associations.

The political motivation is related to mature political consciousness mixed with active social engagement: the person is conscious and concerned that his/her action has both political and social consequences. Both user and volunteer personas correspond to this motivation.

The economic motivation describes the profitable and more opportunistic part, so we created a persona called the *seller*, and another one called the *worker*, specifically referring to Cerf, which aim is even to *"give a job to people"*.

The following sensitizing scenario (Waern *et al.*, 2020) that involves the four personas illustrates the complex and multifaceted issues local communities face.

Leo wants to change his habits and he feels it is time to be more useful to other people: for this reason, for months now he is reading about non-profit organizations and associations aiming to help people and the planet. He has found an online

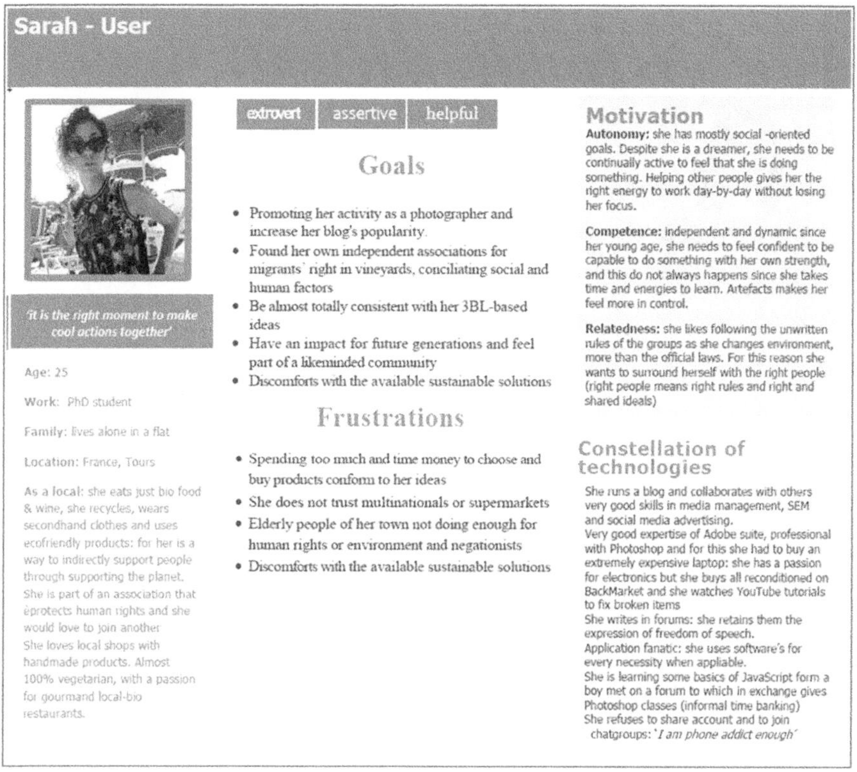

Figure 6.2. Persona user.

article from **Sarah**, a freelance journalist, about a local association where she is a regular customer, dealing with zero-waste products to replace polluting or disposable ones. Since during the past year Leo has been spending a considerable amount of his time looking on the Internet for anything that would have replaced little by little all the tools and items that do not conform to his environmentalist values, Leo decides to visit the website of the association. He finds it very business-oriented; the association owns a few shops that are run half by volunteers, half by workers that sell only local "replacement" products. The Facebook page of the association depicts pictures of a usual workday with some comments of colleagues or regular customers, and posts of the latest news related to the shops and the events. Scrolling down the products' list from the main website, Leo decides to visit the closest shop, and before, he asks to the official Instagram profile of the shop, ran by a volunteer, **Lucas**, if he can bring some old jars (in case they would find it useful for containing zero-packaging food). Lucas asks by SMS to the president of the association if it is possible. The president wants to push the association more and more towards a "green" vision and she specifically enjoys being personally in contact with the associates. The President then replies that yes, they might need as many jars as

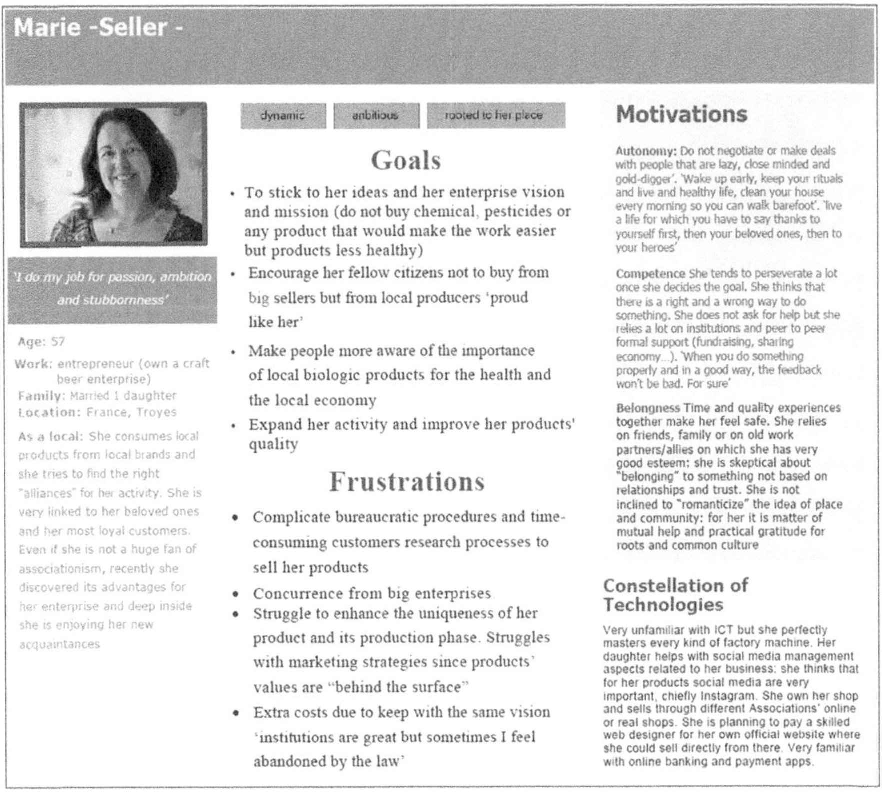

Figure 6.3. Persona seller.

possible: not because there is a plan for them, but who knows which amazing idea those jars would inspire to the associates. For this reason, she decides to share pictures of the jars in the group conversation, asking for suggestions. According to her, every member should share ideas and unused objects. Thus, Leo goes to the shop to give his jars. While visiting the shop, he notices that there is even the possibility to fill paper surveys, positioned right close to the cash register. So before leaving people can write a feedback on a piece of paper and leave it in a big box. He is very impressed by the range of products available, way more than the ones displayed online, so he reports his opinion directly to the shop assistant, Lucas, which encourages him to contact the web developer, who works for an enterprise. Leo decides to talk with the web developer and since he has some web-design skills, he decides to join the volunteers' group just to change some features and some colors of the landing page. He then asks the approval of the Association main runners. Once he has done this, he remembers that he saw some very creative beer bottles and he would like to buy some. He checks the catalog of the main sellers of the Association and he finds **Marie**, a producer specialized in hop cultivation. Leo decides to contact her through the association and he makes a request. Marie responds that

Lucas - Worker -

'I was just looking for a job, but this association is doing so nice things'

Age: 46

Work: association's shop clerk

Family: Re-married, 3 sons

Location: Reims, France

As a local: He is a PlayStation fan and he invites all his friend and their families to join him and his wife: those gatherings became central for his "niche" of friends, which somehow formed a community of interests. Even the associates love spending time together in his place and they often meet there for Association meetings or just for an evening afterwork with good food from the shop's unsold (or expired).
By posting on Instagram and Pinterest his outdoor activities is giving visibility to local landscape and sport attractions. He buys (or brings back) food from the local shop for which he works

creative | generous | Weighty

Goals

- Have an honest and interesting job which ensure him a fair monthly income and keep him passionate
- Found a "repair cafe" atelier in the Association where he works
- Combine work and family life, ensure a future to his sons, but also have the right amount of free time for himself
- Have a lifestyle that respects his ideas and beliefs: since he is working for the Association, he is paying more attention to his choices, especially related to food

Frustrations

- Dealing with monthly expenses and 'special" expenses (especially related to his sons' education or holidays)
- Difficulties to adapt to an ambiguous job market
- Massive digitalization of job and public life: he would like to separate truth from falseness understand the '*messy muddle of things available on internet, what you can do with them and which risks for my kids* '

Motivations

Autonomy he does what he enjoys or likes doing: this is the best way for him to make things possible. He likes planning and achieving with the people he loves. He usually tries to maximize the result/effort rate. He would like not to be pushed by the lack of money (or its fear) when he takes some decisions.

Competence He strongly believe in mutual learning, he let himself contaminate with new ideas, choosing with his insight what to follow and what to abandon. He does not like hard feedbacks unless they are necessary.

Relatedness. Since he suffered bullying at 15, he hates being put on the side and he thinks that sometimes sacrifices should be done to be included. He loves exploring new fields of interest through his beloved ones: sharing a growth path motivates him

Constellation of technologies

He is often online (especially when the shop is empty) consulting Facebook and some websites, usually online illegal film streaming ones. He is a fan of YouTube tutorials and of the updates related to Play Station and games. Once in a moth he plays online role game. He is a very discrete player on the net. However, he does not have real digital competencies and skills: for example, more than once his wife warned him that he believed to a fake new. He borrows and lends games for PlayStation and he occasionally shares his Netflix accounts when some of his friends asks him if they can watch a specific movie.
He shares an Instagram profile with his wife. Since they both are very sporty, they use it mostly have a monothematic profile. Eventually they equipped their mobile for adventurous days out and to film sports without the risk of breaking it. This shared passion lead them to raise money on the "scaled sharing platform" www.patchworkpresent.com to buy a professional GoPro and other useful tools. He came to know about this fundraising website from one associate and he found the idea so genial that he sent the website all over his group chats

Figure 6.4. Persona worker.

it is possible and asks her bottle supplier for an out-of-order. Then, the product is sent to the shops as "reserved for Leo" while the payment already happened online, using the website as an intermediary.

If local communities could benefit from the work of designers, web designers or user researchers, the personas presented should be taken into account when designing artifacts. Indeed, conformity to community's ideals and needs is necessary to move towards platform collectivism, which, as mentioned in Section 2, fosters a sense of community and is more sensible to issues that are important for the members we have interviewed: privacy and participative democracy. Platform collectivism is intimately related to Woodbourne's idea of sharing nicely, in contrast to the "capitalistically platformised sharing". The four personas give a snapshot of the users' motivations and attitudes that might be found in a local community interested in social, economic, and environmental sustainability. Therefore, taking all of them into account when designing the artifacts ecology should allow to align this ecology with users' values and to pursue community and users' motivations.

6.8 Conclusion

In this chapter, we started by discussing sharing in local communities, stating that we are interested in the way an ecology of artifacts could support *sharing nicely*. We then described how two associations with common characteristics (strong sustainability ideology, involvement in the local economy, minded alike users and infrastructures, locality, strong autonomy) share nicely through their community's ecology of artifacts. Our analysis led us to identify four kinds of motivations for sharing nicely: sustainability, sociability, politics, and economy. We then derived four personas and a scenario from these motivations, and explained how we envision the use of personas and scenarios for supporting the design of a community ecology of artifacts that would embrace all the different motivations.

The contribution of this chapter is then twofold: On one side, we shed light on the motivations of participants in grassroots initiatives, understanding what pushes them to share nicely. In the cases that we studied, the ideals related to the protection of the environment and socio-political change co-exist with the need to earn money, get a job, and health improvement through better food. Those findings are consistent with the existing work about the triple bottom line model, which foresees that sustainability can only be realized through the coexistence of three interwoven factors: environment, society and economy (Elkington, 1997). Even if our approach aims at supporting the design of a community ecology of artifacts, the technology is not the focus, nor the "user" is, but we rather pay attention to the diverse motivations of participants. We consider motivations the first thing to address when conducting design research with communities, and throughout the community's life and evolutions. Consequently, the second contribution concerns how to address those motivations in grassroots initiatives that, as said, are strongly ideals and/or needs driven. The personas we propose aim at re-thinking technology both for society and communities, focusing on people instead of platforms (Avram *et al.*, 2019; Srnicek, 2016). Therefore, we offer the personas to anyone who can make the difference in a local context: designers, NGOs, policymakers, and not for profit organizations. On this wavelength, as Ann Light and colleagues (Light *et al.*, 2017) state, design should not accept things to be done "as usual", since we are facing extraordinary times characterized by a "wave of change and uncertainty". CSCW and HCI have their role in changing society and imagine a better future to pursue and realize through design and human-centered technology. *Targeting the right change-makers* is fundamental to start this process.

References

Avram, G., Choi, J.H., Paoli, S.D., Light, A., Lyle, P., and Teli, M. (2019). Repositioning CoDesign in the age of platform capitalism: From sharing to caring. *CoDesign, 15*(3), 185–191. https://doi.org/10.1080/15710882.2019.1638063

Bellotti, V., Ambard, A., Turner, D., Gossmann, C., Demková, K., and Carroll, J.M. (2015). A muddle of models of motivation for using peer-to-peer economy systems. *CHI 2015 – Proceedings of the 33rd Annual CHI Conference on Human Factors in Computing Systems: Crossings*, 1085–1094. https://doi.org/10.1145/2702123.2702272

Belk, R.W. (2014). You are what you can access: Sharing and collaborative consumption online. *Journal of Business Research, 67*(8), 1595–1600.

Bødker, S., Korsgaard, H., and Saad-Sulonen, J. (2016). 'A Farmer, a Place and at least 20 Members': The Development of Artifact Ecologies in Volunteer-based Communities. *Proceedings of the 19th ACM Conference on Computer-Supported Cooperative Work & Social Computing*, 1142–1156. https://doi.org/10.1145/2818048.2820029

Böcker, L. and Meelen, T. (2017). Sharing for people, planet or profit? Analysing motivations for intended sharing economy participation. *Environmental Innovation and Societal Transitions, 23*, 28–39. https://doi.org/10.1016/j.eist.2016.09.004

Braun, V. and Clarke, V. (1 January 2006). Using Thematic Analysis in Psychology. *Qualitative Research in Psychology, 3*, 77–101. https://doi.org/10.1191/1478088706qp063oa.

Carroll, J.M. and Beck, J. (2019). Co-designing platform collectivism. *CoDesign, 15*(3), 272–287. https://doi.org/10.1080/15710882.2019.1631353

Cinderby, S., Haq, G., Cambridge, H., and Lock, K. (2014). *Practical Action to Build Community Resilience: The Good Life Initiative in New Earswick*. 74.

DeWalt, K.M., DeWalt, B.R., and Wayland, C.B. (1998). "Participant Observation." pp. 259–99 in Handbook of Methods in Cultural Anthropology, edited by H.R. Bernard. Walnut Creek, Calif.: AltaMira Press.

Dourish, P. (2010). HCI and Environmental Sustainability: The Politics of Design and the Design of Politics. In *Proceedings of the 8th ACM Conference on Designing Interactive Systems – DIS '10*, 1. Aarhus, Denmark: ACM Press, https://doi.org/10.1145/1858171.1858173.

Elkington, J. (1997). Cannibals with forks. In: The Triple Bottom Line of 21st Century. Capstone Publishing Ltd., Oxford, UK.

Feigin, S., Owens, G., and Goodyear-Smith, F. (2014). Theories of human altruism: A systematic review. *Annals of Neuroscience and Psychology, 1*(1). Retrieved from http://www.vipoa.org/neuropsychology

Hawlitschek, F., Stofberg, N., Teubner, T., Tu, P., and Weinhardt, C. (2018). How Corporate Sharewashing Practices Undermine Consumer Trust. *Sustainability, 10*(8), 2638. https://doi.org/10.3390/su10082638

Jayathilake, H.A.C.K., Jayasinghe-Mudalige, U.K., Perera, L.D.R.D., Gow, G.A., and Waidyanatha, N. (2017). Fostering technology stewardship approach to promote knowledge sharing among farming communities in Sri Lanka. *Tropical Agricultural Research, 28*(3), 238–246. https://doi.org/10.4038/tar.v28i3.8228

Keyes, O., Hoy, J., and Drouhard, M. (2019). Human-Computer Insurrection: Notes on an Anarchist HCI. *Proceedings of the 2019 CHI Conference on Human Factors in Computing Systems*, 1–13. https://doi.org/10.1145/3290605.3300569

Kuutti, K. and Bannon, L.J. (2014). The turn to practice in HCI: Towards a research agenda. *Proceedings of the SIGCHI Conference on Human Factors in Computing Systems*, 3543–3552. https://doi.org/10.1145/2556288.2557111

Lay, J.C. and Hoppmann, C.A. (2015). Altruism and Prosocial Behavior. In N.A. Pachana (Éd.), *Encyclopedia of Geropsychology* (pp. 1–9). Springer Singapore. https://doi.org/10.1007/978-981-287-080-3_69-1

Light, A. and Miskelly, C. (2015). *Sharing Economy vs Sharing Cultures? Designing for Social, Economic and Environmental Good.* 14.

Light, A., Shklovski, I., and Powell, A. (2017). Design for Existential Crisis. *Proceedings of the 2017 CHI Conference Extended Abstracts on Human Factors in Computing Systems*, 722–734. https://doi.org/10.1145/3027063.3052760

López, C. and Farzan, R. (2015). Lend me sugar, I am your neighbor!: A content analysis of online forums for local communities. *Proceedings of the 7th International Conference on Communities and Technologies – C&T '15*, 59–67. https://doi.org/10.1145/2768545.2768558

Manzini, E. (2015). *Design, When Everybody Designs: An Introduction to Design for Social Innovation.* Translated by Rachel Coad. ISBN 978-0-262-02860-8. The MIT Press Cambridge, Massachusetts, London, England.

McMillen, H.L., Campbell, L.K., Svendsen, E.S., Kealiikanakaoleohaililani, K., Francisco, K.S., and Giardina, C.P. (2020). *Biocultural Stewardship, Indigenous and Local Ecological Knowledge, and the Urban Crucible.* https://doi.org/10.5751/ES-11386-250209

Mauss, M. (2002). *The Gift: The Form and Reason for Exchange in Archaic Societies.* Routledge.

Norström, A.V., Cvitanovic, C., Löf, M.F., West, S., Wyborn, C., Balvanera, P., Bednarek, A.T., Bennett, E.M., Biggs, R., de Bremond, A., Campbell, B.M.,

Canadell, J.G., Carpenter, S.R., Folke, C., Fulton, E.A., Gaffney, O., Gelcich, S., Jouffray, J.-B., Leach, M., ... Österblom, H. (2020). Principles for knowledge co-production in sustainability research. *Nature Sustainability, 3*(3), 182–190. https://doi.org/10.1038/s41893-019-0448-2

Ostrom, E. (1990). *Governing the Commons: The Evolution of Institutions for Collective Action.* Cambridge University Press.

Parguel, B., Benoît-Moreau, F., and Larceneux, F. (2011). How Sustainability Ratings Might Deter 'Greenwashing': A Closer Look at Ethical Corporate Communication. *Journal of Business Ethics, 102*(1), 15. https://doi.org/10.1007/s10551-011-0901-2

Poderi, G. and Dittrich, Y. (2018). Participatory design and sustainability: A literature review of PDC proceedings. *Proceedings of the 15th Participatory Design Conference on Short Papers, Situated Actions, Workshops and Tutorial – PDC '18*, 1–5. https://doi.org/10.1145/3210604.3210624

Rossitto, C., Bogdan, C., and Eklundh, K. (2014). Understanding Constellations of Technologies in Use in a Collaborative Nomadic Setting. *Computer Supported Cooperative Work (CSCW).* 23. 10.1007/s10606-013-9196-4.

Ryan, R. and Deci, E. (1 February 2000). Self-Determination Theory and the Facilitation of Intrinsic Motivation, Social Development, and Well-Being. *The American Psychologist 55*, 68–78. https://doi.org/10.1037/0003-066X.55.1.68.

Satchell, C. and Dourish, P. (2009). Beyond the user: Use and non-use in HCI. *Proceedings of the 21st Annual Conference of the Australian Computer-Human Interaction Special Interest Group on Design: Open 24/7 – OZCHI '09*, 9. https://doi.org/10.1145/1738826.1738829

Seland, G. (2006). *System Designer Assessments of Role Play as a Design Method: A Qualitative Study.* 10.

Srnicek, N. (2016). *Platform Capitalism.* John Wiley & Sons.

Steyaert, P., Barzman, M., Billaud, J.-P., Brives, H., Hubert, B., Ollivier, G., and Roche, B. (2007). The role of knowledge and research in facilitating social learning among stakeholders in natural resources management in the French Atlantic coastal wetlands. *Environmental Science & Policy, 10*(6), 537–550. https://doi.org/10.1016/j.envsci.2007.01.012

Teli, M., Di Fiore, A., and D'Andrea, V. (3 April 2017). Computing and the Common: A Case of Participatory Design with Think Tanks. *CoDesign 13*(2), 83–95. https://doi.org/10.1080/15710882.2017.1309439.

Volpato, T., Allian, A., and Nakagawa, E.Y. (2019). Has social sustainability been addressed in software architectures? *Proceedings of the 13th European Conference on Software Architecture – ECS '19 – Volume 2*, 245–249. https://doi.org/10.1145/3344948.3344979

Waern, A., Rajkowska, P., Johansson, K.B., Bac, J., Spence, J., and Løvlie, A.S. (2020). Sensitizing Scenarios: Sensitizing Designer Teams to Theory. *Proceedings of the 2020 CHI Conference on Human Factors in Computing Systems*, 1–13. https://doi.org/10.1145/3313831.3376620

Woodburn, J. (1998). Sharing is not a form of exchange: an analysis of property-sharing in immediate-return hunter – gatherer societies. In Property relations: renewing the anthropological tradition (ed. Hann C.M.), pp. 48–63 Cambridge, UK: Cambridge University Press [Google Scholar].

DOI: 10.1561/9781680838411.ch7

Chapter 7

Towards a Political Definition of the Sharing Economy: Reflections on the Development of a Sharing Economy Initiative Outside of Big Cities

By Hannelore Goyens and Liesbeth Huybrechts

7.1 The Street as a Shared Space

The public space – the street – belongs to all of us. However, during the last century, in many European cities, the mobility system, and more specifically the dominance of the car, has gradually reduced and divided the space for people literally, but also figuratively. There exists a growing uncertainty and disagreement about how to deal with the complex challenge of increasingly busy car roads: their economic and functional necessity is weighed against their disadvantages for social cohesion and ecological balance (Illich, 1974; Gehl, 2010; te Brömmelstroet and Verkade, 2020). The increase of cars has contributed to a perception that today the street divides us instead of connecting us: neighbourhoods are cut by busy car roads, cyclists and pedestrians have to fight for their space on the streets. In multiple projects, the research work of the group Spatial Capacity Building challenges the current mobility system by learning about mobility and how it interacts with other domains of everyday life. The starting point of this learning process is to look for what connects us, **"what we share"** rather than what divides us (Huybrechts *et al.*, 2018): we learn how to reclaim mobility as something we share.

Within the context of the complex North-South Limburg project (Studio NZL, 2019), which focuses on the redesign process of a very busy and important regional connection in a rather rural part of Flanders, called Limburg, we question the current mobility system as something that for years has "divided" the community and its politics, by rediscovering it as a shared space. The busy road divides residential areas, cuts through green spaces and divides between cars, bikes and pedestrians, but also divides people on the question of what position this road should take in the space and how it should be designed.

In this article we will (1) discuss how sharing was a conscious approach to deal with a subject that divides people, and particularly how this is practiced **outside of the big cities**, in a more **rural region**. Indeed, sharing concepts thrive in urban contexts, but have been under-discussed in semi-urban and rural contexts. (2) We elaborate on how we have developed a **platform-methodology to support sharing** in this particular context, in a way that enables to root sharing in a particular space and time. (3) We discuss the **Platform Mobility in the North-South context** and the particular ways in which it operated, to end with a discussion on what sharing means and to embrace sharing as a term that is political, that embraces and respects the diversity of the everyday life in which it is situated and thus does not exclude what and who is often perceived as at the margin.

7.2 Sharing Outside of the Urban Context

In "The Age of Sharing" Nicholas John (2016) dissects the increasingly popular word "sharing" and sharing economy: sharing bikes, food, houses etc. He finds that when we talk about sharing, we point to prosocial behaviours that claim to promote greater openness, trust and understanding between people. He also sees, however, that 'sharing' has not always been associated with these values. In his book he focuses on three spheres of sharing: sharing as the constitutive activity of social media and our technologically mediated lives; sharing as a model for economic behaviour and thus our economic lives as producers and consumers; and sharing as a category of speech that impacts our emotional, interpersonal lives. His book allows us to differentiate between how the concept of sharing is ideally articulated, what we desire from it in creating a sense of community and the actual daily life transactions that people label as sharing, but that are often the victim of the market dynamics and commodification. What he talks less about is that this idea of sharing has an elitist character and the platforms that facilitate sharing involve people who are well-educated and with sufficient incomes (Bársony, 2017).

Indeed, what requires additional investigation is how sharing is rooted in the world we live in. In line with Latour (2018) claim to come down to earth, we are interested in how sharing is performed in daily life, between people, people and animals, people and water, water and plants, etc. This way of looking at sharing reveals a great diversity of sharing practices in the world and provides angles to see how actors that are often marginalised in the sharing economy can become part of it. Coming down to earth also reveals the particularities of sharing in more rural contexts, where people do not live in close proximity, and maybe do not always have access to the technologies needed to participate in sharing economies. Indeed, often sharing economy initiatives emerge in urban contexts. In this case we explored sharing as a basis for sustainable mobility transition in a less urban context, where sharing might emerge in less spontaneous ways (Davidson and Infranca, 2016). Because in these contexts that can be situated in-between the urban and the rural there is a great need for sustainable transitions in mobility, building, working etc., we explored deliberate sharing strategies to shape these transitions. This approach to sharing forms the basis of our platform-methodology, that is focussed on starting from particular situated contexts to build platforms for diverse actors, spread out in space, who can share and act together.

7.3 Approach: A Platform-Methodology to Support Sharing

We discuss a particular design research trajectory that resulted in a sharing economy initiative that we initiated in the complex North-South Limburg project (Studio NZL, 2019). In this context, the team decided not to start from what divides people, but examined "sharing" as a stepping stone for a sustainable mobility transition. We questioned the current mobility system as something that "divides" by learning together about mobility and its interactions with everyday life: how do we think about mobility and its interaction and how do we want to shape it. By researching together "what we share" (Huybrechts *et al.*, 2018), we build on a tradition of participatory design research that looks at "commons" (Berlant, 2016; Gil and Baldwin, 2014; Marttila *et al.*, 2014; Seravalli, 2014; Teli, 2019) and "partial economies" (Avram *et al.*, 2017).

To enable this situated understanding of how sharing economies develop in this context and develop platforms that can support these situated sharing activities, we used history and a design anthropological approach. On the one hand, we traced actualities and histories of sharing through intense observations, interviews and archival work. We built on our experience with historical approaches in participatory design research as earlier developed in Huybrechts *et al.* (2016) and in Zuljevic and Huybrechts (2019) and further develop them for the particular context of the sharing economy. On the other hand, we start designing Platforms – physically and digitally – that support the growth of these initiatives. Because in this case, many institutions played a role in the development process of the sharing economy initiative, we particularly dived into the question of **how can we design platforms that can enable sharing initiatives to develop in contexts, and with actors that are systematically marginalised in the sharing economy initiatives, in close interaction with existing institutions** that drive sustainable mobility transitions?

First of all, we decided to start with detecting **the historical and existing sharing collectives in the field** (often citizen-driven) who already were reclaiming modes of sharing mobility and the street. We define collectives as assemblages of both non human as well as human agents (Latour, 1999) that radically interdepend (Escobar, 2018). We discovered these sharing collectives and their radical interdependencies based on a design anthropological trajectory made of 250 field interviews and even more field observations via photography, drawing and video-portraits as methods. As one of the conclusions of our extensive participation process, it was noticed that the discussion on sharing mobility revolved mainly around two central themes. First, there were groups who organised themselves around mobility in the sense that they wanted to regain their position in "sharing the street", as a space for people who live and work around it. Second, we noticed that in the less organised group of inhabitants, there was only very limited cycling in the area of North-South Limburg, including by children and young people. Often because people didn't dare to bike, didn't know how or didn't own a bike. Therefore, there was a need for collective sharing of bikes, but also of collective sharing of knowledge and practices on biking.

Second, we decided to build on these insights to collaboratively give form to **platforms (see scheme "platform-approach", Figure 7.3) that can support self-organised initiatives with impact on a sustainable mobility transition and that can document how these evolve.** Four platforms were created: on open space, on living, on working and – and this is the one we will discuss in the remainder of this paper – on mobility. In line with the understanding that there is a shared demand for more quality use of bicycles and public transport by children and young people, within the context of their schools and neighborhoods, children and young people became one of the central target groups in the platforms' activities, a group involving also their parents, grandparents, etc.

7.4 Insights: Case Study "North-South Limburg Bicycle Library" as Part of the Platform Mobility in the North-South Context

7.4.1 Historical Research

7.4.1.1 East-West and North-South

As we mentioned, there are two central themes in the history of the North-South connection in Houthalen-Helchteren (HH) – a municipality in the centre of Limburg: the attention for moves from east to west and the (often supralocal) attention from north to south (see Figure 7.1).

In the landscape where the water structure of the Demervallei and the sandy structure of the Kempisch Plateau met, a unique east-west oriented settlement pattern grew on the edges and peaks. These east-west "corridors", or "village routes"

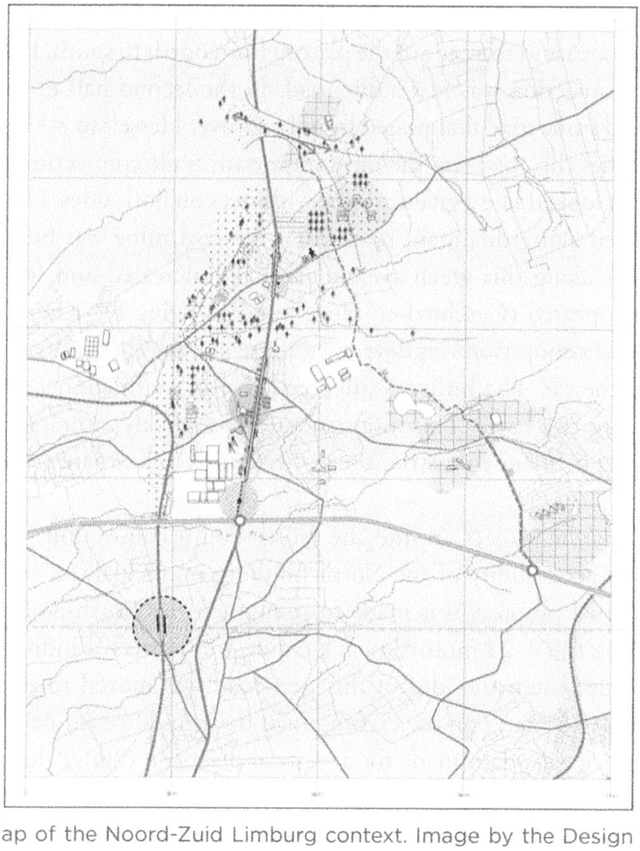

Figure 7.1. Map of the Noord-Zuid Limburg context. Image by the Design team Studio NZL (2019).

(OSA, 2018) formed the historical basis of the local road infrastructure as a connector between living, working and facilities, in dialogue with the landscape of the little valleys. Coal was discovered on the same axis at the beginning of the 20th century, as part of a supra-local movement stretching from the German Ruhr area over Limburg to the French region of Hauts-de-France. The Houthalen mine had a slower start-up and finally went into production in 1939 as the youngest of the seven mines in Limburg. The facilities that supported the exploitation of coal were also part of these "village routes". In the same way, the cité "Meulenberg" was connected with a historical axis parallel to the valley of the Laambeek and leading to the church of Houthalen (Atlas der Buurtwegen, 1841). The miners named this axis "Koolmijnlaan", because it connected the cité with the mining site. Because of its favourable location at the junction between cité and historical network, the Koolmijnlaan created the perfect climate for small independent entrepreneurs. The foreign miners wanted shops, cafes and restaurants like in their own country, so the Koolmijnlaan grew into a colourful and lively trading street.

In the meantime, also a supra-local North-South route steadily grew, crossing both the valley area and the edge of the plateau from north to south. Before the 18th century this connection was just a dirt road. In the second half of this century it became a paved stone road that passed from Liège over Hasselt to 's-Hertogenbosch. For two centuries this road was the most important trade connection in the east of Belgium, but it looked like a green avenue with trees on both sides. Until the 1920s, few people lived along this main road. After the coal mine was built, the houses and companies along this green avenue gradually increased and from the 1950s the first cars appeared (Geschied-en Heemkundige Kring De Klonkviool, 2018). This supra-local connection was developed more intensively thanks to the growing popularity of the car. The hitherto still green North-South connection was paved and divided into two lanes. This infrastructure was quickly saturated in the 1960s and the existing two-lane infrastructure was extended to four lanes, losing its green character.

In the 1970s the project around the North-South connection started, mainly to improve the accessibility of the North Limburg municipalities and companies. In 1971 an initial proposal was made to strengthen the North-South connection via constructing the A-24 motorway as a connector between Eindhoven and Huy on the East side. Later, the idea of this new road was moved to the west of the municipality (1978), later on the existing road that would be tunneled (2000) and finally the choice was again made for a western diversion (2007) that was also put aside in 2017. Characteristic of this debate is that the North-South connection was mainly approached from the perspective of the car, with little attention for the local east-west network that traditionally connected the people, animals and nature elements who inhabit the space. At the end of 2018, with the start of the complex project North-South Limburg, the interdisciplinary research and design team Studio North-South Limburg (Studio NZL) – of which the authors of this text are part – actively reopened up this focus on a car connection to themes such as living and working, open space, entrepreneurship and also mobility in all its aspects (cycling, walking, driving, public transport, etc.). Within the design goal of the project that needs to provide an answer to some of the issues the supra-local North-South connection is confronted with, the research team of the University of Hasselt supports in opening up this perspective by involving people, animals and natural actors that inhabit the surrounding rural context.

7.4.1.2 The district of Meulenberg

In order to deepen our understanding and commitment to this local network, we started working with specific neighbourhoods. We became, among others, more actively involved with the network in Meulenberg, because we were aware of the fact that this neighbourhood had received too little attention and care in recent history. From the seventies onwards we see how Meulenberg is increasingly evolving into an island within the local east-west network in Houthalen-Helchteren, because of the closure of the mines in Limburg and an increasingly busy North-South connection. In 1970, the government threatened to cut back on mining activity, causing miners to go on strike en masse. This context gave birth to the first community work "vzw Buurtopbouwwerk Meulenberg" in 1972 and its own youth work in 1978 (Lingier, 2011). In 1964 the Houthalen mine, the youngest mine in Limburg, had to close its shafts after twenty-five years, because of a lack of coal deposits and unfavourable geological conditions. Fortunately, the mine merged with that of Zolder and a large group of miners and managers were able to keep their jobs: the shafts in Houthalen remained operational for descents and after two years the head offices were put back into operation by management and administration. The Zolder mine was definitively closed in September 1992 as the last mine, both in Limburg and in Belgium (van Doorslaer *et al.*, 2012).

During the mining industry, the mine took good care of its employees: it provided housing embedded in a high quality and green public space, schools, shops, opportunities for relaxation, etc. In return, the mine did not tolerate any disorder, which was meticulously supervised: not pruning the hedge or not taking care of the garden, was punished with a fine. After the closure, this care ceased overnight. For a long time, the municipalities, which had never had to invest in the mining districts, denied the problems raised by the inhabitants. The community workers encouraged the residents to take the initiative themselves, and to confront the municipality. Slowly the confrontation policy of the municipality was transformed into a cooperation policy. At the beginning of the 2000s, community work was encapsulated within the non-profit organisation Rimo Limburg, a sub-contractor of the municipal authorities, and youth work was also encapsulated in the municipality's cross-neighbourhood youth welfare work. This growth of neighbourhood work was obvious during our preparatory fieldwork in the context of the complex North-South Limburg project.

At the same time, Meulenberg was isolated by the increasingly busy North-South axis. Some neighbourhoods and groups were able to defend their position around the North-South. This was the case for many supra-local established platforms and local action groups, who are today also strongly involved in the current complex North-South Limburg project from the start: Beweging.net, Boerenbond, Bond Beter Leefmilieu, Ondernemingsclub, Limburgse Milieukoepel, Unizo Limburg, VKW-Limburg, Voka, etc. We see that these platforms, together with the local community, and supported by local political parties, united over time in a few local action groups: Aktiekomitee "A-24", Aktiekomitee "Grote Baan", action group "Noord-Zuid-Nooit", action group "Om-Nu", action group "Om-U", action group "Om-nee" etc. However, because the various routes that were discussed in the historical debate have had little direct impact on the local network on the east side of Houthalen-Helchteren and Meulenberg, the local community on the east side was not involved from the outset in the North-South Limburg complex project, although the Meulenberg district became strongly isolated from the centre of Houthalen. Gerard, creative coach at vzw L.A.C.H. in Meulenberg (interview, 8 June 2019) explains that his father never let him cross the Grote Baan. According to him and other people we interviewed, the historically grown opposition between the East and West of Houthalen-Helchteren grew over the years. There was a perception of a difference between the rich West, which flourished thanks to the railway station, and the poor East, which became increasingly isolated from the centre by the busy Grote Baan, resulting in even more social deprivation (G. Aerts, interview, 8 June 2019; F. Didden, interview, 31 June 2019; D. Pauli, interview, 22 September 2019).

7.4.2 Platform-Approach: "North-South Limburg Bicycle Library" as Part of a Platform Mobility

In order to understand and design with the collectives that were active and spread out in this rural area who cared for and worked on a more caring approach to mobility which could include the excluded actors and areas, we slowly developed a platform for mobility for the North-South Limburg connection. From our historical overview it became clear that there exist platforms that have been striving for better and more inclusive mobility around North-South for years now, such as union groups, political parties, neighbourhood actions groups and more recently groups like the "fietsersbond (bikers union)", but that they often represent only a part of the community and often fail to reach groups that are lower educated, women, children and youth and migrant communities. Therefore, we developed a platform-approach, namely an approach that aimed for carefully developing historical-aware and situated platforms that support sharing between a larger diversity of groups (see also Botto and Teli, 2017). In order to understand the diversity of actors we continuously mapped and brought together the local network that cared for mobility; we supported groups and people we discovered in the network to design for existing and new opportunities for mobility, we invested in scaling small initiatives – often very distributed in space, because of the less densified character of the municipality – in order for them to be part of stronger and more robust networks. We also communicated these activities to a broader audience in order to inform and strengthen the network.

After an extensive field research Platforms were set up at the end of 2019 as a way to interweave the design trajectory of the "Studio NZL" design team that worked with the more established groups on the North-South, and the participatory trajectory of the UHasselt team that was aiming for including more diverse voices in the process. Today, the design team Studio NZL is one of the many actors – on an equal footing with other actors such as pupils, schools, the municipality, neighbourhood organisations, other professionals, etc. – in driving these platforms. Sometimes in a pulling role, other times in a supporting role, or just as a stakeholder. So we see that the platform has evolved from "a platform that served the environment of the complex project North-South Limburg" to "a platform as a building block for a sustainable society".

In order to answer our research question on how can we design platforms that can enable sharing initiatives to develop in contexts, and with actors that are systematically marginalised in sharing economy initiatives, **in close interaction with existing institutions that drive sustainable mobility transitions?**", we did a bottom-up analysis of how we researched and designed the platform, all the interviews, images produced, conversations during workshops. Based on this, we distinguished five capabilities that supported us and the people involved in building a trusted platform around a shared space as a building block for a sustainable society: networking, coaching, scaling, communicating and imagining. On the basis of the project "North-South Limburg bicycle library" – started from the Platform Mobility – we clarify these five capabilities below.

a. Networking

During the project we were continuously actively present in the field – physically and digitally – in order to collect information on historical and present individuals and groups who cared for their street and their mobility, each in their own way. In the beginning of the project we were mainly physically present in the field, but since March 2019 COVID-19 challenged us to explore new methodologies to reach and engage individuals and groups. Since physically meeting in a large group was no longer possible, we looked for alternatives both digitally and physically through a one-on-one meeting or a doorstep conversation. All these conversations were visualised in a map that grew through time and gave an overview of people, artefacts and groups that were engaged with the mobility theme over time. A small sample of the map (see Figure 7.2) is shown below.

In our interviews and observations on the larger regional scale, but very specifically also in Houthalen-Helchteren, one set of caring relations became apparent: we heard from various target groups their care for learning to cycle. Many parents or grandparents cannot teach their child or grandchild how to ride a bike because they have never learned it themselves or because they have forgotten it over the years. For many newcomers who are staying temporarily in the refugee center 'De Kazerne', the bicycle symbolizes their freedom (I. Martens, interview, 2 September 2019). The bicycle is also an ideal means of transportation for people in socially vulnerable situations. In the De Standaard neighborhood, a group of women are also asking to learn how to cycle. Schools indicate that they would like to use a bicycle more often with students for short trips or day trips (E. Bogaerts, W. Hoebers, C. Pistolas & T. Verheyen, Platform Mobility, 12 December 2019). This is not self-evident. Pupils do not all have a bike or are not allowed to bike to school

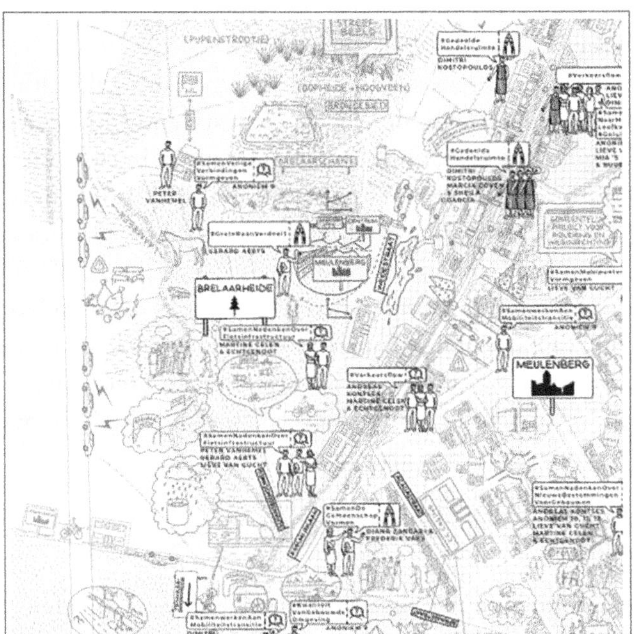

Figure 7.2. Small sample of the map. Image by Stieglitz and Goyens (2019).

on a daily basis, because the route to and from school is perceived as too dangerous by their parents. Teachers also note that students are less able to cycle over the years and more accidents happen during class trips.

When we brought together some of those people we discovered in the network – people who care for learning to cycle and safe cycling – the idea of a bicycle library (de Fietsbieb) grew: a lending point for children's bikes up to 12 years old. The existing Fietsbiebs focus more on the individual, not yet on groups or schools: if you need a bike as a child, you and your parents will choose a bike in the library. The field research showed that there is currently no service where schools or associations can temporarily lend out bicycles. To explore further this idea we met Beweging.net, a social movement network that grew from the Workers Movement connecting 11 organisations (such as women and elderly organisations) that manages most bicycle libraries in the region.

Furthermore, we engaged more actively with the network in the cité Meulenberg, because of the historically detected lack of attention and care throughout the years. In the preparatory fieldwork, we detected many committed citizens and organisations who are sincerely concerned with the future of the neighborhood: vzw RIMO, vzw Stebo, youth well-being Meulenberg, cultural center Casino, vzw L.A.C.H., the elementary school Sprankel, etc We entered this network through a "Live project", a project in which each year we as architecture researchers, together with our master students (interieur)architecture of the UHasselt and with the local community, turn an ongoing research project into a live built intervention, to explore and materialise some deeper insights on location. This intervention made clear how themes such as cycling and qualitative public space already were present in the neighbourhood and could be further explored (Birgit, Brent, Busra, Fadime, Louise, Mahya, Sebnem, Selin, Tom, interview, 16 September 2019).

One of the core actors in the community was situated at the top of Meulenberg: the elementary school Sprankel. Eighty years ago, Isia Isgour, a modernist architect, formed the basis for this inspiring learning environment. He designed an educational kindergarten in the form of separate, comfortable cottages framed as communities for small groups of toddlers. Based on the same pedagogical vision, he also designed an elementary girls' and boys' school, with gymnasium and stage, around a large playground. This former mine school has an interesting excess of green space between the primary school and the nursery school, which at one point was closed off from the public space. The entire school site was fenced off. This once lively green oasis became with time a forgotten piece of green. During the Live project, this forgotten piece of green started to live again: we as researchers and students removed part of the school's fence and literally built a gate welcoming everyone in this green oasis. Equally important, this gate was also conceived as the starting point of a walking route connecting all existing and potential meeting green places in the neighborhood. This intervention ended up boosting the plans of the local community to found a children's campus that today already hosted functions such as the Chiro, the sports hall, pre- and after-school care, the drawing academy and also youth well-being. After the Live project, in September 2020 this site was again expanded with a nursery for babies and toddlers and also a secondary Freinet school and a bicycle library. They also plan a traffic park and a music academy for 2021.

> "The children's campus thus becomes an ideal place for working parents who want to offer their child as many opportunities as possible, without having to ride back and forth. On the traffic-free site, children from the same family can go to daycare, to the Jenaplan kindergarten and elementary school and to the Freinet secondary school. After school these children can circulate independently between daycare, sports hall, drawing or music academy. On Wednesdays and Saturdays sports activities and playground activities are offered. On Sundays there is Chiro. All in one place. This unique concept focuses on the development opportunities and talents of all children and young people," Christos Pistolas tells us. (interview, 26 September 2019).

This network became the basis for building a caring "Platform Mobility" in Meulenberg and by extension around the North-South Limburg road connection, which led to projects such as the bicycle library and a bicycle park to learn to cycle. The school offered a space within the building for the bicycle library that is spatially related to the neighbourhood. Beweging.net started searching connections between the library in development and the recycle centre in the neighbourhood, that brings used and repaired bicycles to the school. When the coordinator of youth well-being became in charge for the library project, the project also became networked with the municipality.

b. Coaching

We visualised the group of people and organisations that were engaged with the mobility theme in the map that gave an overview of people, artefacts and green (see small sample of the map, Figure 7.2). In a first stage, they were research subjects for us as researchers, but later they became trainers for our students and colleagues less familiar with the neighbourhood. At the start of the Live project they showed a group of our master students around in the cité Meulenberg. During the Live project we as researchers, together with our students, kicked off the bicycle library by temporarily building a bicycle training environment in the middle of this play and learning landscape. The children were asked to take their bikes with them so that they could test the temporary bicycle training environment during the physical education lessons and after school hours. After a while, a network of people emerged that started coaching each other in cycling: the school trainers and the (grand)parents gave cycling lessons for small groups of children through the schools, for (grand)parent and (grand)child, for young adults who are new in the country, and so on. Parents came to school earlier and stayed a bit longer while their child played and cycled together with their friends. Christos Pistolas (interview, 26 September 2019), director of the Jenaplan elementary school Sprankel, was already very excited:

> "This project fits in seamlessly with the vision for the future that we as a school have in mind from September '20: an open school community linked to a play and learning landscape that pupils, young people, adults and the elderly can make use of before, during and after school hours. The entire teaching staff now sees the potential of this vision of the future temporarily translated into space".

Together with the local community who already experienced their role as trainers in the network and Beweging.net we then managed to develop a Fietsbieb as the ninth location in Limburg. Slowly the collaboration between the school, Beweging.net and UHasselt/Studio NZL developed further. The idea grew that the temporarily built bicycle training environment could be made sustainable with a more permanent bicycle training park. The more the children's campus and Beweging.net became the core drivers and core trainers in the project, the more the role of Uhasselt/Studio NZL changed into a support in the back-office and finally being the liaison between the bicycle library and the Platform Mobility in which the different smaller initiatives around mobility in the North-South Limburg project are discussed. This platform became a learning environment enhancing collective learning between pupils, (grand)parents, schools, neighborhood organisations, the municipality, spatial professionals, the social movement platform (Beweging.net) and the Flemish government on how a shift in mobility culture can occur.

Within the Platform Mobility, the Fietsbieb project, located at the school, now served as a sharing economy platform in itself within which children, their mothers and fathers, schools, neighbourhood organisations, the municipality and spatial professionals can learn from each other. The coaching role became more and more distributed between multiple actors: groups that set up initiatives (the schools), groups that needed to learn how to cycle (the pupils and their (grand)mothers and (grand)fathers), a regional organisation with experience in the field of Fietsbiebs (Beweging.net).

c. Scaling

By mapping the network of people who care for cycling, we discovered actors and groups on a micro and on a macro level and detected possibilities for new collaborations between them. The most relevant decision in this particular case was that a collaboration was set up with a regional social movement network to set up a bicycle library to make it part of a bigger discourse on mobility in the region (an approach which we have called institutioning in Huybrechts *et al.*, 2018). The choice for the network Beweging.net versus the other network, "Op Wielekes" was it strong relation with the region (Beweging.net runs 10 libraries in the region versus Op Wielekes 1) and its affinity with the more rural and multicultural character of the region, the (women's and elderly and other) groups, rather than its political colour. It was felt that in our region there is a great need for collaboration between small initiatives on regional level, because of the lack of an urban context that can surround the small initiative with a lot of dynamic. While the project was solely on a voluntary basis until January 2019, at the end of 2019 we applied for funding from cera foundation to enable the start-up of this Fietsbieb. Early February 2020 the funding was approved.

The intention was also to consider this Fietsbieb as a pilot project for and by the schools in the larger municipality and the neighbouring municipalities. The district of Meulenberg as a location for the Fietsbieb was interesting on a regional level because no bicycle libraries had yet been started in this region of Limburg. The intention is that in time this Fietsbieb will also have faces (e.g. information stands) at various schools (municipal level) in the vicinity because together with Meulenberg, the entire region also needs a new perspective on cycling. The design process of the Fietsbieb creates a collaboration with primary and secondary schools in which children and young people – often from underprivileged groups – can learn skills to take control and design qualitative shared space, in collaboration with researchers, teachers and students. In this way we want to develop the skills of everyone – especially groups that are often neglected in spatial design processes, such as disadvantaged groups – in order to give them a voice and to develop further in the spatial debate.

Finally, this Fietsbieb with bicycle track as part of a green spot in the neighbourhood, located at the top of the Meulenberg, will also be used strategically as part of an incubator that contributes to the revival of the isolated district of Meulenberg among the general public in order to break through the historically grown contrast between East and West, reinforced by the ever busier Grote Baan.

From the Fietsbieb we thus entered into links with different scale levels via collaboration with the regional social movement running different bicycle libraries, with elementary and secondary schools, with urban planners/designers in the university and the participating agencies, and within the neighbourhood.

d. Communicating

In order to support the diverse voices in the sharing economy platform initiative in development, the communication strategy became more and more decentralised over time. While at the start the mobility Platform mainly used the communication of the complex North-South Limburg project to gather people and communicate the initiatives, over time the different partners took over and the North-South Platform communication diminished. The bicycle library keeps a fixed project page on the website of the North-South project as one of the "short-term wins" of the project. Here, regular updates are published.

Beweging.net offered a fixed method to guide the communication of Fietsbiebs during the start-up. Together with UHasselt, the organisation supported the funding application in the search for a start-up budget. Because the organisation had a large network with many local anchor points, they could efficiently facilitate the search for a possible location, volunteers and children's bikes. Finally, each new Fietsbieb also gets its own page on Facebook and its own email address so that the volunteers can stay in touch with their (new) members and start building their own local network. Also the school's role became more active over time. During the school party of the primary school on Saturday, March 14 2019, and also during the opening of the children's campus in the beginning of the school year 2020–2021, the school wanted to introduce the arrival of the Fietsbieb to the general public. As a result of the measures concerning the Coronavirus, a large party was postponed for an indefinite period of time. Instead, guided tours were organised in small groups in the autumn where children could discover the campus and the library together with their parents. Additionally, the municipality had played an active role in communicating the project, via their monthly magazine and via their online communication.

The launch of the sharing platform for bikes was in the end organised online and not via an event on location. A digital platform was set up by launching a webpage "Fietsbieb Houthalen-Helchteren" on Facebook. All partners could use this Facebook Page to spread both the arrival of the library and the call to volunteers and children's bikes within their own network in order to reach the widest possible target group. The regional newspaper "Het Belang van Limburg" published the arrival of the Fietsbieb as part of the children's campus, as well as a call for volunteers and (children's) bikes. Volunteers who wanted to host the library or repair bikes, could register by email and children's bikes could be brought in on a daily basis to either the elementary school "Sprankel", the future location of the Fietsbieb, or the recycle centre in the neighbourhood. In exchange for offering a children's bike to the library, the children receive a voucher with which they can borrow a bike for a year. In the meantime three technicians have reported themselves as volunteers. The library opened in full capacity in April 2021.

e. Imagining

The whole trajectory was not only focussed on networking, coaching, scaling and communicating what was and what is, but also on what "could be". The bicycle library was potentially present in the past and present network and was strengthened by the above mentioned capabilities, but what made it emerge as a project was the capability of the collective to imagine a future. In this imagination process they looked at the already existing relations in the community and discovered that they had all the resources, when working together, to collaboratively realise a project that could impact the transformation from a car-centred to a more bike-oriented community. What the project put forward was the capacity as a collective to work together concretely on an abstract theme such as 'sustainable mobility culture' for the future, right from the start, based on relations that already exist, but could be strengthened in the community. Imagining a potential project thus served as a driver for the other capabilities: imagining how to strengthen the networking potential, stimulating people to take a coaching role, enabling to scale in more concrete ways (namely, by imagining the role of the bicycle library within the larger sustainable mobility project) and supporting the shift in communication strategy from central to decentralised, from professional to personal.

7.5 Discussion: Towards a Political Definition of the Sharing Economy

In the above process it became clear that tracing the history, studying the present and imagining the future of situated caring relations enabled the emergence of a more political definition and conception of a sharing economy initiative. It became clear that the capability of imagination – as mastered and introduced by the designers involved in the process – became a driver for the different actors to engage in a careful exchange of their capabilities, which allowed existing groups present in the community to self-organise themselves as a sharing collective (as defined with Latour, 1999 and Escobar, 2018 earlier) that in its diversity and radical interdependence could take a step forward in the direction of a more sustainable mobility transition. The process showed that the step from the potential of capabilities in the network and the realisation of an imagined sharing economy initiative, required a collective learning process. This was especially important in a context where the awareness on the potential of sharing economy initiatives was particularly low. The set-up of a platform (van Dijck *et al.*, 2018) that explicitly enables this collective

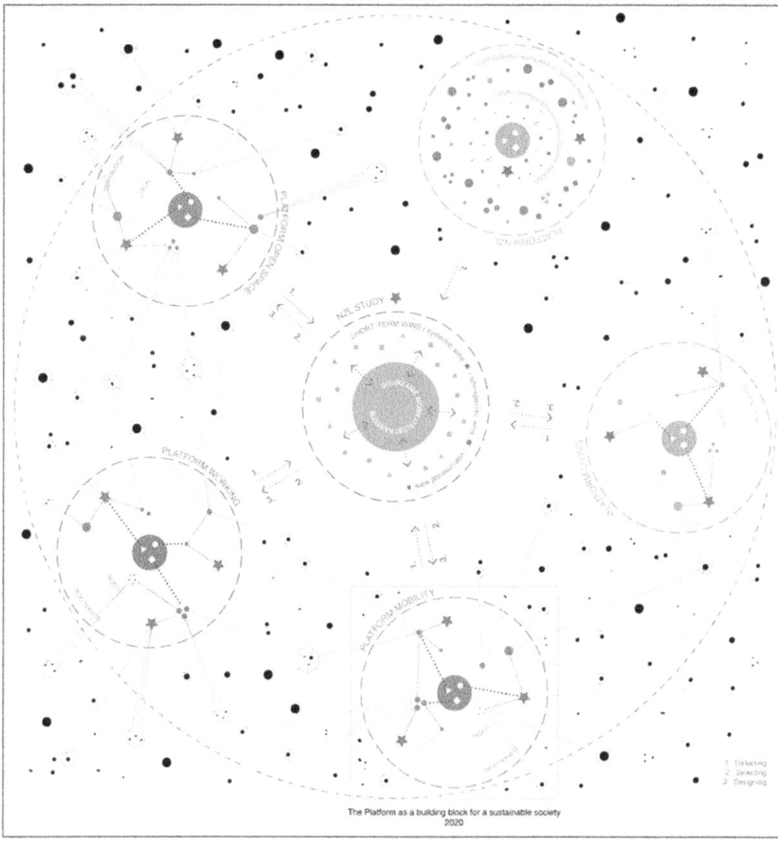

Figure 7.3. Scheme "platform-approach". Image by Goyens (2020).

learning on how we share and can share in the future, was an effective way to debate the ownership and power relations in the sharing initiative under development and to sustain the sharing which we had built on and prototyped for two weeks during the Live Projects.

When tracing back the recent history of the project, we noticed that there were a few aspects that were important to support this collective learning process as a stepping stone from prototype to sustainable sharing economy initiative. More particularly, we distinguish five capabilities that supported us and the people involved in building a trusted platform around a shared space: networking, coaching, scaling, communicating and imagining. The schema above (see Figure 7.3) illustrates this platform-approach.

a. Networking: Platform as part of the world

In the process from a divided to a shared space for a sharing economy initiative to emerge with sensitivity to politics, we entered continuously into situated networks and expanded and sharpened them. This growing network is anything but stable today, but it does have sufficient resilience to collaboratively reflect upon (in this case for young people, often from vulnerable backgrounds) and to take up complex challenges. **By embedding a prototype of the sharing economy initiative in a strategic location in the neighbourhood, a careful connection between people and space in the community is created.** By temporarily imagining this potential project, a stable network is gradually emerging in the neighbourhood that is embedded both locally and supra-locally and forms a basis for building a platform for caring mobility.

b. Coaching: Platform as a space activated by actors

All working methods to create a caring platform were continuously supervised, yet not necessarily by spatial planners and designers, or even participation experts. The strength of the platform is that the coaches of the several projects were present in the social network itself and gradually formed a new network, sustained by a potential project in the field. We detected different types of coaches: the participation professional, the designer, the communication expert, the social professionals (the schools, the unions), the family and friends. All of them have different motivations to engage: the schools were looking for projects that taught their students to think and communicate socially, the trade unions were looking for partners to reconnect their social movements to particular neighbourhoods and the designers were looking for a greater involvement of children and young people, to enhance a diversity of groups in their design process.

c. Scaling: Platform as a political space

Throughout the whole process imagination supported the step from the personal, situated world to the complex and uncertain world of "sustainable mobility transition" and infrastructure plans in the far future and vice versa. A prototype of the sharing economy initiative of the Fietsbieb helped to make this process of scaling up and/or down of the impact of the initiative a step-by-step process, or at least provided handles for it.

d. Communicating: Platform as a sustainable (memory) space

In order to support the diverse voices in the platform, over time the communication strategy became more and more decentralised. While at the start the Platform Mobility mainly uses the communication of the North-South project to gather people and communicate the sharing economy initiative, over time the different partners (Beweging.net, elementary school, municipality, etc.) took over and the North-South Platform took a supportive communication role. Situating the project

Figure 7.4. Picture of the bicycle library in development via small scale interventions, in the context of the Noord-Zuid Limburg project. Image by Goyens (2019).

in the past, present and future of the field supported the shift in communication strategy from central to decentralised.

e. Imagining: Platform as a space for collective imagination

The whole trajectory was continuously focussed not only on networking, coaching, scaling and communicating what was and what is, but also what "could be". The imagination of the bicycle project (see Figure 7.4) in the network brought out the capability of the collective to give form to a situated future. This imagination process served as a driver for the other capabilities: imagining revealed, stabilised and strengthened collectives – in their radical interdependency (Escobar, 2018) – locally and supra-locally. It stimulated people to take a coaching role, it enabled to scale in more concrete ways and supported the shift in communication strategy from central to decentralised, from professional to personal. **The interactions between**

these capabilities, triggered by imagination, contributed to creating the networks between inhabitants and institutions (Huybrechts *et al.*, 2018) that set a sustainable mobility transition in motion.

7.6 Conclusion: Situating the Sharing Economy in Rural Contexts

Based on the idea that sharing is a situated practice (John, 2016) and on our desire to create platforms that facilitate sharing between diverse groups of people, beyond those groups who live in big cities, who are well-educated and with sufficient incomes (Bársony, 2017), we developed a platform methodology (building on e.g. Botto and Teli, 2017). This methodology aimed for rooting the sharing economy initiatives in a history of sharing in the environment and then further developing and tracing how the sharing economy initiative develops in the present and further into the future. This was done via the slow discovery and introduction of new actors, tools and networks via a process of imagination. This process proved to be beneficial for a more political understanding of the development – and the power and ownership relations – of the sharing economy initiative. It allowed the existing "sharing collectives" that were distributed in the rather rural context of the region, with their particular histories and interests (e.g. their own street), to become "collectives that share in time and space": they entered in a learning exchange between each other and other individuals in the discussed study. They exchanged about different traditions in sharing, in what to share and how to share and how this can contribute to a sustainable mobility transition on a larger scale.

From this process we learned that as long as we continue to approach mobility as a theme that divides us (between the urban and the rural, between cultures, between young and old etc), it is difficult to find a ground for sharing economy initiatives that have the potential to include certain marginalised groups and spatial contexts. If we approach mobility and the street as a theme that we share, this opens a new dimension of living together: What do we share in the discussion, how can we gain knowledge about it through time and build a future around it? And how can we do that in an environment that we have built together? The platform methodology slowly took shape via the collaborative articulating, tapping into and developing five capabilities and formed the ground for the development of the discussed sharing economy initiative. While it still requires a lot of research and experimentation, in the case of North-South Limburg it offered both methodological and thematic support in shaping a space of care together in times where many issues, such as mobility, are discussed in polarising ways.

References

Atlas der Buurtwegen van Vlaanderen. (Cartographer). (1841). Atlas der Buurtwegen [Historical map]. Consulted on 22 September 2020, from http://www.geopunt.be/kaart.

Avram, G., Choi, J., De Paoli, S., Light, A., Lyle, P., and Teli, M. (2017). Collaborative Economies. *Proceedings of the 8th International Conference on Communities and Technologies*, 305–307. https://doi.org/10.1145/3083671.3083712

Berlant, L. (2016). The commons: Infrastructures for troubling times. *Environment and Planning D: Society and Space, 34*(3), 393–419. https://doi.org/10.1177/0263775816645989

Bársony, F. (2017). The age of sharing by Nicholas A. John (Cambridge, Polity, 2016), book review. *Corvinus Journal of Sociology and Social Policy, 8*(1), 146–152. https://doi.org/10.14267/cjssp.2017.01.11

Botto, F. and Teli, M. (2017). PIE News. A public design project toward commonfare. *The Journal of Community Informatics, 13*(2).

Davidson, N.M. and Infranca, J.J. (2016). The Sharing Economy as an Urban Phenomenon. *Yale Law & Policy Review, 34*(2), pp. 215–279.

Escobar, A. (2018). *Design for the Pluriverse.* Durham: Duke University Press.

Geschied-en Heemkundige Kring De Klonkviool. (2018). *De Noord-Zuidkeuze in Houthalen-Helchteren.* Houthalen-Helchteren: Geschied- en Heemkundige Kring De Klonkviool.

Gehl, (2010). Cities for People. Washington DC: Island Press.

Gil, N. and Baldwin, C.Y. (2014). Sharing DesignRights: A Commons Approach for Developing Infrastructure (Working Paper No. 14-025). Harvard Business School.

Goyens, H. (2019). Interview with G. Aerts in the context of the complex project Noord-Zuid Limburg. Interviewed on 8 June 2019.

Goyens, H. (2019). Interview with F. Didden in the context of the complex project Noord-Zuid Limburg. Interviewed on 31 June 2019.

Goyens, H. (2019). Interview with I. Martens in the context of the complex project Noord-Zuid Limburg. Interviewed on 2 September 2019.

Goyens, H. (2019). Interview with students Birgit, Brent, Busra, Fadime, Louise, Mahya, Sebnem, Selin, Tom in the context of the live project "Koolmijnlaan". Interviewed on 16 September 2019.

Goyens, H. (2019). Interview with D. Pauli in the context of the complex project Noord-Zuid Limburg. Interviewed on 22 September 2019.

Goyens, H. (2019). Interview with C. Pistolas in the context of the complex project Noord-Zuid Limburg. Interviewed on 26 September 2019.

Huybrechts, L., Benesch, H., and Geib, J. (2018). Co-Design and the public realm. *CoDesign, 13*(3), 145–147. https://doi.org/10.1080/15710882.2017.1355042

Huybrechts, L., Hendriks, N., and Martens, S. (2016). Counterfactual scripting: acknowledging the past as a resource for PD. *PDC '16: Proceedings of the 14th Participatory Design Conference: Full papers – Volume 1*, pp. 111–120. https://doi.org/10.1145/2940299.2940304

Huybrechts, L., Yevchenko, Y., and Palmieri, T. (2018). *In Hasselt, Everyone Shares. An Exploration of Sustainability in Hasselt.* Hasselt: the School, City of Hasselt.

Illich, I. (1974). *Energy and Equity.* London: Marion Boyars.

Image 1. Design team Studio NZL (2019). Map of the Noord-Zuid Limburg context [Illustration].

Image 2. Stieglitz, J. and Goyens, H. (2019). Small sample of the map [Illustration].

Image 3. Goyens, H. (2020). Scheme "the platform-approach" [Illustration].

Image 4. Goyens, H. (2019). Picture of the bicycle library in development via small scale interventions, in the context of the Noord-Zuid Limburg project [Photograph].

Inventaris onroerend erfgoed. (2018, 1 februari). *Steenkoolmijn van Houthalen: Cité Meulenberg.* Consulted on 22 September 2020, van https://inventaris.onroerenderfgoed.be/erfgoedobjecten/122142.

John, N.A. (2016). *The Age of Sharing.* Cambridge: Polity.

Joseph-Jean-François graaf de Ferraris. (Cartographer). (1771–1778). Ferraris kaarten [Historical map]. Consulted on 22 September 2020, from http://www.geopunt.be/kaart.

Latour, B. (1999). *Pandora's Hope: Essays on the Reality of Science Studies.* Cambridge: Harvard University Press.

Latour, B. (2018). *Down to Earth: Politics in the New Climatic Regime.* Cambridge: Polity Press.

Lingier, J. (2011). *De kracht van onderuit: een geschiedenis van het buurtopbouwwerk in de mijnstreek.* Houthalen-Helchteren: Haletra.

Marttila, S., Botero, A., and Saad-Sulonen, J. (2014). Towards commons design in participatory design. *Proceedings of the 13th Participatory Design Conference on Short Papers, Industry Cases, Workshop Descriptions, Doctoral Consortium Papers, and Keynote Abstracts – PDC '14 – volume 2*, 9–12. https://doi.org/10.1145/2662155.2662187

Teli, M., Tonolli, L., Di Fiore, A., and D'Andrea, V. (2019). Computing and the common. Learning from Participatory Design in the age of platform capitalism. Trento, Italy: Università degli Studi di Trento. http://doi.org/10.5281/zenodo.3228359

OSA. (2018). *Oost-West/Noord-Zuid Houthalen-Helchteren.* Leuven: KU Leuven.

Platform Mobility (2019). Workshop with E. Bogaerts, W. Hoebers, C. Pistolas and T. Verheyen in the context of the complex project Noord-Zuid Limburg. Held on 12 December 2019 by Studio NZL.

Seravalli, A. (2014). *Making Commons: Attempts at Composing Prospects in the Opening of Production.* Doctoral Dissertation, Malmö University.

Studio NZL (2019). *Ambitienota Noord-Zuid Limburg.* Brussel: Werkvennootschap.

Team Vlaams Bouwmeester. (2014). *Mijnwijk, Tuinwijk, Inwijk: een actieplan voor Meulenberg-dorp.* Brussel: Team Vlaams Bouwmeester.

te Brömmelstroet, M. and Verkade, T. (2020). *Het recht van de snelste: hoe ons verkeer steeds asocialer werd.* Amsterdam: De Correspondent.

van Dijck, J., Poell, T., and de Waal, M. (2018). *The platform society: Public values in a connective world.* Oxford: Oxford University Press, 226.

van Doorslaer, B., De Rynck, P., *et al.* (2012). *Mijnerfgoed in Limburg. Ondergronds verleden, bovengrondse toekomst. Van Beringen tot Eisden.* Gent: Openbaar Kunstbezit Vlaanderen vzw.

Zuljevic, M. and Huybrechts, L. (2019). (Un)Curating the City: Participatory design and urban heritage. Espoo: Nordes 2019, Who Cares?

DOI: 10.1561/9781680838411.ch8

Chapter 8

Assessing Perspectives and Opportunities of Airbnb Hosts in Albania

By Sindiola Koka, Alba Kruja (Demneri) and Eglantina Hysa

Airbnb nowadays is considered as one of the largest networked hospitality businesses that went from sharing rooms into a very successful business spread worldwide. It is continuously expanding its activity and nowadays it has become a leader of networked hospitality by bringing innovation in business models and increasing the number of tourists due to its low prices. In recent years, Airbnb has grown rapidly worldwide, including Albania. This study focuses on Airbnb hosts in Tirana: how they have adapted to the difficulties and benefits this platform offers, what motivated them, how being Airbnb hosts has affected their lives and where they see the future of this platform in Albania. Following a qualitative research method, semi-structured interviews with 8 Airbnb hosts in Albania were conducted. The findings show that economic benefits along with the social ones lie behind the main motivation on becoming an Airbnb host. In terms of business success, being friendly and kind with the customer, but also professionalism are important. The results suggest that there is a need of building a *local* Airbnb community to connect with other hosts, share experiences and advise each other on providing better services to the guests. Moreover, the sector suffers from a lack of regulatory framework which calls on the attention of government and policy-makers. Knowing the problems, challenges, and benefits of operating under Airbnb would be interesting for policy-makers to be able to support the sector with the necessary regulatory framework.

8.1 Introduction

Airbnb is an online marketplace that aims to connect people who want to rent their apartments, homes, villas with people who are looking for accommodation in different locations and destinations. In most of the cases, Airbnb provides relatively affordable accommodation while at the same time offering hosts a way of earning income from their property, often a second house given for rent. In recent years, excluding the period of travel restrictions related to Covid-19, Airbnb has enjoyed rapid growth and today is considered as the largest accommodation service worldwide (Adamiak, 2019). It is indeed a very successful business, becoming a trend all over the world with an exponentially expanding number of members, competing strongly with the more traditional businesses in the hospitality sector.

However, with the Covid-19 pandemic, the accommodation market has experienced a significant decline, even more than other economic activities. Some preliminary studies on the effects of the current pandemic on the Airbnb market have predicted that there might be a shift of interest from the *Capitalist hosts* (they typically have a mortgage on the space they rent out, and are engaged in short-term renting to make commercial profits) to the *Befriender hosts* (enjoy the social aspect of hosting) and *Ethicist hosts* (they are the true believers in the principle of sharing) (Dolnicar and Zare, 2020). According to Dolnicar and Zare (2020), this is a step toward the original narrative of space sharing among ordinary citizens. The reason for this shift might be also the inability to pay for salaried staff in the case of professional hosts and some hosts using P2P accommodation platforms have decided to exit from these platforms (Farmaki *et al.*, 2020). This decision seems to have been encouraged by hosts' disappointment over the minimal support they received from the platforms and the way the platforms handled the pandemic by encouraging guests to ask for full refunds (Farmaki *et al.*, 2020).

But what has happened in Albania before the Covid-19 andemic? What is the trend of the accommodation market in the specific case of Airbnb? Similar to other countries, Albania has experienced an increase in usage of Airbnb with hosts sharing their apartments, homes, or villas. It should be noted that in Albania, Airbnb has not become a trend yet, as in many EU countries, but still it is spread in the main historical cities, representing cultural heritage, and others touristic cities and villages - especially on the seaside. Some of the main destinations are Tirana, Saranda, Korca, Vlora, Berat, Shkoder, Ksamil, Dhermi. The main contribution of this study lies in the analysis of a city where Airbnb has not yet "taken over" the accommodation market. We look at the perspective of hosts in Tirana—specifically, short-term renter of second properties—considering their motivations and challenges in hosting and more generally, how being an Airbnb host has impacted their lives.

Albania is a developing country rich in cultural heritage, where tourism is a crucial sector for its development contributing to one fifth of its GDP (Kruja, 2012; Kruja and Berberi, 2020; Hysa, 2012; Hysa and Gjergji, 2018; WTTC, 2018). Many researchers in Albania have highlighted that investing in information technology by increasing interaction and communication with the customers is decisive for this sector's development (Berhani and Hysa, 2013; Ferizi and Kruja, 2018; Hysa *et al.*, 2021; Kruja *et al.*, 2019; Noti, 2014). Yet there are no studies investigating the effects of Airbnb in Albania from the tourist or host perspective. In fact, this is the first study conducted on understanding the motivation of Airbnb hosts as well as their challenges and benefits. Additionally, the study contributes pointing to practical implications for government and policymakers working on the necessary regulatory framework. In particular, this chapter addresses the following research questions:

1. What are the challenges and benefits of Airbnb hosts in Albania (Tirana)?
2. What are Airbnb host perceptions of their experiences?
3. What are the motivations influencing Airbnb hosts?

To answer these questions, the chapter has the following structure. The second section is devoted to a literature review on the reasons for the popularity of Airbnb and the host motivations. The following two sections present the situation in Albania and the methodology adopted for this study. The fifth section presents the results, followed by their discussion, mostly in comparative terms. The conclusion highlights implications for entrepreneurs and for policy-makers.

8.2 Related Work

8.2.1 Reasons for the Popularity of Airbnb

Airbnb has shown to be a successful digital platform providing entrepreneurial opportunities and income supplement to hosts as well as extending the availability of choices for guests. Although there were already platforms for short-term rentals such as Craigslist which, initiated in 1995, brings together ads of multiple nature, or Couchsurfing, a non-profit portal initiated in 2004, Airbnb has managed to excel over competitors. Literature in general mentions some evident reasons that have supported the success of Airbnb, such as: price, flexibility, trust, brand, design and usability.

- *Price.* Airbnb's competitiveness with respect to traditional accommodation services was favoured primarily by low rental prices (Benítez-Aurioles, 2018; Gibbs *et al.*, 2018; Möhlmann, 2015; Teubner *et al.*, 2017).
- *Flexibility.* Airbnb is an extremely agile platform, capable of easily adapting to sudden development. In the customer's reviews analysed by Cheng and Jin (2019) "flexibility" was found as one of the most important factors positively affecting Airbnb usage.
- *Trust.* The problem of perceived security has always been the Gordian knot for strategy of Airbnb marketing. To foster trust, the company worked primarily on the platform, by introducing a review and ratings system (Fagerstrøm *et al.*, 2017; Newlands *et al.*, 2019; Ter Huurne *et al.*, 2017). The quality of the photos attached to the ads, the geographical location and access to the portal via Facebook Connect, which increased the information relating to the identity of users, also contributed to strengthening it. To this was added, with subsequent improvements, an insurance coverage against theft and/or damage protecting hosts (Möhlmann, 2015; Rosenfelt, 2014; Tussyadiah, 2015).
- *Brand.* Numerous researches also have explored the customers perception of the Airbnb brand. They found that the branding identity of this company was highly appreciated (Lee and Kim, 2018; Stollery and Jun, 2017; Yang *et al.*, 2018; Yannopoulou *et al.*, 2013).
- *Design and usability.* The design of the site contributed to success of the company, as well as to building trust among the users and to retain them. The platform is user-friendly (Forgacs and Dimanche, 2016; Lee and Kim, 2018; Möhlmann, 2015), with images large and high quality, an easy search system and an intuitive interface that allows a quick comparison of offers, also geographically. There are also simple payment methods, with payment occurring exclusively online. According to the study of Forgacs and Dimanche (2016), Airbnb follows what can be defined as emerging 21st-century business models, based on:n meeting real, existing needs; introducing an effective digital platform; generating a steady revenue stream based on transaction fees; using a cloud-based digital platform; focusing on quality and on user reviews; and on establishing a strong brand, and in scaling up rapidly.

Besides this innovative business model, an association with authentic local travel (Forgacs and Dimanche, 2016) is also seen as a key element differentiating Airbnb from other competitors, such as Booking.com. Airbnb has introduced a set of "host experiences", a unique marketing approach well-recognized by experts and observers (Davis and Hillier, 2019; Monllos, 2016). For instance, Davis and Hillier (2019), describes the campaign launched by Airbnb in 2016 "*Live There*" as a brilliant one, which has intelligently captured the company's brand proposition of travel, communities, people and experiences. "*Welcome to the world of trips*" was another new product offered by Airbnb in 2016, aiming to introduce a new set of experiences for its customers (Monllos, 2016). In fact, a study by Hamari *et al.* (2016) ranked enjoyment, sustainability, and economic benefits to be important in using sharing platforms such as Airbnb. Apart from favourable prices, trust, and utility, Möhlmann (2015) identified other effects such as community belonging and familiarity to be important for such services.

8.2.2 Host Perceptions

While the above studies focus on the reasons affecting Airbnb usage, they failed to differentiate the users into providers (host) and consumers (guests), often privileging the latter. In this section, we look at hosts, and on what is already known about them.

A study from Tibulschi (2017) with hosts in Vienna, older than 30, with a higher number of men hosts, all educated and currently working, pointed out that Airbnb hosts were very happy and motivated to be part of the platform. The accommodation type offered by them was mostly an entire apartment or private room especially for students. Some of the motivational factors for them were economic and social in nature, but they were unsure about the future of tax regulations, as well as the data protection by Airbnb. Furthermore, Calinao *et al.* (2019), with a sample of hosts in Makati City, under the age group of 28 to 37, found that they were economically and socially motivated. They considered Airbnb as a great opportunity for doing business and meeting a lot of new people. They believed that monthly income is one of the most important motivational factors that push them to be a part of Airbnb.

A study by Malazizi *et al.* (2018) focused on risk perceptions of Airbnb hosts in Mediterranean countries by applying structural equation modelling. This study pointed out that host satisfaction is negatively influenced by financial, safety and security risks. Furthermore, the intention to recommend this business is negatively affected by political risk, which refers to political instability within a certain country, criminal and terrorist activities, etc. Surprisingly, psychological risk increases satisfaction and intention to continue to use and recommend Airbnb (Malazizi *et al.*, 2018).

There are, however, also some criticalities and ambiguities in the functionality of Airbnb highlighted in the literature. For instance, Farmaki *et al.* (2019) focus on the "morally irresponsible behaviours" of hosts, because of tax avoidance, guest discrimination and providing misleading property information among others. Considering Airbnb hosts as a 'community of practice' (Farmaki and Kaniadakis, 2018), such behaviours were considered to be harmful for communities, and to require therefore strict regulatory framework (Gurran and Phibbs, 2017). In his study, Blasi (1993) defines 'moral responsibility' as the conduct of moral actions, which means that the hosts' moral identity motivates in a way the moral functioning of such businesses. According to the study of Farmaki *et al.* (2019), some of the Airbnb hosts define themselves as 'professional hosts', those who rent one or multiple listings systematically, and and for whom hosting provides the primary source of income (Lutz and Newlands, 2018); and some define themselves as 'non-professional Airbnb hosts', sharing a single room or their property on an ad-hoc basis. In this case they gain a supplement to their income, and/or use renting for socialising reasons. As the professional hosts are long-term businesses, they try to keep with the standards and are attentive to the moral responsibility. Such a distinction among professional and non-professional hosts is a motivation to comply with the moral responsibilities while offering related services. Finally, it can be agreed that the moral actions and responsibility awareness are important factors to the hosts.

8.3 Airbnb in Albania

Although Airbnb is identified as one of the most successful sharing applications in the Balkans region (Čavalić, 2017), the percentage of usage of this platform is limited there compared to other countries. One of the reasons for this low participation is the high informality in this zone. According to Williams and Horodnic (2017), the owners of hotels and restaurants are significantly competing against unregistered or informal firms. For instance, the percentage of hotels and restaurants registered and affirming to compete with the informal sector[1] is 76% in Bosnia and Herzegovina, 72% in Kosovo, 52% in Serbia, 52% in Macedonia, 45% in Montenegro, and 32% in Albania. This shows that there is a severe struggle between the formal sector and the informal one. According to the report of EY (2017), the estimated size of the shadow economy in Albania is 22% of GDP. Additionally, the percentage of hotels and restaurants that see informality as a major obstacle are 69%, 53 in Bosnia and Herzegovina, 40% in Macedonia, and 19% in Albania (Williams and Horodnic, 2017).

1. Williams and Horodnic (2017) used here the answers to the question "Does this establishment compete against unregistered or informal firms?"

In these last years, Albania is attracting a lot of interest as a newly emerging destination in the international tourist market. The development of information and communication technology (ICT) supporting widespread tools and sharing platforms made such destinations more and more reachable. But again, Albania is considered an emerging market in the field of mobile technologies (Eurostat, 2017). According to Network Readiness Index (NRI), one of the leading global indices on the application and utilization of information and communication technology by citizens, Albania is ranked in 75th place out of 121 economies for year 2019, with a slightly better position when it comes to business use, the 57th place (NRD, 2019).

8.4 Methodology

Primary data were collected by conducting semi-structured interviews with people who are Airbnb hosts in Tirana, renting short-term their second apartments/houses. The interview consisted of 35 open-ended questions based on the research conducted by Tibulschi (2017) in Austria. The interview guide (see Appendix) was aimed to investigate attitudes and opinions of the respondents with respect to how they feel being an Airbnb host in Tirana. A sample size of 8 people was selected through purposive sampling as there are only a limited number of primary data sources who can contribute to this study.

The interviews were conducted through online means (using Face Time, WhatsApp video call) or via phone calls, during April-May 2020, and notes were taken without audio recording. Each interview lasted for approximately 60 min. They were conducted in Albanian language.

During the first section of the interview general information about the hosts was collected. Three men and five women, aged 25–40, participated in this study. They were all working, and they held a master's degree. Most of them were engaged as an Airbnb host for less than a year. Only two of them have been part of Airbnb for 2 years. When considering the accommodation type the interviewees offered, most of them rent out their second home, some of them rent out the apartment in which they live but none of them rent out their private rooms. The ones with an apartment near the centre of Tirana mostly rent out weekly, even twice a week, while the others rent out once a month.

8.5 Results

The presentation of the results follows the substantial sections of the interview guide: the factors that motivated interviewees to become involved in Airbnb hosting and the relative benefits; what does it mean to enact the role of the hosts; how hosting affects daily life and the difficulties that they have faced; how they perceive themselves as hosts and entrepreneurs; the presence or lack of an Airbnb local community; finally, how the hosts are dealing with this new experience, the Airbnb platform regulations, and where do they see Airbnb in the future.

8.5.1 Becoming a Host

One of the main reasons and the most motivated one for the hosts we have interviewed is "economic reason", an opportunity to earn extra money. One of the respondents, a 27 years old woman hosting an apartment in the city centre, points out:

> *"In this difficult world that we are living in, we need money more than ever before, I chose this method to make some extra money."*

They considered Airbnb as an easy way of doing business and most of them have rented out spare apartments that were currently not used by them.

Another reason was "sharing new experiences and cultures". Being part of Airbnb gives you the ability to have connections with different people from different countries. Being in contact with them helped the hosts to learn a lot about different cultures. One of the interviewees, a 30 years old woman hosting an apartment in the city centre, said:

> *"One of the most motivating things is the ability to share experiences and cultures with different people. Once I had guests from Moldavia from which I learned a lot of things I did not know before."*

It should be said that no one of the hosts have thought in the past of renting out their house. Most of them came up with this idea based on suggestions by friends or relatives that live abroad and had information about Airbnb. The others were motivated to be Airbnb hosts in Tirana when they first used Airbnb abroad. One of the interviewees, a 25 years old woman hosting an apartment in the city centre said:

> *"Not really, it started two years ago when I first used an Airbnb abroad. I liked the idea a lot and I started thinking about it. I had an apartment which no one of my family used so I decided to rent it out on Airbnb."*

Bringing to Albania and their lives something that recently became widespread in other countries was not only the random way they became host, but also part of their motivation to do so. Some hosts consider themselves as indirect contributors to the Albanian tourism sector and to its development, and they feel very positive for the future. This issue is reported as below by a 27 years old woman hosting an apartment in the city centre:

> *"I think that small people make small things to make the difference. Albanians have hospitality in their blood and when seeing that something is becoming a trend and becoming more successful every day, I think that everyone will try to open up their hearts and homes for foreign tourists."*

8.5.2 Being a Host

The role of a host is very important given the relevance of the feedback of the customers and, potentially, their willingness to come back. Hosts are expected to have strong communication skills and make sure the customers had a great experience to maximise the chance they come back in the future. Most of the interviewees consider themselves *"social"*, *"friendly"* and *"kind"*. But at the same time, they want to be *"professional"* in the service provided to the customers, as emphasized by a 27 years old woman hosting an apartment in the city centre:

> *"You have to be friendly, welcoming but at the same time make some things clear from the beginning."*

Since they are hostingming guests at their property, they try to make sure that everything inside the apartment remains the same as in the beginning.

They consider the Airbnb experience as a nice one but they also said guests' behaviour may be a cause of stress—so interaction with guests is, once more, central to successfully being a host. This issue is pointed out by a 27 years old woman hosting in a periphery apartment:

> *"It is one of the best experiences, but it can be stressful sometimes. You have to deal with different types of people that sometimes can regret regarding the conditions of the apartment only to get discounts."*

The selection of the guests is considered as one of the hardest parts of hosting. With respect to this, interviewees may be divided into two groups:

(1) There are strict hosts which prefer only families or couples and before selecting the guests they ask the reason of the visit, age, their passport as an ID. Furthermore, one of the hosts said:

> *"When they contact me for the availability of my apartment, the first thing I do before I reply to them, I search them on social media and get some information."*

(2) On the other hand, there are hosts who do not select guests at all. The only thing that matters to them is the money they earned. As one of them, a 30 years old man hosting in a periphery apartment said,

> *"I do not practice the method of the guests' selection. When my apartment is available, I answer them immediately because at the end the only thing that is important for me is the money I am earning from them."*

8.5.3 Influence on Daily Life

All the hosts that were interviewed had other jobs as primary, with hosting as a part-time job. This means they now have more responsibilities in their daily life. They must be precise with their guests and have time management skills to combine their other job and hosting, with occasional difficulties in managing time, particularly for check-in/check-out time as they must be present at the apartment. Some of the hosts have decided to do the check-in/check-out during the hours they are not working or during their break at work. To most of the hosts, however, the visitors usually came during the weekend when they were free from work and in any case, hosts are motivated by profits, so they tend to find a way to combine their work with hosting. One of the respondents, a 35 years old man hosting in a periphery apartment, emphasizes:

> *"It does not really take that much time. It is like a leisure time to me. To add, there is profit that motivates me."*

Being an Airbnb host has affected hosts' daily life also in relational terms and by opening their horizons. They met many new people from different countries and different cultures, and they think they learned a lot about people behaviours, as the Airbnb experience has made them understand people's needs from a host perspective. It should be highlighted that Airbnb has changed a lot the hosts' view of tourism in Tirana. They did not know that guests from all over the world visit Tirana and Albania and one of the respondents, a 30 years old woman hosting an apartment in the city centre, was very surprised by that:

> *"I did not know that the city of Tirana was visited by many visitors coming from all over the world. As I said before, once I had visitors from Moldavia and I was very surprised".*

Furthermore, seeing the city from a host perspective, made them very detail-oriented when it comes to what a tourist wants to see. This is pointed out by a 25 years old woman hosting an apartment in the city centre as:

> *"I realized all the sort of visitors that Tirana gets, there are a lot of people that want to know a lot about the places in Albania and the history behind it. I believe that we, from the tourism point of view especially in Tirana, have a lot of things that we should change to make the city more attractive and point out more about the history, and also offer much more services."*

Although interviewees are renting out apartments that were unusable by them hence being part of Airbnb has not affected their own personal space, they experience some costs to maintain the apartment as compared to an earlier time. The interviewed 25 years old woman hosting in the apartment in the city centre Airbnb expresses it as:

> *"For starters, it is a bit costly to maintain the apartment compared to what was before. But as I said this is only the first year. I am positive that the market will have a shift and Airbnb will become much more popular. With the number of visitors increasing, the earnings will be higher, and these costs will be considered negligible."*

Moreover, they spend more time on improving their apartments to maximally meet their guests' expectations and having good feedback from them.

8.5.4 Identity of Entrepreneur

When it comes to the entrepreneurial identity of these hosts, which is something they recently developed, it should be highlighted that they are all risk takers since Airbnb is not very popular in Albania. They feel their hard work and being passionate about hosting on Airbnb helped them grow and become successful, to the point they become an inspiration for other people. A 30 years old woman hosting in an apartment in the city centre, said:

> *"They see me as a successful micro entrepreneur and some people get inspired to do the same thing I do with their extra rooms or apartments."*

Many believe their success is strongly related to the way they perform the host role. They are all very committed to fulfilling the guests' expectations and receiving great feedback from them. They described themselves as *"caring"*, *"social"*, *"friendly"* and always willing to help their guests with everything they needed. Moreover, their strong communication skills have played a key role in receiving good feedback.

8.5.5 Airbnb Local Community

Airbnb Community is best defined as a place which connects hosts to share their hosting stories and experience so they are able to get updates as well as valuable suggestions from each other. Airbnb has its own global community centre which is a very organized and active place where hosts can get every information they need through questions and answers sessions. Some countries have their own Airbnb Community where they organize meetings, share ideas and experiences, know, and support each other. On the other hand, there are countries in which hosts only use pages on social media like Facebook or Instagram to share information.

When looking at Airbnb in Tirana, the hosts had no information if a community exists, but they would like to be part of it. A 35 years old man and a 27 years old woman hosting periphery apartments expressed it as below:

> *"I do not have information about a community in my city, but I guess there is not. I would like to be a part of one because it would boost and improve our job. I think being in an Airbnb community would be a great way to share suggestions with each other."*
>
> *"As far as I know in Tirana an Airbnb Community does not exist. It is a Facebook Page named Airbnb Albania, but has not been updated since 2018."*

8.5.6 Future Perspectives and the Role of Regulations

Although time-management difficulties, interviewees are enjoying hosting a lot, especially the part of socializing, meeting the guests and the good feedback they get from them. Therefore, when it come to the future, they intend to further invest in this sector by increasing the number of rooms/apartments they offered as well as developing the hosting service further, as pointed out by two hosts, a 35 years old man and a 39 years old one, both hosting in periphery apartments:

> *"At first I just gave it a try. I did not expect too much and did not plan to go for so many years with Airbnb either. But now I am thinking of increasing the number of apartments to rent out as well as starting to rent out villas."*
>
> *"I do not see anything different in the future of Airbnb, but my future I think it will be with more apartments not only for the tourists but also for the students who can share only rooms."*

This would be very helpful for students, especially those who come from other cities to finish their studies in Tirana. In general, the hosts plan a career in tourism, see Airbnb as a great opportunity and are willing to invest in this sector.

However, the lack of a proper regulation risks severely hindering this future, particularly when single individuals start to rent multiple properties. All the interviewees think that this kind of business is done illegally in Tirana. The legal framework is not complete. They agree that having regulations ensuring the guests security and comfortability would positively affect their hosting activity.

8.6 Discussion

Airbnb has shown to be a successful digital platform providing entrepreneurial opportunities and income supplement to hosts as well as extending the availability of choices for guests. This research is of importance as it looks at Airbnb hosts in a developing country. Through this study we focused on the analysis of a case like the city of Tirana, where Airbnb has not yet "taken over" the accommodation market. Even though previous studies were conducted mainly in more economically developed countries, the results of the current study show not many differences on the motivation to become an Airbnb host by the interviewees, their view of the role of the host, daily life, and entrepreneurial identity. The findings show that economic benefits along with the social ones lie behind the main motivation on becoming an Airbnb host. This confirms the results by Calinao *et al.* (2019), who emphasized monthly income as the most important motivational factor, and by Hamari *et al.* (2016), who included sustainability as a motivational factor besides the economic and enjoyment benefits.

When comparing Tibulschi (2017) findings in Vienna with Tirana results (Table 8.1), the main difference between the two cities stays on the Airbnb community and its perspectives. Community belonging and familiarity were defined to be crucial for these kinds of services by Möhlmann (2015). While there exists a consolidated community in Vienna, in Tirana it doesn't exist as Airbnb is new to the city. The interviewees sustained the idea that the existence of a community of hosts would help them share experiences and provide better services to their customers. The hosts of Airbnb in Vienna had information about the Airbnb community and also a local person has been hired in the position of Community Organizer, officially opened by Airbnb (Tibulschi, 2017). They also had their own Facebook page of Airbnb and organized meetings with the hosts. On the other side, interviewees of Tirana feel enthusiastic about the future of Airbnb, while the Vienna interviewees feel insecure because of legal regulations and taxations. Hosts in Tirana agree that the legal framework is not complete and has a lot of absences. Furthermore, the hosts in Tirana rent out their second home while in Vienna most of them shared the entire apartment and private rooms. Detailed information on the two cities experiences and perspectives comparison is provided in the table below.

Table 8.1. Vienna and Tirana study results comparison.

General Profile	Vienna	Tirana
Demographic	5 male and 1 female Older than 30 Educated Some working/Some not	3 male and 5 females 25–40 years old Master's degree Currently working
Engagement with Airbnb	From 2013	From 2018
Accommodation type offered	Entire apartment/Private Room	Second home
Motivation to becoming a host	Economic Reasons Social Reasons	Economic Reasons Social Reasons
The role of a host	Easy going Flexible Friendly Nice Strict Professional	Social Kind Friendly Professional Strict
Influence on daily life and well being	Improved life quality More organized More disciplined	Improved life quality More responsible Time-management skills
Identity of entrepreneur and social identity	Airbnb persons Good career	Hard working Passionate Successful
Airbnb Community	Very important	Have no information
Airbnb Regulations and Perspectives	– Hosts were anxious about taxation regulations. – Contributors in tourism and in promoting different locations to the visitors. – Unsure about the future due to regulations.	– this kind of business is done illegally. – Indirect contributors in tourism. – Motivated for the future.

8.7 Conclusions

Through eight semi-structured interviews with Airbnb hosts of Tirana, we understood that they consider Airbnb an easy way of doing business—where the costs are low, the risk is not high (since they are not investing a large amount of money), and unused resources become fruitful—and as an opportunity for developing an entrepreneurial identity. In doing so, the hosts also gained entrepreneurial skills. They are now more responsible, more organized and increased their time-management skills by having to combine current work and hosting.

Another factor that motivated them was the sharing of new experiences and cultures. During the time of hosting, they learned a lot about the culture of their visitors coming from different countries all over the world. Being in touch with new people, hosting taught them a lot with regards to people behaviour.

On the other hand, difficulties for hosts appear to be of two kinds. On one hand, the lack of a local Airbnb community and the information sharing thereof. Even if social media play an important role in advertising and in getting information about different things, it should be said that the Airbnb pages opened on Facebook or Instagram are not updated. Having social media updated and being able to give information to all possible guests, would help the hosts to increase the number of their visitors. Moreover, building an Airbnb community would help in sharing experiences and provide better services to the guests.

On the other hand, the interviewees lament a lack of regulations. Hosts think that this kind of business is done illegally in Tirana and they are not subject to taxation. They considered themselves indirect contributors in tourism and they feel very motivated for the future of Airbnb in Tirana, but the lack of a regulative framework may hinder such a future. This calls on the attention of government and policy-makers.

Appendix

Interview guide

General:

(1) For how long have you been a host for Airbnb?

(2) How do you rent out (room (s), whole apartment, other second home)

(3) How often do you rent out a room?

(4) Do you also use Airbnb for yourself?

Motivations to becoming a host:

(5) Why did you decide to become a host?
(6) What are the motivations for you to be an Airbnb host?
(7) Have you always wanted to rent out your house?
(8) Do you have another job or jobs besides your hosting?

The role of a hosts, i.e., experience, expectations and competencies:

(9) If you had to describe yourself as a host, how would that be? (i.e., strict rules, try to socialize, try to also be authentic as Airbnb states)
(10) How do you position yourself as a host? Any strategies applied?
(11) Do you feel like you have the competencies of being a good host?
(12) If you had to describe the experience in general of being a host, how would that look like?
(13) How do you select visitors and why so?
(14) How much hours per week on average you spend on being a host?
(15) Which activity is the most time consuming of hosting?
(16) Which part of hosting do you enjoy the most and which one the least?
(17) How do you see the future (i.e., do you worry about the future?, or any plans)

Influence on daily life – well-being:

(18) Did you had any expectations of becoming a host, and looking back, do you feel like your expectations are met?
(19) In which way do you feel like your life has changed from the moment you became a host?
(20) Are there also any positive and/or negative consequences (burdens or costs) after becoming a host? What kind and how does this impact you and-or your family?
(21) How do you perceive the sharing of your apartment interferes with your own personal space and time and from any family members? How does this affect you?
(22) How do you combine being a host with other domains in your life? And how does this work out? How do you feel about that?

Identity of entrepreneur and social identity:

(23) Have you always aspired to become an independent–(micro) entrepreneur?
(24) How do you perceive yourself as a host?

(25) Is your host role a part of yourself or it dominates your overall identity?
(26) How would you like others to see you? – How do you think others perceive you?

Airbnb community:

(27) Is there a community of other hosts in the city? Do you feel a part of this community and how? Is the community supportive or competitive?
(28) Do you feel any social pressure of being a popular host in the community? And how does that affect you?
(29) How important is the host community for you?
(30) How do you feel to be a part of the Airbnb community in general? Is there any pressure coming from Airbnb to behave in a certain way? Do you feel happy that Airbnb facilities your hosting?

Other: Airbnb in general:

(31) How concerned are you with the regulations of Airbnb: legal versus illegal issues?
(32) Do you also rent out your apartment to other platforms besides Airbnb?
(33) In which way has your role as a host changed your view of tourism in the city?
(34) How do you think your role as a host but also the Airbnb practices in general has changed your neighborhood and your perception/attitudes of contributing to tourism?
(35) What would you like to see different in the future with Airbnb? How do you think your future will look like?
(36) Any other comments.........................

Source: (Tibulschi, 2017)

References

Adamiak, C. (2019). Current state and development of Airbnb accommodation offer in 167 countries. *Current Issues in Tourism*, 1–19.

Benítez-Aurioles, B. (2018). "Why are flexible booking policies priced negatively?". *Tourism Management*, *67*, 312–325.

Berhani, R. and Hysa, E. (2013). The Economy of Albania Today and then: The Drivers to Growth. In *The 4th International Conference on European Studies* (p. 598).

Blasi, A. (1993). The development of identity: Some implications for moral functioning. In: G.G. Noam and T.E. Wren (Eds.). *The Moral Self.* Cambridge, MA: The MIT Press (pp. 99–122).

Calinao, R.J., Atenta, N., and Riofloriodo, N. (2019). Entrepreneurial Motivations of the Airbnb Business in Makati City. 102–112.

Čavalić, A. (2017). Sharing Economy in Western Balkans: Potential for Rural Development.

Cheng, M. and Jin, X. (2019). "What do Airbnb users care about? An analysis of online review comments". *International Journal of Hospitality Management, 76,* 58–70.

Davis, B. and Hillier, L. (2019). 10 examples of great Airbnb marketing creative. *Econsultancy,* https://econsultancy.com/blog/68225-10-examples-of-great-airbnb-marketing-creative/ (accessed 20 November, 2020).

Dolnicar, S. and Zare, S. (2020). COVID19 and Airbnb–Disrupting the disruptor. *Annals of Tourism Research.*

Eurostat (2017). *Enlargement Countries Information and Communication Technologies – Statistic Explained.* Available online: http://ec.europa.eu/eurostat/statistics-explained/index.php/Enlargement_countries__information_and_communication_technology_statistics.

EY (2017). *Reducing the Shadow Economy in Albania through Electronic Payments,* https://assets.ey.com/content/dam/ey-sites/ey-com/en_pl/topics/eat/pdf/03/ey-albania-shadow-economy-study-report.pdf (accessed 1 December, 2020).

Fagerstrøm, A., Pawar, S., Sigurdsson, V., Foxall, G.R., and Yani-de-Soriano, M. (2017). That personal profile image might jeopardize your rental opportunity! On the relative impact of the seller's facial expressions upon buying behavior on Airbnb™. *Computers in Human Behavior, 72,* 123–131.

Farmaki, A., and Kaniadakis, A. (2018). *Responsibility in the Sharing Economy. Proceedings of the 28th CAUTHE Conference, Newcastle Business School,* The University of Newcastle, Australia (pp. 553–556).

Farmaki, A., Miguel, C., Drotarova, M.H., Aleksiæ, A., Časni, A.Č., and Efthymiadou, F. (2020). Impacts of Covid-19 on peer-to-peer accommodation platforms: Host perceptions and responses. *International Journal of Hospitality Management, 91,* 102663.

Farmaki, A., Stergiou, D., and Kaniadakis, A. (2019). Self-perceptions of Airbnb hosts' responsibility: a moral identity perspective. *Journal of Sustainable Tourism,* 1–21.

Ferizi, V. and Kruja, A.D. (2018). Coastline Hospitality Industry Performance, Challenges, and Opportunities: Evidence From Durres Coastline. In D. Batabyal. *Managing Sustainable Tourism Resources* (pp. 14–38). IGI Global.

Forgacs, G. and Dimanche, F. (2016). "Revenue challenges for hotels in the sharing economy: facing the Airbnb menace". *Journal of Revenue and Pricing Management, 15*(6), 509–515.

Gibbs, C., Guttentag, D., Gretzel, U., Morton, J., and Goodwill, A. (2018). "Pricing in the sharing economy: a hedonic pricing model applied to Airbnb listings". *Journal of Travel and Tourism Marketing, 35*(1), 46–56.

Gurran, N. and Phibbs, P. (2017). When tourists move in: How should urban planners respond to Airbnb? *Journal of the American Planning Association, 83*(1): 80–92.

Guttentag, D. (2013, 12). Airbnb: Disruptive innovation and the rise of an informal tourism accommodation sector. *Current Issues in Tourism*, 1–23.

Hamari, J., Sjöklint, M., and Ukkonen, A. (2016). The sharing economy: Why people participate in collaborative consumption. *Journal of the Association for Information Science and Technology, 67*(9), 2047–2059.

Hysa, E. (2012). Influence of Tourism Sector In Albanian GDP: Estimation Using Multiple Regression Method. *Revista de turism-studii si cercetari in turism*, (13).

Hysa, E. and Gjergji, E. (2018). The Long Run Relationship Between Tourism And Economic Growth In Western Balkan Countries: A Panel Co-Integration Analysis. *Revista de turism-studii si cercetari in turism/Journal of tourism-studies and research in tourism, 25*(25), 1–3.

Hysa, E., Kruja, A., and Shiko, V. (2021). Current and Prospective Expansion of Sharing Economy in Albania. In *Country Reports 2019*, 12–44.

Kruja, A.D., and Berberi, E. (2020). Tourism and Handicraft Industry: Opportunities and Challenges of Operating in the Albanian Market. In I. Chirino-Klevans. *Cases on Global Leadership in the Contemporary Economy* (pp. 119–136). IGI Global.

Kruja, A.D., Hysa, X., Duman, T., and Tafaj, A. (2019). Adoption of Software as a Service (Saas) in Small and Medium-Sized Hotels in Tirana. *Enlightening Tourism, 9*(2), 137–167.

Kruja, A. (2012). The Impact of Tourism Sector Development in the Albanian Economy. *Economia. Seria Management*, 204–218.

Lee, S. and Kim, D.Y. (2018). "Brand personality of Airbnb: application of user involvement and gender differences". *Journal of Travel and Tourism Marketing, 35*(1), 32–45.

Malazizi, N., Alipour, H., and Olya, H. (2018). Risk Perceptions of Airbnb Hosts: Evidence from a Mediterranean Island. *Sustainability*, 10. 10.3390/su10051349.

Monllos, K. (2016). *Airbnb's CMO Wants to Redefine Experiential Marketing with the Company's New Offering*. Adweek, http://www.adweek.com/news/advertising-branding/airbnbs-cmo-wants-redefine-experiential-marketing-companys-new-offering-174717 (accessed 20 November, 2020).

Möhlmann, M. (2015). Collaborative consumption: determinants of satisfaction and the likelihood of using a sharing economy option again. *Journal of Consumer Behaviour, 14*(3), 193–207.

Newlands, G., Lutz, C., and Fieseler, C. (2019). *The Conditioning Function of Rating Mechanisms for Consumers in the Sharing Economy.* Internet Research, *29*(5), 1090–1108.

Noti, E. (2014). *Ndikimi i teknologjise se informacionit ne sipermarrjet turistike ne Shqiperi.* Tirana: Tirana University.

NRD (2019). *Insight Report "The Network Readiness Report: Toward a Future-Ready Society"*, Eds. Dutta, S. and Lanvin, B., Portulans Institute.

Rosenfelt, R. (2014, August). *Going for Global.* from https://www.youtube.com/watch?v=U3oICoTN1-M

Stollery, A. and Jun, S.H. (2017). "The antecedents of perceived value in the Airbnb context". *Asia Pacific Journal of Innovation and Entrepreneurship, 11*(3), 391–404.

Ter Huurne, M., Ronteltap, A., Corten, R., and Buskens, V. (2017). Antecedents of trust in the sharing economy: A systematic review. *Journal of Consumer Behaviour, 16*(6), 485–498.

Teubner, T., Hawlitschek, F., and Dann, D. (2017). "Price determinants on Airbnb: how reputation pays off in the sharing economy". *Journal of Self-Governance and Management Economics, 5*(4), 53–80.

Tibulschi, L. (2017, October 19). *The Motivations of Airbnb Hosts.* from https://www.modul.ac.at/uploads/files/Theses/Bachelor/undergrad_2017/Thesis_131178_TIBULSCHI_Ludmila.pdf

Tussyadiah, I.P. (2015). An exploratory study on drivers and deterrents of collaborative consumption in travel. In *Information and Communication Technologies in Tourism 2015* (pp. 817–830). Springer, Cham.

Williams, C.C. and Horodnic, I.A. (2017). Regulating the sharing economy to prevent the growth of the informal sector in the hospitality industry. *International Journal of Contemporary Hospitality Management.*

WTTC (2018). *Travel & Tourism Economic Impact 2018: Albania.* World Travel & Tourism Council.

Xiang, Z., Schwartz, Z., Gerdes, J., and Uysal, M. (2014). What can big data and text analytics tell us about hotel guest experience and satisfaction? *International Journal of Hospitality Management,* 120–130.

Yang, S.B., Lee, H., Lee, K., and Koo, C. (2018). "The application of Aristotle's rhetorical theory to the sharing economy: an empirical study of Airbnb". *Journal of Travel and Tourism Marketing, 35*(7), 938–957.

Yannopoulou, N., Moufahim, M., and Bian, X. (2013). "User-generated brands and social media: couchsurfing and Airbnb". *Contemporary Management Research*, *9*(1), 85–90.

Zervas, G., Proserpio, D., and Byers, J. (2017). *The Rise of the Sharing Economy: Estimating the Impact of Airbnb on the Hotel Industry*. 687–705.

DOI: 10.1561/9781680838411.ch9

Chapter 9

The Open Source Platform Federation Business Model

By Justin Larner

9.1 Introduction

The digital collaborative economy has led to new forms of working, where workers perform a range of tasks that include transport, household repairs, information work and domestic service. These workers are matched with their temporary employers through software platforms such as Uber, Airbnb or Amazon Mechanical Turk, where (Kenney and Zysman, 2016, p. 61) note that these 'digital platforms ... shape the terms on which participants interact with one another', giving power to the corporations that operate them through a centralized form of governance (Gol *et al.*, 2019). The platform economy has been promoted as enabling collaboration and sharing, but its real value to firms is in being able to lower the transaction costs of accessing goods, services and particularly workers (Drahokoupil and Fabo, 2016).

The platform economy has been seen as a continuation of the Industrial Revolution, where technology enabled centralization of the means of production (Fuchs, 2014), characterized by extractive rather than generative ownership of firms (Bauwens and Niaros, 2017). Platform firms deliver a value proposition to the customer, who benefit from services such as transport at a reduced cost (Kenney and Zysman, 2016). These workers are a vital resource to the platform firm, but are classed as independent contractors, with an uncertain income as a result. The Frankfurt Paper on Platform-Based Work (2016, p. 2) notes that workers as independent contractors in the digital platform economy are 'typically excluded from the legal and social protections established for employees over the last hundred years', and that 'worker organizing has for decades been correlated with the economic well-being of working people' (p. 6), calling for a 'co-operative turn', 'in which workers, clients, platform operators, investors, policy makers, and worker organizations work together to improve outcomes for all stakeholders' (p. 3).

Workers are creating Internet-based forums to share knowledge and experience as a form of collective action (Fabo *et al.*, 2017), such as Ride Share Drivers United (2021), and in some cases researchers have set up forums (Irani and Silberman, 2013) that are now run by their worker community (Turkopticon, 2021). These forums can benefit workers who use them, but do not directly change power relations between those workers and platform operators. Platform economy workers are now aligning with unions, creating guild-like organizations and worker-led platform cooperatives (Vandaele, 2018), which have had some success in niche markets (Scholz, 2016), but 'face many challenges in competing with established companies' (Healy *et al.*, 2017, p. 241). There is thus potential for other alternative business and organizational models in the collaborative economy.

This chapter first introduces the potential for alternative collaborative economy business models in the next section, then the following section presents a methodology where annotated portfolio techniques (Gaver, 2012; Gaver and Bowers, 2012) can enable business model designs to be elicited from ethnographic data. Annotated portfolios are a method used in human-computer interaction design that can bring together a number of artifacts and identify the aspects that are common among them through text annotations. The technique could thus be applicable to the design of business models, where the artifacts are ethnographic data from engagement with business founders and workers, annotated as a portfolio of potential business model design elements. In this chapter, ethnographic data on the author's interactions with two case study business founders and their stakeholders are regarded as a collection of artifacts and annotated in a similar manner. These annotations reveal the Open Source Platform Federation as a collaborative economy business model with a balance of power between founders and members.

9.2 The Potential for Alternative Business Models in the Collaborative Economy

Business models are generally framed in terms of value creation and capture (Shafer *et al.*, 2005; Zott *et al.*, 2011), also in terms of how the firm gains revenue from this value (Chesbrough, 2010). This conceptualization of a business model is reinforced by (Teece, 2010, p. 179) who offers the definition, 'A business model articulates the logic, the data and other evidence that support a value proposition for the customer, and a viable structure of revenues and costs for the enterprise delivering that value'. Other authors take a more resource-based view of business models, Tapscott *et al.* (2000) frame a business model in terms of how the firm uses both internal and external resources to create value that is differentiated from its competitors, while DaSilva and Trkman (2014) argue that resources in themselves do not offer any value to customers: this value is gained through transactions using those resources. Taking this more transactional view of business models highlights that the business model 'is centered on a focal firm, but its boundaries are wider than those of the firm' (Zott *et al.*, 2011, p. 1020), where transactions take place across the boundary. The concept of a focal firm and wider stakeholder network proved key to framing how the organizations that are documented in this chapter operated.

Another key dimension of business models is ownership, and how power is exerted through ownership (Bauwens and Niaros, 2017), where (Kenney and Zysman, 2016, p. 66) ask the questions, 'Who owns or controls the platform?', 'How is value created' and 'Who captures the value?'. Business models are thus a useful concept to frame power relations in the collaborative economy. It is thus important to find business models that can offer more of a balance of power between workers and firms. The next section introduces a research project that aimed to explore the potential for designing new business models in two case studies using annotated portfolio techniques to analyze ethnographic data.

9.3 An Ethnographic Approach to Business Model Design

Ethnography is a research methodology that aims to gain a deep understanding of the experience of individuals and groups in their context through techniques such as participant observation (Silverman, 2007). Ethnography has also been used in participatory and other forms of action research, where it can contribute to reflection that leads to action (Cassell and Johnson, 2006). Action research and participatory design are both methodologies that encourage participation by stakeholders to make real-world change (Foth and Axup, 2006).

Changing the situation of platform economy workers from an existing to a desired state implies a design process, where design is 'concerned with how things ought to be, with devising artifacts to attain goals' (Simon, 1969, p. 59), thus design is about an 'inquiry into the ideal' focusing on what is desirable but 'not-yet-real' (Nelson and Stolterman, 2012, p. 35). Research methods that can promote change and bring about the not-yet-real include action research and design research (Cole *et al.*, 2005). The research tradition of Computer Supported Cooperative Work (CSCW) is helpful here, where early writings highlight the problem of 'representations in work and system design' (Suchman, 1995, p. 63), and in presenting the outcome of ethnographic research in a way that makes sense to system designers (Hughes *et al.*, 1994). Business models are in themselves a representation of how an organization strategically manages value creation and capture, and particularly in the collaborative economy are implemented as computing systems. Ethnographic data can contribute to the design of computing systems for cooperative work, implying that this form of data can contribute to the design of business models as well.

The critical design approach advocated by (Bardzell and Bardzell, 2013, p. 3304) can be helpful as a technique that can offer insight into existing social structures through creating new ones that promote 'social change, from the present to a hoped-for future that is attainable but not immediately within reach'. Building on the critical design approach, (Barab *et al.*, 2004, p. 254), introduce the concept of critical design ethnography, which they see as 'a process that sits at the intersection of participatory action research, critical ethnography, and socially responsive instructional design'. Implementing critical design ethnography starts with understanding cultural context through rich description, as with other forms of ethnography, then making commitments to social change which are expressed in a design for potential action, which can be generalized beyond the original ethnographic context (Barab *et al.*, 2004). The methodological challenge is how ethnographic data can be represented in a form that enables business model design. Annotated portfolios are introduced in the next section as an analysis technique that could help meet this challenge.

9.4 Annotated Portfolios as a Business Model Design Method

Annotated Portfolios as a Design Methodology

Annotated portfolios were originally developed in the context of classroom assessment to enable students and teachers to collaborate more effectively (Yancy, 1992), then were first used in a design context to develop clinical treatment strategies in mental health (Lavori and Dawson, 1998). Annotated portfolios were then re-introduced in the context of human-computer interaction as a method that could bring together a number of artifacts and identify the common aspects among them through text annotations (Gaver, 2012; Bowers, 2012; Gaver and Bowers, 2012). Annotated portfolios can thus be helpful in bridging the gap between research and design, where the 'essence of research is to produce knowledge, and the essence of design is to produce artifacts' (Löwgren, 2013, p. 30) as a level of abstraction between the 'ultimate particular' (Nelson and Stolterman, 2012, p. 27) of each artifact and a more generalizable level of knowledge.

In design terms, annotated portfolios can contribute to producing knowledge of 'what ought to be' (Gaver, 2012, p. 42), or a future change. (Bowers, 2012, p. 71) highlights that any 'material form can be considered for an annotated portfolio including an illustrated monograph, a scientific paper, a curated exhibition and so forth', which implies that ethnographic data could be annotated as a portfolio. Despite this potential, there has been relatively little interest in annotated portfolios of ethnographic data as a design technique, with only a few examples in the literature.

Annotated portfolios were used by Hobye *et al.* (2013) to abstract key qualities from five design case studies, while Kelliher and Byrne (2015) extended the use of annotated portfolios to produce indexing data from documentation of a large public event. Taking a somewhat different approach, Frauenberger *et al.* (2016) linked annotated portfolios with Actor-Network Theory where portfolios can enable a shared understanding of the activities and discourses that form a design process. In contrast, annotated portfolios enabled Barrass (2016) to offer a more general account of the knowledge embodied in a single artifact, the Hypertension Singing Bowl.

Moving closer to the case study methodology documented in this chapter, Sauerwein *et al.* (2018) used annotated portfolios in qualitative analysis of interview data, adding to the five-step analysis method offered by McCracken (1988), where repeating the process with multiple designs enabled pattern recognition through comparing each annotated design. Analyzing qualitative interview data using annotated portfolios implies that analysis of ethnographic data in a design context could also use annotated portfolios. The technique could thus be applicable to the design of business models, where the artifacts are ethnographic data from engagement with business founders and workers, annotated as a portfolio of potential business model design elements.

Annotated Portfolios as a Framework for Business Model Design

Analysis of ethnographic data on engagement with business founders and workers as an annotated portfolio can bring out the latent desiderata that include functionality, practicalities, motivation for designing, potential products and services, aesthetics and performance measures (Nelson and Stolterman, 2012). Gaver and Bowers (2012) also offer a similar set of categories of choices that influence the design of an artifact, including functionality; aesthetics; practicalities; the motivation for designing it; the people for whom it is intended; and sociopolitical concerns.

A useful perspective for business model design is that a business model can be considered 'as a material object, as a scale model of the new venture' (Doganova and Eyquem-Renault, 2009, p. 1568) that can simulate its real-world application (DaSilva and Trkman, 2014), thus a business model design will need to contain all the components that would be found in the business itself. Drawing on the recent literature on business models (Al-Debei and Avison, 2010; Nenonen and Storbacka, 2010; Zott *et al.*, 2011; Fielt, 2013; Spieth and Schneider, 2016; Wirtz *et al.*, 2016), these components can be identified as: personal factors; resources; opportunities; stakeholders; value creation and capture; strategy; boundaries; structure; activities; customers; revenue and costs; and profit. Business model design can also contribute to organizational design, where Stanford (2007) offers the relevant parameters of culture; systems; structure; people; performance measures and processes; products and services; and operating context.

These elements of desiderata, design choices, business model components and organizational design parameters are summarized in Table 9.1 below.

These categories can be combined to create a framework for business model design from the perspective of the business founder or founders:

- Culture;
- Aesthetics;
- Personal factors;
- Motivation for designing;
- People for whom it is intended;
- Sociopolitical concerns;
- Operating context;
- Opportunities;
- Customers;
- Potential products and services;
- Functionality;
- Practicalities;
- Boundaries;
- Strategy;
- Activities;
- Resources;
- Value creation and capture;

Table 9.1. Business model annotated portfolio analysis components.

Author(s)	Business Model Design Element
Desiderata (Nelson and Stolterman, 2012).	• Functionality • Practicalities • Motivation for designing • Potential products and services • Aesthetics • Performance measures
Design choices (Gaver and Bowers, 2012).	• Functionality • Aesthetics • Practicalities • Motivation for designing • People for whom it is intended • Sociopolitical concerns
Business model components (Al-Debei and Avison, 2010; Nenonen and Storbacka, 2010; Zott *et al.*, 2011; Fielt, 2013; Spieth and Schneider, 2016; Wirtz *et al.*, 2016).	• Personal factors • Resources • Opportunities • Stakeholders • Value creation and capture • Strategy • Boundaries • Structure • Activities • Customers • Revenue and costs • Profit
Organizational design parameters (Stanford, 2007).	• Culture • Systems • Structure • People • Performance measures and processes • Products and services • Operating context

- Revenues and costs;
- Profit;
- Performance measures and processes;
- Systems;
- Stakeholders;
- Structure.

The framework can enable the design of business models through guiding annotation of data from engagement with entrepreneurs and business founders. The case studies described in the following section show how annotated portfolio techniques enabled the design of a networked business model that could operate in the collaborative platform economy.

9.5 Case Studies: Applying Annotated Portfolio Techniques to Business Models

Background to the Case Studies

Case studies of two networked micro-enterprises in the North West of England from 2013–2016 illustrate how annotated portfolios can help elicit latent business model designs in ethnographic data. The two case studies aimed to explore how a business founder built their business with their stakeholders. During the author's engagement with founders and other stakeholders in the case study organizations, the open source guild business model (Larner, 2013) was offered as a starting point for discussion. In the open source guild model, the medieval guilds, as a pre-industrial form of network organization (Deakin, 2006), are used as a metaphor, comparing the role of the software architect to the masters in the medieval guilds (Larner *et al.*, 2017). In the case studies, the open source guild model acted as a design provocation (Bardzell and Bardzell, 2013), and also functioned as a boundary object (Star and Griesemer, 1989), by offering a new perspective on doing business which provoked reflection by both founders. The two case study organizations were 3rd Way Coop and the Ethical Small Traders Association (ESTA).

3rd Way Coop

3rd Way Coop was founded to develop and produce Super Insulation, a product that could revolutionize how buildings are insulated and help deal with energy poverty worldwide. The founder developed a multilevel organizational structure, including a network of apprentices who would be trained to install Super Insulation. 3rd Way's founder was interested in the open source guild model, as it had the potential to solve their organizational dilemma concerning patents, investors and values, where some intellectual property could be made available to people working their way out of poverty, but those of a high commercial value could be held in trust for social benefit. The author became involved with 3rd Way Coop as an informal mentor from November 2013, holding unstructured interviews and discussions with 3rd Way Coop's founder.

Ethical Small Traders Association (ESTA)

ESTA developed from its founder's experience of problems with decision-making in the UK voluntary sector, aiming to move away from this sector by creating a business association that promoted economic success through personal development. As a loose network of members, the organizational structure of ESTA connected with the open source guild model, in particular that the founding micro-business admits members to the association at its discretion based on shared values, in this case, that personal development can contribute to business success. The author was involved with ESTA from June 2014 as a member, involved in meetings of ESTA members, also holding unstructured interviews and discussions with the association's founder.

Analysis of Case Study Data as Annotated Portfolios

Case study data from the author's engagement with the founders of 3rd Way Coop and ESTA included notes of meetings with the founder and other stakeholders, the author's email correspondence with each founder and organizational documents that the founders made available to the author. The time frame for data collection was from the start of the author's involvement with each organization until October 2016. Each founder granted permission for their retrospective data to be used in the study, which took place under ethical approval from Lancaster University.

The framework for business model annotated portfolio analysis presented in the previous section was used to analyze case study data. After organizing the data, NVivo software was used to structure it for annotation as a portfolio of business model designs. A summary of this analysis is below, illustrated by relevant quotes from email correspondence with each founder and notes of meetings with them.

3rd Way Coop Annotated Portfolio Analysis

Culture

In an email to the author in November 2013, the founder of 3rd Way Coop emphasized that that the "The Trustees need to be Values Aligned senior people to encourage, develop and support Values Based Leadership throughout the Company and all operations/franchises etc to reform/develop our personal values and hence corporate values (made up of the sum of our personal values)".

Aesthetics

In emails in 2013, 3rd Way's founder highlighted that the firm was founded as an "unofficial "spin out" of Initiatives of Change as the thinking has been encouraged and developed by relating with this international community", clarifying that Initiatives of Change are "one of the leading international communities working in this field with 80 years track record".

Personal factors

The founder's vision for 3rd Way Coop was to make lasting changes to society by alleviating energy poverty, in a February 2015 email he highlighted that it was "realistic to think of such energy efficiency and distributed generation as a viable model to banish UK energy poverty to history".

Motivation for designing

3rd Way Coop took a view of value beyond the economic dimension, in initial emails to the author in 2013, he explained that this goal had become focused on Super Insulation as "the only solution that can lift the home occupier out of long term energy poverty".

People for whom it is intended

In the author's initial meeting in November 2013, 3rd Way Coop's founder highlighted that he had been working since 2010 to alleviate energy poverty through an extensive portfolio of technologies mainly in sustainable energy.

Sociopolitical concerns

In the author's initial meeting in 2013, 3rd Way's founder explained how he had worked for many years to alleviate poverty through applying engineering expertise, and in a follow-up email added that he had "networked with engineers etc for many years to overcome technical barriers preventing substantial progress to deliver UN Millennium Devt Goals/alleviation of poverty".

Operating context

In a 2013 email, the founder of 3rd Way Coop also mentioned their involvement for the past 18 years with Initiatives of Change, who "are one of the leading international communities working in this field with 80 years track record. This is very much a "counselling" community where we can come together, share experiences and encourage/strengthen each other", but was "under resourced and does not have the abilities to materially support the 3rd Way initiative". The founder's intention was to build on this involvement in that "IofC will be a benefactor of the Trust with a view to contribute to efforts to make up this shortfall in future".

Opportunities

In an email in 2013, 3rd Way's founder pointed out that 3rd Way's Super Insulation product can insulate to "passive house standards with an expected heating energy cost (and carbon) saving of around 75–90%" offering a significant business opportunity.

Customers

In an email in early 2015, the founder of 3rd Way Coop had developed a franchising strategy, where the customers were "international machine builders/installers (part of the Mondragon co-operative) interested in working with us to build and install the production systems".

Potential products and services

In a November 2013 email, 3rd Way's founder commented that "The technologies are being taken to market in the form of spin out co-operatives", while in December 2013 he highlighted the potential for installation of Super Insulation as being able to promote skills development, "installers who have gone from no skills and being unemployable to skilled under the Guilds program could be good peer mentors".

Functionality

Further thinking on 3rd Way's structure was expressed in emails by its founder in December 2013: "I have presented it in stages of evolution which hopefully will end up with 3rd Way being a part of an Internet enabled International Academy for Sustainable Development to accelerate delivery of UN Millennium Devt Goals". Then in January 2014: "Also to development of the franchising model so those who do not have the abilities/skills to develop new technologies but want to work to take the benefits to society – including those traditionally marginalised – can do so as a franchisee … I feel this fits very well with the creative commons and open source guilds – responsibly working to serve and build the common good".

Practicalities

Previous work with the Forum for Human Security enabled 3rd Way's founder to identify in an email in 2016 that "one of the main barriers to sustainable socio-economic development is education and especially vocational education", and also the "importance of peer to peer mentoring, quoting that 95% of successful entrepreneurs (leaders) state that peer to peer mentoring was the most important part of their success".

Boundaries

In the author's initial meeting with the founder of 3rd Way Coop in November 2013, he highlighted that the "Super Insulation patents are the core intellectual property and must be protected against exploitation by putting them in a trust", and that he wished to create a Multi Stakeholder Industrial Provident Society that has operational responsibility for producing Super Insulation.

Strategy

In an email in early 2015, the founder of 3rd Way Coop had "developed the production systems so it is modular and may be readily taken to scale as a Social Enterprise providing employment in manufacturing and many more jobs in installation", including working with Mondragon co-operatives.

Activities

By February 2015, 3rd Way's founder was able to reflect on leadership in an email, highlighting the importance of "helping others to overcome the barriers that prevent them from working their way out of poverty. Certainly the leadership team must have such a heart and hence aspirations/commitment".

Resources

In an email to the author in early 2014, the founder of 3rd Way Coop highlighted the problems with gaining investment, "The innovation/entrepreneurial process to identify a need and develop an innovative solution to that need and develop a business to reliably meet that need is very demanding and high risk".

Value creation and capture

In 2013 emails, 3rd Way's founder highlighted that the Super Insulation "patents are potentially of considerable commercial value and are to be placed in a Trust on behalf of those working their way out of poverty", refusing offers from commercial investors "who will take 30% control". By 2015, this thinking had developed, to propose that intellectual property would be "licensed or franchised – ideally within the Social Sector/Co-operative Sector" to housebuilders by the Trust.

Revenues and costs

In a February 2015 email, 3rd Way's founder emphasized the social benefits of Super Insulation, "I am more than happy to share these developments and the business model with the Social Sector to support efforts to alleviate energy poverty, create employment in this sector and generate income streams/ assets".

Profit

3rd Way's founder further highlighted in a February 2015 email that "It is only right that the technologies are owned by a Trust that is legally committed to such aims and implemented primarily through the Social Sector such that profits from meeting commercial market needs may be used to support and accelerate efforts".

Performance measures and processes

In 2013 3rd Way's founder reflected in an email that "As we grow we will inevitably recruit people who only know "business as usual" … hence the need for a Trust to appoint and remove directors".

Systems

In a June 2016 email, 3rd Way's founder considered systems to implement franchising, "I am also interested in evaluating an Open Source ERP system which appears to offer additional benefits".

Stakeholders

In an email in November 2013, 3rd Way's founder set out the organisation's developing structure: "The manufacturing companies need to be franchises or subsidiaries. The energy assessors and installers need to be Guilds. Master Assessors or Installers (responsible for standards, quality assurance and training) could be employed by 3rd Way … Some new technologies to support people working their way out of poverty need to be available under Open Source. Those with high commercial value need to patented and the patents held in trust. Also possible structure for the Foundation which essentially is a vocational training facility focused on developing Values Base Leadership together with audits of training needs etc. Funds come from royalties on patents".

Structure

Further discussions and reflections highlighted the role of the trust, in an August 2014 email 3rd Way's founder said: "I believe the 3rd Way model governed by the Joy at Work Foundation where operational directors are appointed and removed by the Foundation goes a long way towards overcoming these problems".

3rd Way Coop's Business Model Design

The annotated portfolio analysis for 3rd Way Coop reveals the design elements of:

- The Joy at Work Foundation protects Super Insulation intellectual property by holding it in trust.
- The beneficiary of the Joy at Work Foundation is Initiatives of Change, an international charity.
- The Joy at Work Foundation operates according to develop Values Based Leadership.
- The Joy at Work Foundation owns a Multi-Stakeholder Industrial and Provident Society (IPS) that takes operational responsibility for producing Super Insulation.
- The IPS licenses intellectual property to manufacturing firms through franchising, which can operate as co-operatives or other form of social enterprise.
- The IPS creates an installer guild network, that admits members based on shared values and vision.
- Housebuilding customers recruit Super Insulation installers from guild network.
- The installer guild network creates an open source commons of knowledge and experience on installing Super Insulation.
- The installer guild network promotes peer mentoring between members to mutual benefit.

The annotated portfolio analysis reveals 3rd Way Coop's business model design that is presented in Figure 9.1 below. In the 3rd Way network, the patents for Super Insulation are protected through a trust, initially with the founder in control, then a group of trustees, with the ultimate beneficiary of this trust being Initiatives of Change. Placing the patents in trust deals with the problem of investment that the founder had identified, as ethical investors can invest in the trust. Commercial investors can invest in subsidiary manufacturing firms that make Super Insulation under license from the trust. In the installer guild network, Super Insulation installers draw on a commons of knowledge, skills and experience, creating an organic social structure (Vargo and Lusch, 2016), that can be viewed as an assemblage of institutions (Wieland *et al.*, 2017). The guild network also enables peer mentoring of installers.

ESTA Annotated Portfolio Analysis

Culture

In a June 2014 email, the founder of ESTA highlighted that the Association was "an attempt to create a local culture of organisational and business cross-collaboration in the gradual creation of a local sustainable infrastructure".

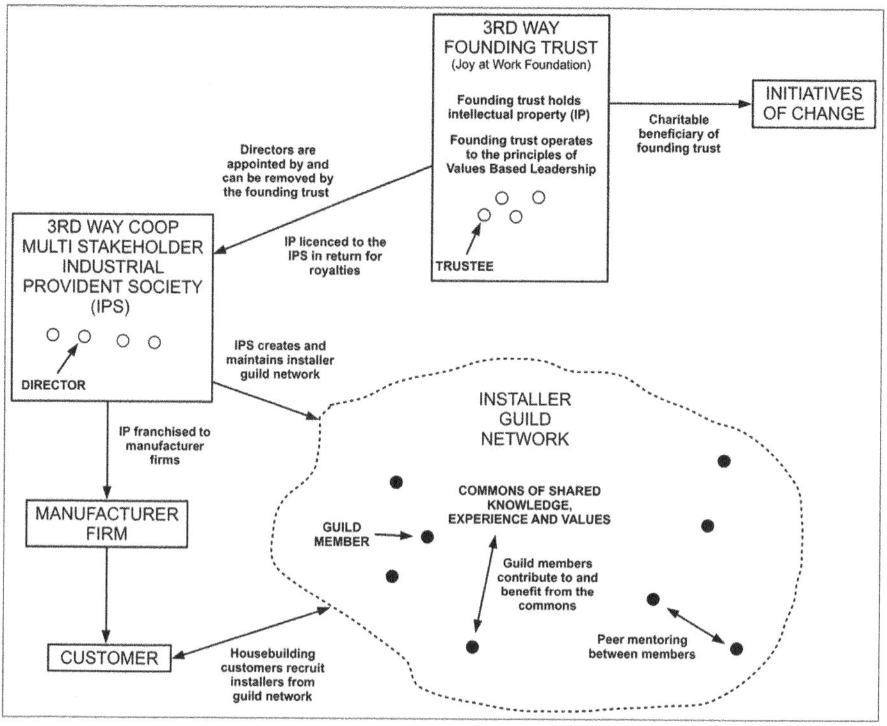

Figure 9.1. 3rd Way Coop's business model design.

Aesthetics

In a discussion with ESTA's founder in December 2014, he likened the network to fungi that are joined by tendrils in the soil, that what appear to be individual plants can act as one organism. In the same way, in ESTA individual micro-businesses are joined by the Association and are able to act together.

Personal factors

The author's first meeting with the founder of ESTA in May 2014 highlighted the importance of personal fulfillment as being compatible with economic success, that ESTA works to "promote intrinsic value in a way which assists people to earn a living".

Motivation for designing

In the author's first few meetings in mid-2014, ESTA's founder made it clear that the existing UK voluntary sector had problems with governance, in particular that marginalized people in society are attracted to organizations that promote open decision-making as they can have a voice as stakeholder and "make a difference", however these organizations had "confused equality at the level of human relationships with equality of decision-making", where the latter implies competence.

People for whom it is intended

In the author's first meeting with ESTA's founder in May 2014, he pointed out that ESTA's core value was the "fourth bottom line" which highlighted how they can contribute to the personal development of members, both individually and through collaboration and co-operation. In a meeting with the founder and other ESTA members in July 2014, he stressed that this was "how I want to work". In the same meeting, the founder highlighted that in the future, ESTA could become "open source", keeping within 100–200 members, and "breaking bits off" if it gets larger.

Sociopolitical concerns

In a January 2016 ESTA Consortium meeting, the founder pointed out that psychopaths would find it hard to manipulate the network, there's "nothing for them to get hold of, nothing here".

Operating context

Underpinning the work of ESTA is its aspiration to a "utopian sustainable future" of a finite but sustainable Earth and infinite capacity for inner growth and development, contrasting with the current dystopian present.

Opportunities

In a meeting with ESTA's founder in 2014, he highlighted how the idea for ESTA originated from experience with Transition Towns in Totnes, which led to developing the "Clockwork Orange bottom-up model which became central to their development".

Customers

In the case of ESTA, on one level, members are customers of the CIC at the core of the network, as it derives its income from membership of the Association. In a July 2014 ESTA meeting its founder highlighted that he was seeking to "intensify trade within ESTA", creating a "hub", where members could become each other's customers.

Potential products and services

In a December 2015 ESTA consortium meeting, ESTA's founder highlighted that "creating a sellable product, an elevator pitch would be a good start" followed by creating a narrative of the skills the consortium can offer. In an email to the ESTA Consortium group in August 2016, he highlighted that "I genuinely do believe that between us we have a huge range of capacities, techniques and resources to offer and also that people and organisations need ways of engaging in genuinely new and more effective ways in order to identify and avoid making the same old limiting mistakes".

Functionality

In the first meeting of the author with ESTA's founder in 2014, he described how the network was established to "promote intrinsic value in a way which assists people to earn a living", and to promote "not just developing the business but the self as an individual".

Practicalities

In 2016 the thinking of ESTA's founder had developed, in an email in October 2016 he proposed that "ESTA itself is a living organisational experiment in how to better integrate and support many active players in becoming effective agents for internal and external capacity building".

Boundaries

In the first meeting with ESTA's founder, he made the connection between ESTA and the open source guild model, where members are admitted to ESTA at the founder's discretion, based on shared values and a commitment to personal development, in this respect the founder acts as a "benevolent dictator", similar to the role of open source software architect. The commons in ESTA is one of interests, stories of personal development.

Strategy

At a ESTA meeting in July 2014 on the theme of "Where ESTA can go next", the founder pointed out that "ESTA is becoming more strategic", with "longer-term relationship-building", moving on from their initial focus on networking. By January 2016, ESTA's founder was now "seeking to: Develop an ESTA-branded Strategic Development Consultancy Consortium that provides help, support and training for groups, organisations, businesses and individuals to assist with structural organisational, personal and inter-personal development".

Activities

In a November 2014 email ESTA founder highlighted that "ESTA is pioneering a new approach to dynamic, active, cross-sectoral and responsive organisational structures, underpinned by reference to the quadruple bottom line". By July 2015, this overall aim had developed a more practical application, "There are moves afoot within ESTA to create our own Professional Consultancy Network with a joined up approach to collaboration, referring on and enhancing the profile of the services that we offer". ESTA's founder further highlighted the importance of the consortium approach in an email in November 2015, "The imperative is that the tendency to commission public service contracts are going to larger and larger organisations and the benefits to local communities are less and less visible".

Resources

In a meeting with ESTA's founder in December 2014, he pointed out that an ESTA member organization had a plan to gain resources, where other ESTA members could buy shares in the enterprise as it becomes established (possibly using the Community Shares model) or give it gifts (which can be given to a business, it doesn't have to be a charity). This allows the enterprise to become established without having to obtain outward investment (such as bank loans or charitable funding, both of which come with their own agenda).

Value creation and capture

During the period of the author's involvement, ESTA became increasingly focused on value capture in the local community. A partnership with Lancaster University in 2013 resulted in the BARTER project, which aimed to keep money local through a discount scheme. ESTA's founder developed the idea further in August 2015 to create the Food Loop Game, where he described in an email to ESTA members that the "aim of the game is to collectively put as much money into the local food economy as possible and to encourage those food businesses we spend with to pass our money on within our local food economy so that it effectively gets spent twice (or more)". By August 2016, he was able to demonstrate that "we have evidenced £11,341 of spending with local food and drink retailers. This has been further spent 1.3 times, creating an additional £15,501 in the local food economy".

Revenues and costs

In a December 2015 consortium meeting, ESTA's founder highlighted that the overall process of the consortium could be one that makes money through business and generates a surplus that could then fund projects to benefit people with disabilities. Income generation could be from charging a fee for successful business through the consortium.

Profit

In the first meeting with ESTA's founder in May 2014, he highlighted that ESTA is moving towards becoming viable financially, the membership fees are providing some income for ESTA CIC as the founding micro-business.

Performance measures and processes

At a members meeting in July 2014, ESTA's founder highlighted that an audit system was being trialed with a few members, that could be rolled out to all members. In a July 2015 meeting in Lancaster, he further highlighted that some members are electing themselves out of ESTA in an organic process when they realize it's not for them, which he saw as a sign of a healthy organization.

Systems

ESTA's founder pointed out in a December 2014 meeting that ESTA was like a "metabolic process" at a social level, where waste becomes resources. An example of this process in action was where a local shop became a centre for recycling Jiffy bags, once they had a room-full they were able to sell them in bulk to a specialist recycler.

Stakeholders

ESTA was set up to deal with personal issues and needs in the context of business, applying the concept of the fourth bottom line as personal development being the key to economic success. In a presentation made by ESTA's founder and members to Lancaster's Minister for Parliament in October 2016, its founder placed ESTA in a new "personal meaning sector".

Structure

In the author's first meeting in May 2014, ESTA's founder was "wary of anything that becomes a legal structure", such structures could be subverted or taken over and in any case will suffer the problems of being governed by a committee. In an ESTA consortium meeting in January 2016 ESTA's founder explained how he is now re-structuring ESTA to have two levels of Basic and Associate members. The basic membership is for a year, and offers access to networking (the benefits people are getting now but with increased emphasis on business promotion). This group is "self-maintaining". After the first year, they can become an associate member after undertaking an audit, additional benefits include "sector-specific" groups, based on businesses who know each other.

ESTA's Business Model Design

The annotated portfolio analysis for ESTA reveals the design elements of:

- ESTA Community Interest Company (CIC) as the founder of the ESTA membership association admits members based on the shared values of a commitment to personal development.

- As a Community Interest Company, ESTA CIC has a beneficiary charity under UK company law.
- The membership association does not have a legal structure.
- Membership fees from the association are an income for the CIC.
- The membership association promotes interaction between members to mutual benefit.
- Members often self select out of the association when they realize it's not for them.
- There are two levels of Basic and Associate members.
- Groups of members can trade and offer mutual support.
- The founder creates consortia of members that can trade with larger customers, creating a surplus that could be used for charitable work.
- Trade between ESTA members and other local businesses through the Money Loop System creates a local multiplier effect, capturing local value.

ESTA's business model design is presented in Figure 9.2 below. In the case of ESTA, the name of the association is registered as Community Interest Company, which prevents appropriation. Surrounding this hard frame (Bate *et al.*, 2000) of a formal institution is an organic social structure (Vargo and Lusch, 2016), based on shared values, but with no legal existence. The members of the association are

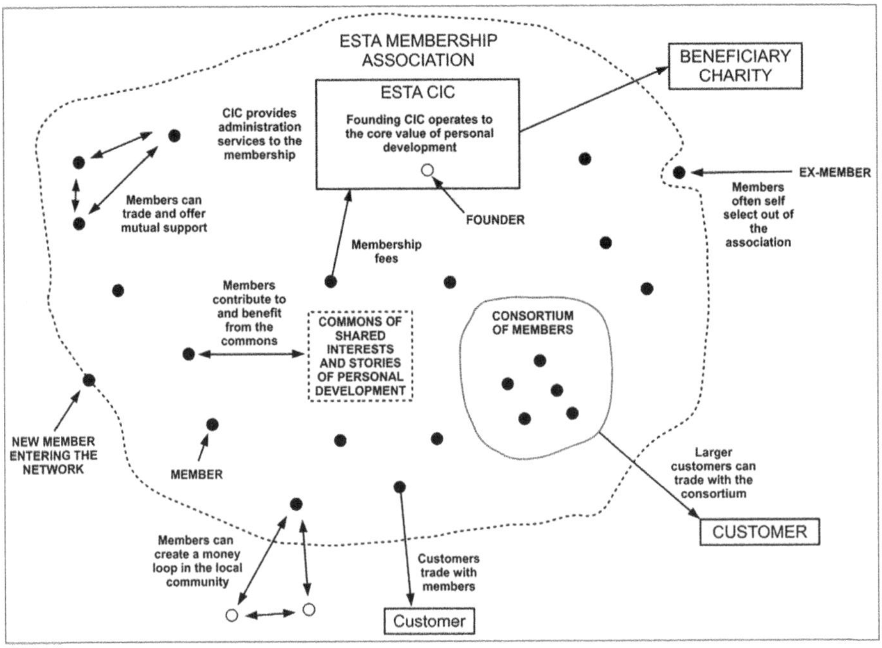

Figure 9.2. ESTA's business model design.

specifically not members of the CIC, and have no direct influence on it. Each ESTA member is a micro-business, therefore the association can be viewed as an assemblage of institutions (Wieland *et al.*, 2017), including consortia within the association. As a whole, ESTA could be viewed as a networked form of business model (Kornberger, 2017) that functions as a boundary object (Jensen, 2013). By having a minimal core organizational structure, the founder of ESTA can minimize governance costs (Santos and Eisenhardt, 2005).

These specific business model designs for 3rd Way Coop and ESTA revealed by the annotated portfolio analysis are then the basis for a more generalizable business model. This model is developed by considering the common factors and unique aspects of each case study business model, and also how each case study business model offers a power-balance between founders and other stakeholders.

Discussion: Common Factors and Power-balances in the Two Case Study Organizations

Common factors in the case study organisations

Both founders are seeking to transform how they do business, in the case of 3rd Way it is about how to use intellectual property of high commercial value to benefit society. In the case of ESTA the transformation is how to create a member network focused on personal development as key to economic success. Working with a community of stakeholders through an open source-like process can enable a transformation through practice (Ahokangas and Myllykoski, 2014).

Each founder wished to create a community rooted in their knowledge and experience (Ahokangas and Myllykoski, 2014). In the case of 3rd Way Coop the community is around creating and installing Super Insulation, and in the case of ESTA the community is around personal development in a business context. In both cases, the community around the focal organization enables exploring a larger market (Doganova and Eyquem-Renault, 2009). 3rd Way proposes to coordinate a network of installers of Super Insulation, where their business model will incorporate being a licensee. ESTA coordinates a network of business members, where their business model will incorporate being able to work with other members in consortia within the network. In both cases, the founder can gain competitive advantage from sourcing resources and competencies (Kornberger, 2017) from their member network.

Both founders adopted a legal form as the core of their business which enabled them to protect key intellectual property. In the case of 3rd Way, the patents for Super Insulation are protected through a trust, initially with the founder in control, then passing control to a group of trustees. In the case of ESTA, the name of the association is registered as a Community Interest Company (CIC), which prevents appropriation of their name. In this respect, ESTA operates like an open source project, where a name or trademark, such as Linux, enables the software architect to protect the boundary of the project (Weber, 2004).

The founding organization can capture value from their wider stakeholder community for the benefit of both the organization and its network, particularly financial value (Al-Debei and Avison, 2010), which then contributes to profits (Teece, 2010). In the case of 3rd Way Coop, the founding trust can capture value from its network of Super Insulation installers through licensing fees. In the case of ESTA, the founding Community Interest Company can capture value from its membership network through fees and through working jointly with them. In both cases, the membership network creates an organic social structure (Vargo and Lusch, 2016), based on shared values, which surrounds the hard frame (Bate *et al.*, 2000) of a formal institution.

The common factors and unique aspects of each case study business model are summarized in Table 9.2.

Case study organizations as power-balanced structures

Considering each case study organization as a 'power-balanced structure' (Moore, 2005, p. 673), in 3rd Way Coop the balance of power is held by the Founding Trust which licenses intellectual property to manufacturing firms. However, the installer guild network in developing their commons of shared knowledge and experience could exert considerable power in how they interact with the Trust and with house-building firms. The minimal structure of ESTA's membership network is a power-balanced structure, in that members can form independent consortia which could leave the network if they disagree with the founder's vision. However, leaving the network would mean losing the advantages of working to a common vision as part of ESTA. A balance of power is a key aspect of the Open Source Platform Federation business model, which is discussed in the next section.

9.6 The Open Source Platform Federation Business Model

This chapter first introduced the problem of power relations in the digital collaborative economy, highlighting how existing business models can exert power over workers through a centralized form of governance (Gol *et al.*, 2019). Alternative business models have been developed that aim to address these power imbalances, including platform cooperatives. Platform cooperatives operate as a form of federation, which are 'formed when two or more actors join in creating a common unit to promote common interests on contracted issues while keeping autonomy

Table 9.2. Common factors and unique aspects of each case study organization's business model.

3rd Way Business Model	ESTA Business Model	Factors that Contribute to New Business Model
Founding trust	Founding CIC	Founding non-profit organization
Values Based Leadership	Personal development values	Values based
Beneficiary charity that holds the founding trust to account	Beneficiary charity that the founding CIC is accountable to	Beneficiary charity that holds the founding non-profit organization to account
Franchising of intellectual property to manufacturing firms via an Industrial and Provident Society (IPS)		Members can benefit from founder's intellectual property as part of the network
Installer guild network	Membership club network	Membership network
Peer support within the installer network	Peer support within the membership network	Peer support within the membership network
Income from manufacturing firms.	Income from membership	Founding non-profit organization gains income from membership network
	Consortia of members to trade with larger customers	Consortia of members
	Promotion of trading in the local economy through a "money loop"	Trading in the local economy to create a "money loop"

on others' (Johnstad, 1997, p. 48). In particular, cooperatives are membership owned organizations which limit the power of leaders by election or by consensus decision-making by the membership (Spear, 2004), creating a balance of power. In the collaborative economy however, such decentralized forms of governance, while enabling more direct communication between workers and other stakeholders, also 'poses serious control and coordination challenges and creates demands for laborious consensus-based decision making' (Gol *et al.*, 2019, p. 176).

The Open Source Platform Federation business model presented here is also a form of federation (Johnstad, 1997), where the founding group recruits a group of members who share their common vision. This model can potentially offer a power-balanced structure through a similar mechanism to how power relations operate in open source software production. Power in open source 'at least to start, belongs to the person who generates the idea and articulates the core values behind the project', but 'as the community takes shape, its very openness moves power away from the leader and toward the followers' (Weber, 2004, p. 260). It is in this tension between the leader and their followers in open source where the power-balance lies, uniquely combining centralized decision-making with a power-balanced structure, thus having the potential to overcome the issue identified by Gol *et al.* (2019) of problems with decentralized governance.

In the Open Source Platform Federation model, the founder can retain some power by putting intellectual property in trust, adopting a legal document to do so that has attached symbolic functions (Searle, 1998). This document then gives the enterprise legitimacy and independence in wider society. What keeps members within the network is being able to access the licensed intellectual property and the opportunities they gain to both benefit from and contribute to the commons, however, groups of members can create their own network. In open source software, this is known as "forking", but this rarely happens in practice, as the benefits of remaining within the existing software project outweigh the benefits of leaving with the source code and setting up a new one (Weber, 2004). This balance of power could potentially enable a membership organization to operate without the traditional forms of representative democracy generally found in such organizations. The Open Source Platform Federation model is shown in Figure 9.3.

In this model, the founder recruits one or more individuals who share their values to form a founding trust or other non-profit organizational structure, to hold core intellectual property. In turn, the founding non-profit has a beneficial charitable organization to which it is accountable. Prospective members are attracted to the founder's vision of a shared desired future, on gaining membership they gain the right to use the founding IP and to access the commons of knowledge and experience. Members can trade with other members based on their shared values, which can include forming consortia to enable meeting the needs of larger customers.

Potential contribution to policy and practice

The trend towards the platform economy has been a concern for policy makers, affecting not only working conditions and employment rights but also having implications for wider issues in society such as welfare systems (Drahokoupil and Piasna, 2017). Mechanisms to deal with these issues have included regulation (Kenney and Zysman, 2016; Fabo *et al.*, 2017; Graham and Woodcock, 2018), and developing alternative worker-led business models, particularly platform cooperatives (Scholz, 2017; Martin *et al.*, 2017; Vandaele, 2018). The Open Source

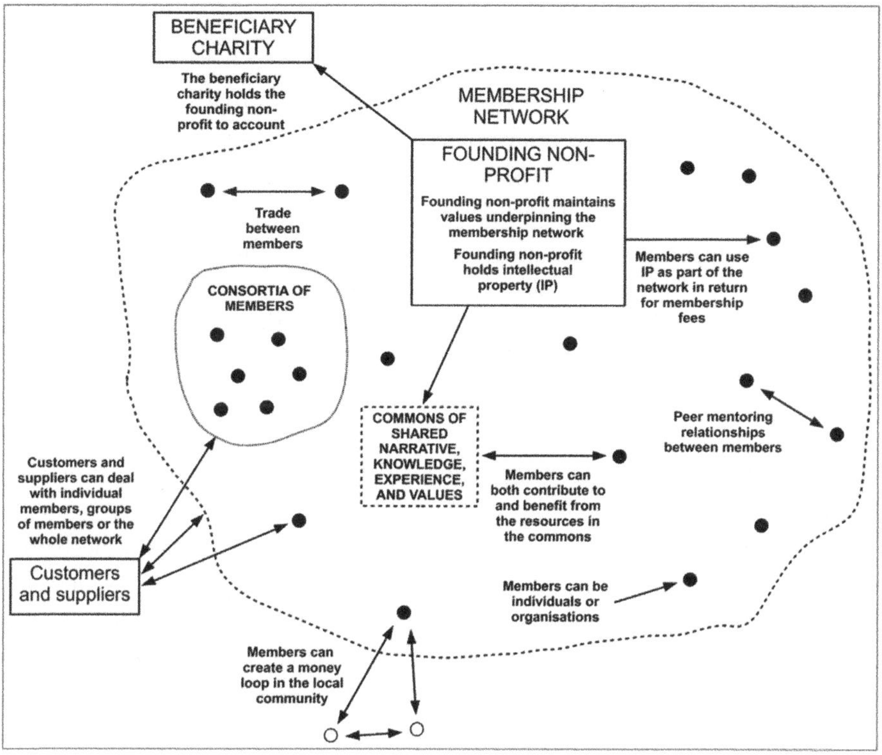

Figure 9.3. The Open Source Platform Federation business model.

Platform Federation model could be helpful to policy makers, as it offers an alternative to platform cooperatives and other models that aim to secure greater worker power. As a power-balanced structure, the Open Source Platform Federation business model could help address some of the societal issues identified by Kenney and Zysman (2016), particularly inequality of workers in the platform economy.

The Open Source Platform Federation model can address the issue highlighted by (Martin *et al.*, 2017, p. 1396), that the academic literature has 'yet to address novel forms of platform governance (democratic or otherwise) emerging within the sharing economy'. Entrepreneurs could also find this model helpful, as it can enable them to build their business through a membership network in which members share their vision and can also benefit from being part of the network. It is this value capture for the benefit of the membership that is a key benefit of this form of business model to workers, together with the balance of power they have with the business owner. However, organizations based on this model will not be owned by their membership so may not be seen as true cooperatives, despite the balance of power between the founding organization and their wider membership.

9.7 Conclusion

This chapter first introduced the potential for creating alternative business models in the digital collaborative economy, then presented how annotated portfolios can be used to enable a founder's wishes and desires for their emerging business to be articulated as a business model design. The chapter then presented two case studies of emergent micro-businesses who wished to use the platform economy to create a networked business model for more than financial benefit. Using annotated portfolio techniques as a framework to analyze ethnographic data gathered from engagement with each founder enabled their wishes and desires to be articulated as business models designs, rooted in their knowledge and experience in a particular domain. The common factors and power-balances in the case study business model designs were then abstracted as the Open Source Platform Federation business model.

The Open Source Platform Federation business model can offer a mechanism for an entrepreneur to build a business with others who share their values, first through creating a founding trust or other non-profit organization, then through establishing a wider membership network. Members within the network can benefit from being able to both access and add to a commons of knowledge and experience. As a business model, the founding organization sets the boundary of the network and captures value from that network for the benefit of members as well the founder. Entrepreneurs could find this model helpful, as it can enable them to build their business through a membership network in which members share their vision and can also benefit from being part of the network. This model has its limitations, however, in that it may not be as effective if the founder wishes to create an organization for purely economic gain, or who do not wish to engage with a membership network.

Acknowledgements

I wish to thank the other chapter authors who provided very valuable input into the development of this chapter during the mutual collaborative review process. This chapter is developed from a presentation to the Ethnographies of Collaborative Economi(es) Conference, Edinburgh, October 2019, I wish to acknowledge the helpful comments of two anonymous reviewers on the conference paper which helped strengthen the arguments made. I also wish to thank the time and input that the founders of 3rd Way Coop and Ethical Small Traders Association gave over several years to this project. The research that underpins this chapter was funded by the UK Engineering and Physical Sciences Research Council through the Digital

Economy Programme (RCUK Grant EP/G037582/1), which supports the High-Wire Centre for Doctoral Training.

References

Ahokangas, P. and Myllykoski, J. (2014). The practice of creating and transforming a business model. *Journal of Business Models, 2*(1), 6–18.

Al-Debei, M.M. and Avison, D. (2010). Developing a unified framework of the business model concept. *European Journal of Information Systems, 19*(3), 359–376.

Barab, S.A., Thomas, M.K., Dodge, T., Squire, K., and Newell, M. (2004). Critical Design Ethnography: Designing for Change. *Anthropology and Education Quarterly, 35*(2), 254–268.

Bardzell, J. and Bardzell, S. (2013). What is "critical" about critical design? In *Proceedings of the SIGCHI Conference on Human Factors in Computing Systems (CHI '13)*, 3297–3306. New York: ACM.

Barrass, S. (2016). An annotated portfolio of research through design in acoustic sonification. *Leonardo, 49*(1), 72–73.

Bate, P., Khan, R., and Pyle, A.J. (2000). Culturally sensitive structuring: An action research-based approach to organization development and design. *Public Administration Quarterly, 23*(4), 445–470.

Bauwens, M. and Niaros, V. (2017). The emergence of peer production: challenges and opportunities for labour and unions. http://www.etui.org/Publications2/Policy-Briefs/European-Economic-Employment-and-Social-Policy/Collective-labour-law-under-attack-how-anti-crisis-measures-dismantle-workers-collective-rights accessed 8 April 2021.

Bowers, J. (2012). The Logic of Annotated Portfolios: Communicating the Value of Research Through Design. In *Proceedings of the Designing Interactive Systems Conference (DIS 2012)*, 68–77. New York: ACM.

Cassell, C. and Johnson, P. (2006). Action research: Explaining the diversity. *Human Relations, 59*(6), 783–814.

Chesbrough, H. (2010). Business Model Innovation: Opportunities and Barriers. *Long Range Planning, 43*(2–3), 354–363.

Cole, R., Purao, S., Rossi, M., and Sein, M. (2005). Being proactive: where action research meets design research. International Conference on Information Systems 2005, Proceedings Paper 27.

DaSilva, C.M. and Trkman, P. (2014). Business Model: What It Is and What It Is Not. *Long Range Planning, 47*(6), 379–389.

Deakin, S. (2006). The Return of the Guild?: Network Relations in Historical Perspective. Working paper No. 322, Centre for Business Research, University of

Cambridge. https://pdfs.semanticscholar.org/dfe3/05fb135a1504148a28e81564f1999dcc6f27.pdf accessed 7 April 2019.

Doganova, L. and Eyquem-Renault, M. (2009). What do business models do?: Innovation devices in technology entrepreneurship. *Research Policy*, *38*(10), 1559–1570.

Drahokoupil, J. and Fabo, B. (2016). The platform economy and the disruption of the employment relationship. ETUI Policy Brief N° 5.

Drahokoupil, J. and Piasna, A. (2017). Work in the Platform Economy: Beyond Lower Transaction Costs. *Intereconomics*, *52*(6), 335–340.

Fabo, B., Karanovic, J., and Dukova, K. (2017). In search of an adequate European policy response to the platform economy. *Transfer*, *23*(2), 163–175.

Fielt, E. (2013). Conceptualising Business Models: Definitions, Frameworks and Classifications. *Journal of Business Models*, *1*(1), 85–105.

Foth, M. and Axup, J. (2006). Participatory design and action research: Identical twins or synergetic pair? In *Expanding Boundaries in Design: Proceedings of the Ninth Participatory Design Conference 2006* (Vol. 2), 93–96.

Frankfurt Paper (2016). Frankfurt Paper on Platform-Based Work. https://www.igmetall.de/download/docs_20161214_Frankfurt_Paper_on_Platform_Based_Work_EN_b939ef89f7e5f3a639cd6a1a930feffd8f55cecb.pdf accessed 28 April 2021.

Frauenberger, C., Makhaeva, J., and Spiel, K. (2016). Designing smart objects with autistic children: Four design exposès. In *Proceedings of the 2016 ACM Annual Conference on Human Factors in Computing Systems (CHI '16)*, 130–139. New York: ACM.

Fuchs, C. (2014). Critique of the Political Economy of Informational Capitalism and Social Media, in Fuchs, C., and Sandoval, M. (Eds.), Critique, Social Media and the Information Society, 51–65. New York: Routledge.

Gaver, W. (2012). What should we expect from research through design? In *Proceedings of the 2012 ACM Annual Conference on Human Factors in Computing Systems (CHI '12)*, 937–946. New York: ACM.

Gaver, B. and Bowers, J. (2012). Annotated Portfolios. *Interactions*, *19*(4), 40–49.

Gol, E.S., Stein, M.K., and Avital, M. (2019). Crowdwork platform governance toward organizational value creation. *The Journal of Strategic Information Systems*, *28*(2), 175–195.

Graham, M. and Woodcock, J. (2018). Towards a fairer platform economy: introducing the Fairwork Foundation. *Alternate Routes*, *29*, 242–253.

Healy, J., Nicholson, D., and Pekarek, A. (2017). Should we take the gig economy seriously? *Labour & Industry: A Journal of the Social and Economic Relations of Work*, *27*(3), 232–248.

Hobye, M., Padfield, N., and Löwgren, J. (2013). Designing social play through interpersonal touch: An annotated portfolio. In *Proceedings of Nordes 2013: Experiments in Design Research*, 366–369.

Hughes, J., King, V., Rodden, T., and Andersen, H. (1994). Moving out from the control room: ethnography in system design. In *Proceedings of the 1994 ACM Conference on Computer Supported Cooperative Work*, 429–439. New York: ACM.

Irani, L.C. and Silberman, M.S. (2013). Turkopticon: Interrupting Worker Invisibility in Amazon Mechanical Turk. In *Proceedings of the SIGCHI Conference on Human Factors in Computing Systems (CHI '13)*, 611–620. New York: ACM.

Jensen, A.B. (2013). Do we need one business model definition? *Journal of Business Models*, *1*(1), 61–84.

Johnstad, T. (1997). Co-operatives and Federations. *Journal of Co-operative Studies*, *30*(89), 48–60.

Kelliher, A. and Byrne, D. (2015). Design futures in action: Documenting experiential futures for participatory audiences. *Futures*, *70*, 36–47.

Kenney, M. and Zysman, J. (2016). The Rise of the Platform Economy. *Issues in Science and Technology*, *32*(3), 61–69.

Kornberger, M. (2017). The visible hand and the crowd: Analyzing organization design in distributed innovation systems. *Strategic Organization*, *15*(2), 174–193.

Larner, J. (2013). Open source guilds: enabling micro businesses to create a sustainable community of practice? In *Proceedings of the November 2013 Digital Economy Conference*, Salford, UK.

Larner, J., Cheverst, K., MacDonald, M., Hoile, C., and Soutar, A. (2017). The open source guild: creating more sustainable enterprise? *Journal of Management Development*, *36*(1), 71–80.

Lavori, P.W. and Dawson, R. (1998). Developing and Comparing Treatment Strategies: An Annotated Portfolio of Designs. *Psychopharmacology Bulletin*, *34*(1), 13–18.

Löwgren, J. (2013). Annotated Portfolios and Other Forms of Intermediate-Level Knowledge. Interactions January + February 2013, 30–34.

Martin, C.J., Upham, P., and Klapper, R. (2017). Democratising platform governance in the sharing economy: An analytical framework and initial empirical insights. *Journal of Cleaner Production*, *166*, 1395–1406.

McCracken, G. (1988). The Long Interview. London: Sage.

Moore, G. (2005). Corporate Character: Modern Virtue Ethics and the Virtuous Corporation. *Business Ethics Quarterly*, *15*(4), 659–685.

Nelson, H.G. and Stolterman, E. (2012). The Design Way (Second Edition). London: MIT.

Nenonen, S. and Storbacka, K. (2010). Business model design: conceptualizing networked value co-creation. *International Journal of Quality and Service Sciences, 2*(1), 43–59.

Ride Share Drivers United (2021). Mission Statement. https://ridesharedriversunited.com/mission-statement accessed 8 April 2021.

Santos, F.M. and Eisenhardt, K.M. (2005). Organizational Boundaries and Theories of Organization. *Organization Science, 16*(5), 491–508.

Sauerwein, M., Bakker, C., and Balkenende, R. (2018). Annotated Portfolios as a Method to Analyse Interviews. In *Proceedings of DRS 2018*, 1148–1158.

Scholz, T. (2016). Platform Cooperativism. Challenging the Corporate Sharing Economy. New York: Rosa Luxemburg Foundation.

Scholz, T. (2017). Uberworked and Underpaid: How workers are disrupting the digital economy. Cambridge: Polity Press.

Searle, J.R. (1998). Mind, Language and Society: Philosophy in the Real World. New York: Basic Books.

Shafer, S.M., Smith, H.J., and Linder, J.C. (2005). The power of business models. *Business Horizons, 48*(3), 199–207.

Silverman, D. (2007). A Very Short, Fairly Interesting and Reasonably Cheap Book about Qualitative Research. London: Sage.

Simon, H.A. (1969). The Sciences of the Artificial. Cambridge, MA: MIT Press.

Spear, R. (2004). Governance in democratic member-based organisations. *Annals of Public and Cooperative Economics, 75*(1), 33–59.

Spieth, P. and Schneider, S. (2016). Business model innovativeness: designing a formative measure for business model innovation. *Journal of Business Economics, 86*(6), 671–696.

Stanford, N. (2007). Guide to Organisation Design: Creating High-performing and Adaptable Enterprises. London: Profile Books.

Star, S.L. and Griesemer, J.R. (1989). Institutional Ecology, "Translations" and Boundary Objects: Amateurs and Professionals in Berkeley's Museum of Vertebrate Zoology, 1907-39. *Social Studies of Science, 19*(3), 387–420.

Suchman, L. (1995). Making work visible. *Communications of the ACM*, 38(9), 56–64.

Tapscott, D., Lowy, A., and Ticoll, D. (2000). Digital Capital: Harnessing the Power of Business Webs. Boston, MA: Harvard Business School Press.

Teece, D.J. (2010). Business Models, Business Strategy and Innovation. *Long Range Planning, 43*(2–3), 172–194.

Turkopticon (2021). Turkopticon. https://turkopticon.ucsd.edu accessed 8 April 2021.

Vandaele, K. (2018). Will trade unions survive in the platform economy? Working Paper 2018.05. Brussels: European Trade Union Institute.

Vargo, S.L. and Lusch, R.F. (2016). Institutions and axioms: an extension and update of service-dominant logic. *Journal of the Academy of Marketing Science*, *44*(1), 5–23.

Weber, S. (2004). The Success of Open Source. Harvard: Harvard University Press.

Wieland, H., Hartmann, N.N., and Vargo, S.L. (2017). Business models as service strategy. *Journal of the Academy of Marketing Science*, *45*(6), 925–943.

Wirtz, B.W., Pistoia, A., Ullrich, S., and Göttel, V. (2016). Business Models: Origin, Development and Future Research Perspectives. *Long Range Planning*, *49*(1), 36–54.

Yancy, K. (1992). Portfolios in the Writing Classroom: An Introduction. Urbana: National Council of Teachers.

Zott, C., Amit, R., and Massa, L. (2011). The Business Model: Recent Developments and Future Research. *Journal of Management*, *37*(4), 1019–1042.

DOI: 10.1561/9781680838411.ch10

Chapter 10

Legal and Organizational Aspects of Labour Relations in the Collaborative Economy

By Kosjenka Dumančić, Ivana Načinović Braje and Ana Aleksić

The app-enabled collaborative economy has grown dramatically in recent years and initiated some fundamental changes in work and employment. This chapter examines the differences in worker-employer relationship in traditional jobs that are not facilitated by the digital platforms and new jobs that are growing rapidly due to the use of digital platforms and are based on short-term, on demand, labor contracting. We specifically look at features of the work enabled by the digital platforms. The chapter emphasizes changes in labour driven by the digital platforms and features of work being the result of these platforms. Main legal issues associated with the position of the gig worker, employment status and employment contract for workers in the collaborative economy are also discussed, including the vision of future labour law rules and regulation.

10.1 Introduction

The world of work has changed tremendously in recent decades. At the moment, as some argue, we are living in the 4th industrial revolution driven by digitalisation (World Economic Forum, 2016). The changes we are witnessing today are introducing numerous uncertainties into existing social relations, which is especially evident as a part of new platform-based business models that can be referred as part of the "collaborative economy" (see Introduction, this volume).

The collaborative economy has brought not only economic changes but has also changed the nature of work and labour markets. The number of jobs created or intermediated by a platform-based collaborative economy has been constantly increasing in the last years. In North America and Western Europe approximately 150 million workers have left traditional and stable jobs to become independent contractors by 2018 (Petriglieri *et al.*, 2018). The U.S. Bureau of Labor Statistics reported in 2017 that 55 million people in the U.S. are "gig workers" which accounts for approximately 34 percent of the U.S. workforce, projected to increase to 43 percent in 2020 (ILO, 2020). In Europe, the proportion of the population having done at least some crowd work differs between countries, ranging from 9% in Germany and the UK to a high of 22% in Italy (Huws *et al.*, 2017; Pesole *et al.*, 2018).

Platforms have introduced changes in the structure of work, in the relations between employers and employees, in the skills and knowledge required from employees, but also new perspectives and practice in labour law. There are many "grey areas" when discussing the legal status of workers in a collaborative economy. Many consolidated forms of employer-employee relations have been disrupted in less than a decade. An increasing number of companies conduct their core businesses through outsourced workers and assets owned by these workers (e.g. Uber). Such workers often significantly differ from the "traditional employee". Although this new type of working relationship has its roots in the idea of flexible and often part-time job facilitated by the platforms, in the meantime this has become the only job for many of these workers. For example, only ride-sharing apps, primarily Lyft and Uber, have grown from almost nothing in 2012 to over a million drivers in the U.S. in 2018 (Oyer, 2020). Thus, in the context of exponential growth of jobs provided by online platforms, on-demand work has become an important social phenomenon.

Platform work is difficult to measure since the identification of such work as a form of employment requires not only that someone has done some work via platforms in a particular period, but also the regularity, intensity and significance of that work (Urzi Brancati *et al.*, 2019). Accordingly, several aspects of work should be considered: locus (online or in person), regularity over a specific reference period, time allocated in the reference period and income earned. Based on these criteria, work via platforms can be classified as: **regular job, secondary job** (employment is somewhat regular but lesser work hours) or **sporadic** (infrequent and inconsequential in terms of time or income).

The concept of employment as we know it today was developed in the beginning of 20th century when, during Fordism, the full-time permanent employment, with stable income, labour and social security protection became the norm for standard employment relationship (Schoukens and Barrio, 2017). The Fordist model of full-time employment in combination with applying technology to the production process, namely the assembly line, enabled growing worker productivity and higher wages (Collier *et al.*, 2017). During the following decades, and thanks to workers' unionism and mobilizations, workers' rights were placed under special public protection via the state laws and the differences between the role of worker and that of employer were legally defined. The concept of strict division of roles has been retained to this day, and it is further defined by the national laws, and at the EU level protected by special legal norms contained in secondary law relating to social rights.

With the emergence of digital platforms and work performed through their mediation, the division of roles between workers and employers has been modified and the traditional role of the employer, as defined by the legal systems, often lost. The newly created digital labour market consists of freelancers (Degryse, 2016), which is actually more similar to what is defined in EU law as self-employment. According to Eurofound, self-employed persons are those who work in their own business, professional practice or farm for the purpose of earning a profit (Pedersini and Coletto, 2009). In the U.S. legal framework, such employees that do not receive protection as salaried employees and have freedom to choose how and whether to perform work are called independent contractors (Nerinckx, 2016). Workers included in the new working arrangements of the collaborative economy have been called by different names: on-demand, crowd workers, platform workers, independent and flexible contractors and quite often gig workers. The latter term comes from the employment of musicians that were hired to play for a particular set or performance and is now used broadly to describe workers hired on the spot for a particular task (Friedman, 2014).

With the development of new forms of work, questions have been raised with regard to protection of such workers and the application of labour law. Labour relations are part of the EU social agenda and therefore encompassed by the EU legislation and harmonized at the EU level. Despite the differences between various national legislations, a set of minimum standards has been established at the supranational level (Barnard and Deakin, 2002; Schiek, 2017). With the rise of different "informal" forms of working relationships with digital platform as an intermediary, the relationship between worker and employer is widely reinterpreted. Furthermore, the informal character of the collaborative economy has opened up questions on workers' rights and welfare provisions connected with employment status, such as working hours, social rights, health protection, pensions, annual leave and right to collective bargaining and representation. According to Valenduc and Vendramin (2016), the new forms of work gave rise also to questions such as: what are the options for ensuring freedom to move between employment and self-employment while retaining the necessary degree of security; how to balance private and professional life and how to manage time; and, in case of very fluid boundaries between the roles of employer and contractor, what basis should be used to create a legal framework for this type of contractual relations. All these organizational and legal aspects will be further explored in the following sections.

10.2 Organizational Structure of the Work Enabled by the Digital Platforms

New types of companies that have emerged as a part of the collaborative economy, i.e., platform-based companies, connect requestors with individual service providers. These digital platforms and new technological companies thus, in theory, just match a client (demand) with a worker who will perform the task (supply). By doing so, the platform actually operates a two-sided market; it connects providers and customers, but charges both sides for its services (Isaevoli *et al.*, 2018). Even though digital platforms actually directly match clients and workers to perform the task, companies owning these platforms are declaring themselves as databases, not as employers (Todolí-Signes, 2017a).

The new division of labour implemented by the digital platform and "on-demand economy" could be compared to a *virtual* assembly line (Todolí-Signes, 2017b). Technology enabled new division of tasks; tasks are simplified and compensated by piece rate, as opposed to time-based pay that dominated in the traditional employment. Therefore, platform's organizational practices are not genuinely novel, just "old wine in old bottles" (Stanford, 2017; Vandaele, 2018), or as Degryse (2016) calls it, digital Taylorism as only small tasks are offered to workers. As argued by Todolí-Signes (2017b), internet-based digital connectivity allows platforms to have large pools of workers waiting for a customer's request, where these workers will be hired only when someone orders a product or service, and dismissed (or stop getting paid) immediately afterwards. The concept of the digital collaborative economy therefore uses the idea of "zero marginal costs" as digitized services

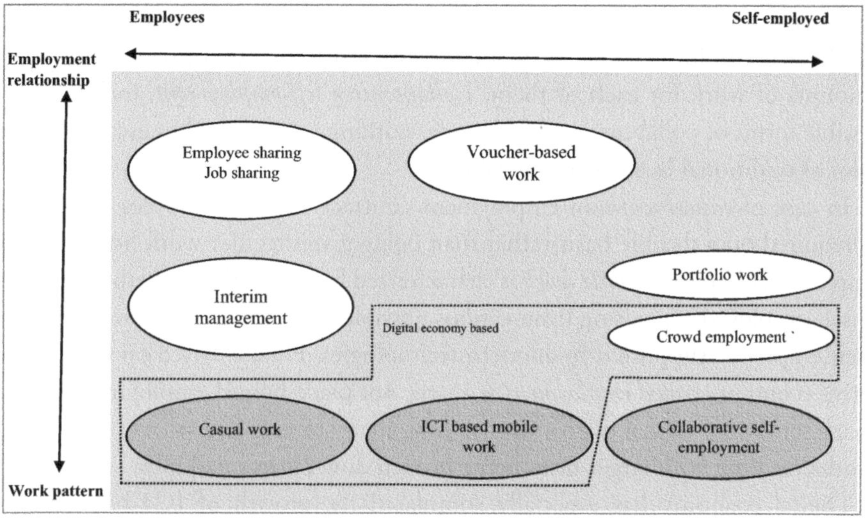

Figure 10.1. Classification of new forms of employment.

Source: Eurofound (2015). *New forms of employment*, Publications Office of the European Union, Luxembourg, p. 8.

and goods may be possessed without necessarily being owned and used on a non-exclusive basis (Valenduc and Vendramin, 2016). The technology of the platform can thus be thought of not as making the worker more productive in the actual production process, but rather as making the market more efficient by lowering transaction costs (Collier *et al.*, 2017).

Modification of the traditional one-to-one relationship between worker and employer in the past decade has resulted with several new forms of employment (Eurofound, 2015). Based on the (1) relationship between employer and employee (ranging from classic employment to self-employment options) and (2) the way work is conducted (ranging from traditional working arrangements to fully new work patterns) quite a few new forms of employment have been recognized in the European Union, as shown in Figure 10.1.

Employee sharing is seen in cases when an individual worker is jointly hired by a group of employers (who are not clients of a traditional temporary work agency), and where such workers rotate between the different companies. In contrast to this is *job sharing*, in which a single employer hires two or more workers to jointly fill a specific job. *Voucher-based work* is a form of employment in which the relationship and related payment is based on a voucher rather than an employment contract. In most cases, these workers then have a status somewhere between employees and self-employed (Eurofound, 2015).

Interim management describes situations in which a worker, usually a highly skilled expert, is hired for a temporary period of time by an employer, often to conduct a specific project or solve a specific problem. Similarly, *portfolio work* done by the self-employed refers to work for numerous clients, providing just small amounts of work for each of them. *Collaborative self-employment*, includes more flexible forms of collaboration (such as co-working spaces) used to escape the confines of traditional business partnerships.

In case of *casual work* an employment contract allows employees to be called as required on a flexible basis rather than being given regular work hours by their employer. *ICT-based mobile work* is characterized by the worker (whether employee or self-employed) operating from various possible locations outside the premises of their employer, supported by modern technologies. Finally, based on the collaborative economy, *crowd employment* is a new, not place-bound employment option, characterized by virtual platforms matching many buyers and sellers of services or products, often with larger tasks being broken down into small jobs.

Digital economy has especially stimulated the growth of ICT-based mobile workers, crowd working and – in certain respects – casual work (Valenduc and Vendramin, 2016). Still, even platform work can be classified into several categories, according to several criteria (see Table 10.1). Most broadly, platform work can be divided as online or offline work (Vandaele, 2018). Online work can take micro (repetitive micro-tasks, e.g. click-work like data entry, content tagging or interpretation, or finding information), and macro form (requires professional knowledge and competences, e.g. graphic design work, web and software development, editing and translation, etc.). The latter is usually associated with placeless crowd workers. Unlike online work, offline work is typically demanded by apps, it is usually place and time on-demand work, performed locally (Todolí-Signes, 2017b). Common to all of these forms of contingent or on-call labour is piece-based compensation, the requirement that workers provide their own capital equipment and digital

Table 10.1. Comparison of two main types of platform work.

Crowdsourcing (Crowd Work)	On-demand (Gig Work)
Online work	Offline work
Global workforce	Local workforce
Remote delivery of electronically transmittable services	Delivery of the services is physical and/or requires direct interaction
Difficult to control and regulate	Possibility to control and regulate
Infinite number of workers and clients	Place-based and geographically limited work.
Payment for work is decentralized and irregular	Payment for work is irregular but centralized and prices are set by the platform

Source: Authors' work.

intermediation necessary to commission the work, deliver it to the final customer or facilitate payment (Stanford, 2017). Interestingly, only the work demanded by apps is the result of modern technology, while all others are old practices, not some novelties implemented by the digital economy (Valenduc and Vendramin, 2016; Stanford, 2017).

Another important distinction in platform work can be referred to as crowdsourcing (crowd work) vs. on-demand work (Collier *et al.*, 2017; Eurofound, 2015; World Bank, 2015; Greenhouse, 2015; Berg and De Stefano, 2015; De Stefano, 2016). Crowd work is arranged for and fulfilled remotely and online on virtual platforms by workers in response to on-line calls and potentially involving people from all over the world (e.g. ClickWorker or Amazon Mechanical Turk). In such job structure a vendor can recruit an on-call contingent worker from a different geographical area for a quick or instant job task, which potentially raises serious problems regarding anti-discrimination law, workplace health and safety and social arbitrage. The second type, on-demand work, is fulfilled in person in the physical world (e.g. Uber) and therefore locally, offline (Aloisi, 2016; De Stefano, 2016). As a consequence of this distinction, crowd work platforms construct a potentially global labour market that integrates high- and low-wage economies, whereas on-demand platforms construct a local market and although such a platform can expand to many localities and organizes separate local markets, it is generally easier to control and regulate (Collier *et al.*, 2017). What links crowd-work and work-on-demand via apps is, at a first glance, the enabling role of technology and the common business model (Graham and Shaw, 2017). The significance of this relationship stands in the content of the three-sided contractual relationship between platform, requester and worker (often defined partners, rabbits etc.), or the two-sided contracts where one is concluded between the platform and the requester and the other between the platform and the worker.

The combination of the different concrete elements of the employment relationships leads to legal concerns regarding rights, obligations and liabilities of the workers as well as employers (Risak and Warter, 2015), particularly due to the fact that it is not well-defined *who* is the employer.

Platform supported work is often epitomized with terms such as gig or task to indicate a new type of employment in which labour protection and employment regulations are assumed not to apply by default. Workers that are called by a click of a mouse could be seen by customers as simply an extension of the on-line platform or IT device (De Stefano, 2016). Nevertheless, as would have been expected from a more traditional organization, such platform-based companies want to keep a high level of quality when providing their services. The platform thus has to ensure that workers provide a fair service to its clients even though platforms themselves keep classifying their workers as independent contractors, or self-employed workers

(Todolí-Signes, 2017a). In work on-demand, jobs related to traditional work activities such as transport, delivery, cleaning, forms of clerical work etc. are offered and assigned through mobile apps. The businesses running these apps intervene in setting minimum quality standards of service and in the selection and management of the workforce (De Stefano, 2016).

Workers in the collaborative economy depend upon reviews of past activities as these might have consequences on gig workers ability to work or earn in the future with a particular app or to accede to better paying jobs on crowd-sourcing platforms (De Stefano, 2016). The technology-enabled possibility of receiving instant feedback and rates of workers performance replaces a formal human supervisor and traditional performance reviews (Kuhn and Maleki, 2017), ensures flexibility and keeps organizations lean. Workers with superior ratings gain better positions and poor performing workers are penalized or removed from the platform, meaning that technology and platform architecture impacts their working experience (Jabagi *et al.*, 2019).

Gig work, which is a central part of the business model of most platforms, takes advantage of the technology efficiency. The technology of labour platforms achieves efficiency by, in effect, shifting the balance between the gig and the search (Collier *et al.*, 2017). To solve this puzzle of gig work, one of the first steps is to determine whether online platform workers somehow remain within the organisational field of a company and are functioning under its control (Todolí-Signes, 2017a) since the basic relationship of worker and employer is the relationship between company and worker. In this new relationship where platforms are connecting service providers and service users, one of the key issues is whether the workers have the status of employees or independent contractors (Cherry, 2016). It is therefore necessary to define the function of workers over platforms – whether they are employees, self-employed or undeclared workers, and also, is it possible to enforce some social regulations on these workers (Degryse, 2016).

10.3 Changes of Labour Concept Driven by the Digital Platforms

When analysing the concept of work and the relationship between employer and worker, there is still a strong expectation that work should provide an income, secure employment, meaningful activity and social bonds. However, an open question is the extent to which the new forms of employment and work in the digital collaborative economy will be up to this task (De Stefano, 2016).

Employment arrangements and work itself in the platform economy are characterized by instability. Millions of jobs are offered by platforms but for on-demand

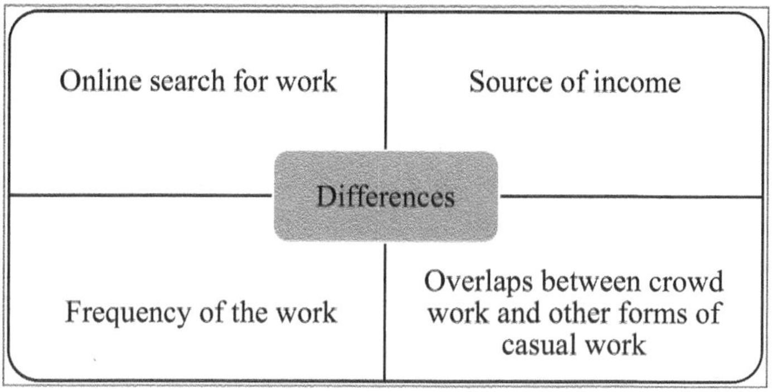

Figure 10.2. Differences between crowd work and traditional work.
Source: Adapted from Huws *et al.* (2017).

workers no one is permanent and thus stable. In the beginning, the idea of giving the possibility of work for everyone was very attractive, but with time the reality turns out to be different. This system gives workers more freedom and flexibility, with the possibility to choose when to work and how to work on the one side, but gives no security or control, on the other side. Work through digital platforms can bring numerous benefits, economically but also socially, with marginalized groups such as unemployed or geographically isolated ones (Johnston and Land-Kazlauskas, 2018) that can more easily enter the labour market and find jobs. A range of relevant variables distinguishes these "new" workers from more traditional ones has been summarized by Huws *et al.* (2017) (see Figure 10.2).

The worker that is a part of the "virtual community" is often named as the "partner", with all the rights and obligations that are based on the relation between himself/herself and the platform and himself/herself and the requester of the service. It is often described that his/her position is that of a self-employee, and the worker has to take care for his/her own social protection (unemployment, retirement pension, occupational sickness provision), work health and safety protection (Degryse, 2016). All forementioned is the justification why gig work in not considered as a standard job but rather as a certain business model or model of entrepreneurship, and as such there is no legal or regulatory support that can be applied to gig workers as employees (Degryse, 2016).

Work facilitated by the platform was often considered as a "side job" performed by individuals who had spare time or wanted to earn an extra income beside their regular job. Katz and Krueger (2019) calculate a 50 percent increase from 2005–2015 in the number of individuals using alternative work as their primary work, accounting for total net employment growth in the U.S. economy. Even though one of the main reasons for such a huge raise of platform work is certainly the recession that began in 2008, this period also corresponds with the rapid development of digitalization. According to the Study of the JP Morgan Chase Institute (2016), most participants do not use the platform as their primary source of income. Platform income can be considered as an additional compensation during work in employer-worker relationship, compensation for insufficient retirement pensions or "early" unemployment or it can be a way to limit or cope with student debt; and it can be an opportunity for flexible income while launching a start-up (Collier *et al.*, 2017). Nevertheless, this income cannot compensate for numerous other relationships that are raising in the relation between worker and employer.

The growth in platform work contributes to the trend toward part-time, short duration, and low-wage jobs, raising questions about how the issues will be institutionalized. Work through a platform is a form of flexible employment that is available to a worker "between", "around", or "in addition to" other traditional jobs that have disappeared, or are themselves irregular or flexible, and sometimes inadequate sources of income (Collier *et al.*, 2017). Research shows that for most Europeans, crowd work represents only a supplement to the main income, so it generates less than 10% of their total income (Huws *et al.*, 2017). Nevertheless, there is a small minority (ranging from 3% in Austria and Germany to 12% in Switzerland) for whom crowd work provides the only source of income (Huws *et al.*, 2017).

The payment for the work provided via the platform can be thought of two parts: (1) the unpaid work of looking for a gig worker and making a contract, and (2) the paid work of fulfilling the contract. The remuneration from the paid production component must cover the unpaid search component. The technology of the platform makes the first of these more efficient, but does not affect the productivity of the worker during the second. Thus, wages rise by working more per time period, assuming the same rate of **remuneration** for the paid gig/contract work. Maintaining this wage rate may be a particular challenge on crowd work platforms, which put pressure on wage rates by globally integrating high and low-wage labor markets and where compensation is usually based on a piece rate system (Collier *et al.*, 2017). Alternative to this is contest (competitive) crowdsourcing, where only the first ones to complete the task successfully are offered compensation by the requester (Todolí-Signes, 2017b).

When analysing payment for the work enabled by the platforms, an additional unique feature is that it is not negotiated between the platform and the worker. That is only one of the key differences when comparing the traditional worker-employee relationship and platform-worker-requestor relationship. Workers can set their own rate and offer their labour at a stated price only with a limited number of platforms. The requester then chooses among workers on the basis of this offer, combined with other information about worker experience, qualifications and ratings by past requesters on the platform. On other platforms, the requester lists a task or project at a set price, and workers decide if they want to apply. In these cases, the platform does not control wage rates, though it might indirectly affect them by constructing a larger market of workers and requesters (Collier *et al.*, 2017).

Platforms are creating a parallel labour market that is ultra-flexible (Degryse, 2016). In terms of **working conditions**, this form of employment is adaptable, based on the obligation contract and not the labour contract, with no standards that are guaranteed by the labour law such as: wage standards, standards relating to working hours, workplace, training, access to trade unions, or collective action. Platform workers do not have paid breaks, lunchtime, vacation, and sick leave, which were considered part of the standard working conditions (Collier *et al.*, 2017). Platform work can augment health and safety risks, reduce social rights, blur boundaries between work and private life and can lack transparency with respect to surveillance practices, rating system or job allocation (Vandaele, 2018). Institutional rights of platform workers are almost non-existent due to their status of independent contractors/self-employed. Exceptions, as in the case of France where platform workers have the right to unionize and take industrial actions, are rare (Vandaele, 2018).

Control is an important characteristic of the platform work (Collier *et al.*, 2017). Based on the criterion of the nature of control provided by the platform, the decision on employee status may be made. Most platforms sustain that workers are not employees but rather independent contractors who maintain autonomy in their work, particularly over the number of hours they work and the decision to accept a gig. In this situation, the contract will apply, and the relationship will be regarded as a contract on obligation. In other situations, where the platform performs control over conditions of work it would be possible to apply the labour law rules since the platform could be treated as an employer. Some authors in the literature emphasize that, even if we acknowledge that there has been a decrease in dependency between the worker and the platform, this does not mean that workers have become self-employed. Every job consists of a certain level of freedom which itself does not mean that the worker acts as an independent contractor (Sprage, 2015).

10.4 Some Insights for Regulation

Workers in the gig-economy are often classified as independent contractors (De Stefano, 2016; Stewart and Stanford, 2017) which excludes them from the usual worker-employer relationship and applicable labour law. This allows shedding not only potential vicarious liabilities and insurance obligations towards customers but also a vast series of duties connected to employment laws and labour protection (Rogers, 2015 as cited in De Stefano, 2016). The appearance of this new way of work and doing business has given rise to two fundamental questions in the labour market. The first concerns a legal issue, questioning whether the traditional legal concept of an "employee" is still valid in connection with this new way of working. The second question concerns policy, examining whether there is any need to extend the scope of labour law protection, disconnecting subordinate work from such protection (Todolí-Signes, 2017a).

Traditionally, a self-employed worker is considered to be a person who works directly for the market, i.e. someone who offers his/her services to one or more companies without becoming part of them. Self-employed people are owners of their own organization and have the independence needed to choose whether to accept a task. They own the tools and materials needed for the work and bear all business risks (Barroso, 2005). New types of workers – working through an online platform, owning the tools and materials needed for the work, choosing when (schedule freedom), for how long (freedom of working hours) and whether to perform the work – would therefore seem to fit more into the self-employed category and less into a traditional employment relationship (Todolí-Signes, 2017a). However, we have seen how this is debatable.

Classification given by the platforms is often in contradiction with the legal tradition and existing regulation. There are situations where the labour platforms fit into the existing regulatory regime regarding employees (Collier *et al.*, 2017). If labour platforms do not fit, then the question is: should the definition be expanded and new categories be devised, or should new forms of social protection independent of employment be adopted? (Ibid). As a result of the debate on occupational safety and health protection for the digital platform workers, some EU countries introduced regulations for the online platform economy. According to the Garben (2017), there are different types of approach to the regulation: applicable or extension of the existing regulation to online platform work; devising a new category of work (between employment and self-employment); decoupling the application of existing regulations from employment; introducing new sector-specific regulatory regimes for online platform work. The analysis made on selected countries (France, UK, Denmark, Sweden, Finland, Netherlands, Belgium and Ireland) shows that the huge debate was caused in the taxi industry which raises the question of the protection of the Uber taxi drivers. That caused the adoption of new rules regarding the regulation of that sector. In other industries, the initiatives exist but they are in realisation (Ibid).

What also has to be considered is whether the protection required for new workers is the same (or different) to the traditional protection guaranteed to subordinated work. In this sense, a new type of legal protection may be needed (Todolí-Signes, 2017a). Some platforms operate as digital versions of the employment agency while others exert more control over the conditions of work (Collier *et al.*, 2017). The status of platforms needs to be clarified in order to determine whether they can be regarded as temporary work agencies and expect to comply with the same regulations. **The degree of control over work can be used for differentiating between the labour platforms.**

There are situations in which the platform is not only intermediary between the worker and the requestor (client) but is managed by a company and the worker is the employee of the company. In this situation, often the platform has a role to create the service, make it available and often generate the need for the service provided. This was particularly discussed in the Court of Justice of the European Union cases Uber, and lately Airbnb, where the main qualification of the service itself was the distinction between the "information society service" and service in the meaning of the EU law (ECJ cases 526/15 Uber Belgium BVBA v Taxi Radio Bruxellois NV [2016] ECLI:EU:C:2016:830; Case 434/15 Asociación Profesional Elite Taxi v Uber Systems Spain, SL [2017] ECLI:EU:C:2017:981; Case 320/16 Criminal proceedings against Uber France [2018] ECLI:EU:C:2018:221 and Case 390/18 Criminal proceedings against X [2019] ECLI:EU:C:2019:1112 (Airbnb Ireland). In both situations' platforms act as intermediaries between client and self-employed professionals. The difference between the platforms is whether it also creates the service and makes it available or/and also generates the need for that particular service. Based on that difference the ECJ gave its judgment. This also has to be taken into account when discussing the legal responsibility for the work which depends on which of the above situations is involved. The responsibility depends on the service provider that is why it is important to understand who is the service provider, is that the platform or someone else. Service provider has the responsibility for the damage if it occurs and liability for the service or product provided (De Stefano, 2016).

In the collaborative and sharing economy framework, workers do not have working hours and are able to offer their activities on apps and platforms whenever they want. The gig-economy may enable them to benefit from job opportunities that they might not be able to access otherwise and on a flexible-schedule basis, allowing matching work with other working, family-related, study or leisure activities. This may enhance the possibilities of moonlighting (working the second job after the normal working hours) and for jobs offered in the virtual world, it can provide the opportunity to earn some income to people that are home-bound for various possible reasons, for example for disabilities. This flexibility on the workers side is often assumed to equate the undisputable flexibility the gig-economy affords to business (De Stefano, 2016). The other side of this flexibility is the mix of the private and professional life, visible in the pressure to work as much as possible, with little or no control on the maximum working hours per day, week or even month. More generally, there is a lack of protection of the workers private and family life.

The worker protection system is tightly connected with the social security protection (Daugareilh *et al.*, 2019). Some have suggested that the social protection model of the future should move away from the current emphasis on professional activities (Risak and Dullinger, 2018) and instead be focused on individuals throughout their lifetime, with a view to making career trajectories more secure in a word of discontinuity (Aloisi, 2016).

A possible solution to fill the regulation gap affecting the gig-economy would be to introduce an intermediate category between employee and self-employee in jurisdictions where it does not exist and cover workers with some limited form of labour protection (De Stefano, 2016). Nevertheless, to apply one of these two existing categories can cause two problems. On the one hand, a legal consequence of court's rulings that establish that platform workers should be treated as employees would make it necessary to apply all the labour rules to them. However, some of these rules just do not fit this new business model. Courts, faced with this sort of decision, would be unable to choose which rules would apply or which new solutions could be better. On the other hand, solution based on a court ruling would mean that, up to that moment, companies would have been misclassifying workers as self-employed (Berg *et al.*, 2018), i.e. they would be subject to sanctions and other legal liabilities for breaking the law until that time. For both reasons, some authors think it would be better to find a legislative solution. In a matter of legal policy, there should be an open debate over which kind of protection such workers should enjoy (Davis, 2015 as cited in Todolí-Signes, 2017a).

Apps explicitly specify that the relationship between the persons executing work and the business running the platform or the app will be one of self-employment. These kinds of clauses are quite frequent in personal service agreements as individuals engaging persons to execute tasks may seek to avoid costs and regulation associated with employment (Aloisi, 2016). These clauses are perfectly legitimate when the classification of the relationship between the parties corresponds to the reality of the transaction i.e. when the person hired fully preserves her autonomy in the actual execution of the task. If this is not the case, the relationship could be reclassified as one of employment whereby the determination of the existence of an employment relationship is to be guided by the fact relating to the actual performance of work and not on the basis of how the parties described the relationship. Classifying workers as independent contractors is a very frequent business practice in the gig-economy. In some cases, these companies may recur to provisions in their agreements that go beyond the ordinary extent of independent-contractor clauses. A crowd work company, for instance, represents that the platform only "provides a venue for third-party" (Valenduc and Vendramin, 2016; Todolí-Signes, 2017b).

Regulators are at crossroads: on the one hand, innovation in the platform economy should not be stifled by excessive and outdated regulation; on the other hand, there is a real need to protect service providers from "platform capitalism" (Berg *et al.*, 2018; Srnicek, 2016) and also the users of these services from fraud, liability and unskilled service providers (Ranchorda, 2015).

The question is, can these collaborative economy practices be qualified as innovations worth protecting and encouraging? Second, should the regulation of these practices serve the same goals as the existing rules for equivalent commercial services? Third, how can regulation keep up with the evolving nature of these innovative practices?

Even if the employment relationship can be interpreted to fit the new type of workers, this does not mean that the protection needed by both the new and the old types is the same. Rules protecting working conditions do not fully match the new business model, one of the main characteristics of which is working time flexibility. Workers are allowed to choose when and for how long they wish to work, a concept distant from traditional regulations on working hours, schedules, compulsory rest periods and holidays. Fixed salaries and minimum wages seem difficult to fit into a business model where a worker can also choose how long he/she is going to work (Fisher, 2015; Weber and Turcios, 2015:12 as cited in Todolí-Signes, 2017a). Moreover, regulations about a pool of on-call workers or a preferential right to work in on-demand jobs seem incompatible with a business which lets clients choose a specific worker, as clients will select the worker they prefer based on the public evaluations. The application of collective bargaining also has its difficulties. Bargaining entities are hard to establish on an online platform where it is unlikely that the number of workers is known, and they can work for different platforms at the same time (Felstiner 2011: 183–185 as cited in Todolí-Signes, 2017a). Moreover, in a business where workers do not know each other, mutual trust to agree on union representatives is doubtful (Salehi *et al.*, 2015:1621 as cited in Todolí-Signes, 2017a). Lastly, under current regulations a company has to pay for any expenses incurred by an employee. However, in a business allowing freedom to the employee (on how to perform the job) this would seem unfair to the company (Todolí-Signes, 2017a).

A special labour relationship should be applied to those who perform work offline through a specific online platform, as described in this article. In this specific work, basic labour rights should be protected, without impeding the normal development of the industry. More specifically, the regulations should ensure fair representative procedures to allow self-regulation through collective agreements. The following questions should be included: Instructions given by platform owners or customers, freedom of schedules and working hours, freedom to work on more than one platform, employee's liability for damage, medium wage, reimbursement of expenses, subsidiary labour law (Todolí-Signes, 2017a).

10.5 Conclusion

As shown in the course of this chapter, digital transformation has eroded standard employment relationships and introduced numerous new working arrangements. This chapter provides insight into legal and organizational aspects of labour brought by the digital collaborative economy. From an organizational point of view, the chapter analyzed the differences between the traditional and new forms of work, as well as specifics of platform-related work. From a labour law perspective, the goal was to identify the main issues related to the status of worker in cases when people are working as service providers with the platform as an intermediary, and to provide some insights for future labour law regulation.

Digital platforms have both created new labour markets and transformed some old ones. Still, even with these changes, the old challenges and politics of work did not disappear; they have just taken on new forms (Graham and Shaw, 2017) and existing patterns of employment relations persist (Collier *et al.*, 2017). The issue is therefore how to regulate the new labour relations in a manner consistent with European legal framework and general values when in come to workers' rights and protection.

References

Aloisi, A. (2016). Commoditized workers: Case study research on labor law issues arising from a set of "on-demand/gig economy" platforms. *Comparative Labor Law and Policy Journal, 37*(3), 620–653.

Barnard, C. and Deakin, S. (2002). Negative and positive harmonization of labor law in the European Union. *Columbia Journal of European Law, 389*, 410–411.

Barroso, M.R. (2005). Trabajo autónomo y trabajo subordinado. Delimitación, análisis y propuestas de reforma. *Estudios Financieros, Revista de Trabajo y Seguridad Social: Comentarios, Casos Prácticos : Recursos Humanos, 273*, 71–144.

Becker, S. (2019). *Digital Structural Change and the Welfare State in the 21st Century.* EU Monitor. Deutsche Bank Research. Available at https://www.dbresearch.com/PROD/RPS_EN-PROD/PROD0000000000489434/Digital_structural_change_and_the_welfare_state_in.pdf?undefined&realload=3rKotA3qV0yRpeJkxE0ZwCGlkzIzVaOd6jpdiHU8/x5MlIocrJrljywlbxzwm3lljUX3A2yCnMASl7t7ioUPkw== accessed 24.11.2020.

Berg, J. and De Stefano, V. (2015). *Regulating Work in the "Gig-economy"*. Available at http://iloblog.org/2015/07/10/regulating-work-in-the-gig-economy/ accessed 23.11.2020.

Berg, J., Furrer, M., Harmon, E., Rani, U., and Six Silberman, M. (2018). *Digital Labour Platforms and the Future of Work. Towards Decent Work in the Online World.* International labour office, Geneva.

Case Court of Justice of the European Union 390/18 AIRBNB Ireland EU:C:2019:336.

Case Court of Justice of the European Union 526/15 *Uber Belgium BVBA v Taxi Radio Bruxellois NV* [2016] ECLI:EU:C:2016:830.

Cherry, M.A. (2016). Beyond misclassification: The digital transformation of work. *Comparative Labor Law & Policy Journal, 37*(3), 544–577.

Collier, R.B., Dubal, V.B., and Carter, C. (2017). *Labor Platforms and Gig Work: The Failure to Regulate.* Research Paper No. 251, Legal Studies Research Paper Series, IRLE, University of California, CA.

Daugareilh, I., Degryse, C., and Pochet, P. (eds.) (2019). The platform economy and social law: Key issues in comparative perspective, Working Paper 2019.10, European Trade Union Institute.

De Groen, W.P. and Maselli, I. (2016). *The Impact of the Collaborative Economy on the Labour Market*, Centre for European Policy Studies (CEPS), Brussels, Belgium.

De Stefano, V. (2016). *The Rise of the "Just in Time Workforce": On-demand Work, Crowdwork and Labour Protection in the "Gig-economy"*, Conditions of Work and Employment Series no. 71., International Labour Organisation, Geneva.

Degryse, C. (2016). *Digitalisation of the Economy and its Impact on Labour Markets.* Working paper 2016.02, European Trade Union Institute, Brussels, Belgium.

Eurofound (2015). *New Forms of Employment*, Publications Office of the European Union, Luxembourg.

Fisher, D. (2015). Uber says drivers oppose lawsuit that would make them employees. Forbes. Available at: http://www.forbes.com/sites/danielfisher/2015/07/09/uber-says-its-drivers-want-no-part-of-lawsuit-to-make-them-employees/#1e45d4b1387d

Fisk, C. (2016). *Hollywood Writers and the Gig Economy*, Legal Studies Research Paper Series, No. 2016-55, School of Law, University of California, CA.

Friedman, G. (2014). Workers without employers: shadow corporations and the rise of the gig economy. *Review of Keynesian Economics, 2*(2), 171–188.

Garben, S. (2017). Protecting workers in the Online Platform Economy: An overview of regulatory and policy developments in the EU. *European Agency for Safety and Health at Work,* available at https://osha.europa.eu/en/publications/protecting-workers-online-platform-economy-overview-regulatory-and-policy-developments accessed 2.1.2021.

Graham, M. and Shaw, J. (2017). *Towards Another World of Gig Work. In Towards a Fairer Gig Economy*, Meatspace Press, London, UK.

Greenhouse, S. (2015). Uber: On the road to nowhere. *The American Prospect*, Available at http://prospect.org/article/road-nowhere-3, accessed 23.11.2020.

Greve, B. (2019). The digital economy and the future of European welfare states. *International Social Security Review, 72*(3), 79–94.

Huws, U., Spencer, N.H., Syrdal, D.S., and Holts, K. (2017). *Work in the European Gig Economy*. Available at: https://uhra.herts.ac.uk/bitstream/handle/2299/19922/Huws_U._Spencer_N.H._Syrdal_D.S._Holt_K._2017_.pdf, accessed 20.11.2020.

ILO (2020). *Helping the Gig Economy Work Better for Gig Workers*. Available at https://www.ilo.org/washington/WCMS_642303/lang--en/index.htm, accessed 28.10.2020.

Isaevoli, G., Michelini, L., Grieco, C., and Principato, L. (2018). Mapping the sharing economy: a two-sided markets perspective. *Sinergie Italian Journal of Management, 38*(106), 181–201.

Jabagi, N., Croteau, A.-M., Audebrand, L.K., and Marsan, J. (2019). Gig-workers' motivation: thinking beyond carrots and sticks. *Journal of Managerial Psychology, 34*(4), 192–213.

Johnston, H. and Land-Kazlauskas, C. (2018). *Organizing On-demand: Representation, Voice, and Collective Bargaining in the Gig Economy*. Conditions of work and employment series, 94. International Labour Office, Geneva.

JPMorgan and Chase (2016). *Paychecks, Paydays, and the Online Platform Economy*. Available at https://institute.jpmorganchase.com/content/dam/jpmc/jpmorgan-chase-and-co/institute/pdf/jpmc-institute-volatility-2-report.pdf, accessed 23.11.2020.

Katz, L.F. and Krueger, A.B. (2019). The rise and nature of alternative work arrangements in the United States, 1995–2015. *ILR Review, 72*(2), 382–416.

Kuhn, K.M. and Maleki, A. (2017). Micro-entrepreneurs, dependent contractors, and instaserfs: understanding online labor platform workforces. *The Academy of Management Perspectives, 31*(3), 183–200.

Nerinckx, S. (2016). The 'Uberization' of the labour market: some thoughts from an employment law perspective on the collaborative economy. *ERA Forum, 17*, 245–265.

Oyer, P. (2020). *The Gig Economy*, IZA World of Labor. Available at https://wol.iza.org/uploads/articles/523/pdfs/the-gig-economy.pdf, accessed 28.10.2020.

Pedersini, R., Coletto, D. (2009). *Self-employed Workers: Industrial Relations and Working Conditions*, European Foundation for the Improvement of Living and Working Conditions, Dublin, Ireland.

Pesole, A., Urzí Brancati, M.C., Fernández-Macías, E., Biagi, F., and González Vázquez, I. (2018). *Platform Workers in Europe*, EUR 29275 EN, Publications Office of the European Union, Luxembourg.

Petriglieri, G., Ashford, J.S., and Wrzesniewski, A. (2018). Thriving in the Gig economy. *Harvard Business Review, 96*, 140–143.

Ranchorda, S. (2015). *Does Sharing Mean Caring? Regulating Innovation in the Sharing Economy.* Research paper series, no. 06/2015, Tilburg Law School Legal Studies, Tilburg, Netherlands.

Risak, M. and Dullinger, T. (2018). *The Concept of "Worker" in EU Law Status Quo and Potential for Change,* Report 140, European Trade Union Institute.

Risak, M. and Warter, J. (2015). *Legal Strategies Towards Fair Conditions in the Virtual Sweatshop.* Paper presented at the IV Regulating for Decent Work Conference, ILO, July 8–10, 2015, Available at http://www.rdw2015.org/download, accessed 28.10.2020.

Schiek, D. (2017). Comparing labour laws in the EU Internal Market – a social actor perspective. *International Journal of Comparative labour Law and Industrial Relations, 33*(1), 171–194.

Schoukens, P. and Barrio, A. (2017). The changing concept of work: When does typical work become atypical? *European Labour Law Journal, 8*(4), 306–332.

Sprage, R. (2015). Worker (mis)classification in the sharing economy: Square pegs trying to fit in round holes. *Journal of Labor & Employment Law, 31*, 53–76.

Srnicek, N. (2016). *Platform Capitalism.* Polity press. Oxford, UK.

Stanford, J. (2017). The resurgence of gig work: Historical and theoretical perspectives. *The Economic and Labour Relations Review, 28*(3), 382–401.

Stewart, A. and Stanford, J. (2017). Regulating work in the gig economy: What are the options? *Economic and Labour Relations Review, 28*(3), 420–437.

Todolí Signcs, A. (2017a). The 'gig economy': Employee, self-employed or the need for a special employment regulation? *Transfer: European Review of Labour and Research, 23*(2), 193–205.

Todolí-Signes, A. (2017b). The end of the subordinate worker? Collaborative economy, on-demand economy, gig economy, and the crowdworkers' need for protection. *International Journal of Comparative Labour Law and Industrial Relations, 33*(2), 241–268.

Urzi Brancati, M.C., Pesole, A., and Fernandez Macias, E. (2019). *Digital Labour Platforms in Europe: Numbers, Profiles, and Employment Status of Platform Workers*, Publications Office of the European Union, Luxembourg.

Valenduc, G. and Vendramin, P. (2016). *Work in the Digital Economy: Sorting the Old From the New*, Working Paper Series No. 2016.03, European Trade Union Institute, Brussels, Belgium.

Vandaele, K. (2018). *Will Trade Unions Survive in the Platform Economy? Emerging Patterns of Platform Workers' Collective Voice and Representation in Europe.* Working Paper Series No. 2018.05, European Trade Union Institute, Brussels, Belgium.

Weber, P.M. and Turcios, E.N. (2015). Uber hits a speed bump in California: Labor commissioner rules driver is an employee. *Employee Benefit Plan Review, 70*(3), 3–13.

World Bank (2015). *Regulating the Gig Economy.* Available at http://www.worldbank.org/en/news/feature/2015/12/22/regulating-the-gig-economy accessed 18.09.2020.

World Economic Forum (2016). *The Fourth Industrial Revolution: What it Means, How to Respond*, Available at https://www.weforum.org/agenda/2016/01/the-fourth-industrial-revolution-what-it-means-and-how-to-respond/ accessed 15.06.2020.

DOI: 10.1561/9781680838411.ch11

Chapter 11

Mapping the Impact: Assessment Methodologies and Policy Implications of the Collaborative and Sharing Economy

By Venere Stefania Sanna and Laura Michelini

In this chapter the authors examine different methodologies for assessing the diverse impact of the Collaborative and Sharing Economies (henceforth, CSE) on many aspects of society, then make policy recommendations based on the outcomes of these assessments. The chapter briefly describes the multifarious CSE landscape, then goes on to enumerate the major frameworks currently used to evaluate and assess CSE impacts. It notes that there is no single methodology that can fully capture the wide variety of impacts, but identifies two methods as the most useful: multidimensional assessment, and Theory of Change (ToC) framework analysis. The chapter then elaborates on the benefits and drawbacks of each methodology, before suggesting the use of both methodologies in a mixed format. The authors undertake a brief literature review to examine recent approaches to impact assessment, and focus in on what those approaches have revealed from the points of view of social value, environmental impact, economy, and political impact, concluding with findings about 'rebound effects' in terms of indirect behavioural and other changes caused by CSE initiatives. The chapter maps out a suggested comprehensive inventory for multidimensional assessment, then describes a methodology for assessing the CSE and its impacts via the ToC Framework. It concludes with determining the implications on policy of the impact assessments. This section views policy through the lens of implications derived from the general impact of the CSE, and then via the impact assessment methods and tools derived from the ToC framework. In the first part it assesses considerations that need to be taken by policymakers due to CSE impacts on the market, government, the workforce, consumers, and the environment, and makes recommendations on each. In terms of impact assessment, it recommends more long-term and strategic actions to cope with the ongoing disruption caused by the CSE. In conclusion it recommends a robust and adaptive approach to assessing the impacts caused by flourishing and inevitable growth of the CSE.

11.1 Introduction

Assessing the impact of Collaborative and Sharing Economy practices ('CSE practices' or 'CSE initiatives') is not an easy task, especially because of the difficulties in identifying qualitative and quantitative metrics for such an assessment, due to the heterogeneity of the domain of activities, as CSE practices and/or platforms operate in different industries (food, hospitality, mobility, money, etc.), to the wide range of interested parties – such as academics, practitioners, entrepreneurs' regulatory agencies, policy and program developers – that may have different needs and purposes for using impact assessment methodologies, to the geographical level at which CSE initiatives operate, and so on.

CSE initiatives can be analysed using different methodologies and frameworks. Over the past few decades, a plethora of impact assessment approaches has been developed. The wide range of impacts of CSE initiatives can cover different areas and can be analysed from distinct perspectives, using qualitative and/or quantitative methods, or a combination of the two.

This complexity has resulted in the development of many different techniques. This contribution aims to: (i) identify the main areas of impact of the Collaborative and Sharing Economy (CSE), (ii) explore some methods for evaluating the wide range of impacts of CSE practices that researchers and practitioners involved in the COST Action "From Sharing to Caring" encountered during their work, and (iii) provide information and policy recommendations for decision-making on the main open issue relating to the impact of the CSE.

From the vibrant debate and the growing body of literature around the possible definitions of the "collaborative economy" and "sharing economy" (Botsman and Rogers, 2010; Bardhi and Eckhardt, 2012; Botsman, 2013, 2014; Slee, 2015; Frenken, 2017), we are aware that these terms are not synonymous, and that each has a specific connotation, meaning, and potential outcomes and impacts (see Introduction, this volume).

Generally, these forms of economy involve three categories of actor: (i) peers or providers that share (often underutilized) assets, resources, time and/or skills; (ii) users of these; and (iii) intermediaries that connect – mostly via an online platform – providers with users and that facilitate transactions between them ('collaborative platforms' which allow 'access rather than ownership').

Building on these premises, this contribution does not aim to provide any new definition and/or classification of the "collaborative economy" or "sharing economy", but instead brings together the expertise and research experience of a transdisciplinary team to synthesise some of the evidence, insights and critical reflections about qualitative and quantitative methods of measuring and assessing the wide range of impacts of Collaborative and Sharing Economy (CSE) practices.

The need to evaluate impacts of the CSE arises from the disruptive nature of many CSE initiatives and platforms, and their potential for harm as well as benefit. The biggest CSE platforms are owned and operated multi-nationally, and the underlying digital technologies may still be considered fairly new. For regulation of such novel platforms – particularly in the fields of privacy, labour rights, and environmental protection – a clear understanding of the impacts of CSEs is vital, and it is towards this comprehension that this chapter strives.

Adopting a mixed-method approach, the techniques discussed in this chapter cover four main areas of impact: social, economic, political, and environmental. It presents two different analytic perspectives, based on two approaches: (i) a multidimensional assessment based on a set of selected indicators, and (ii) the 'Theory of Change' method.

For the first, a set of indicators is proposed, for the multi-dimensional assessment of a wide range of CSE practice impacts across several countries, active in a variety of industries and at different geographical levels. The CSE practices may be either platform-based or 'offline'. The indicators can be selected case-by-case, in one or multiple areas of impact, and can be used with different evaluation methodologies such as multi-criteria analysis, benchmark analysis, cost-benefit analysis, performance assessment system, etc. By contrast, a new framework of assessment was applied using the 'Theory of Change' by considering the wider organization level of the CSE platform as a "unit of analysis", in order to evaluate the impact of CSE platforms at industry level.

Finally, the contribution reflects on the policy implications of the mapped areas of impact and proposed assessment methods, and suggests some recommendations about the most problematic open issues.

11.2 Methodological Framework

The growth of Information and Communication Technology (ICT) solutions to mediate transactions between providers and users has allowed online platforms to become globally dominant intermediaries for the CSE. Nevertheless, in recent years, and increasingly in times of deep socio-economic crisis, there has also been a flourishing of CSE practices that are not powered by such technology – such as new forms of volunteering and community-led initiatives, and community and collective ownership models.

The wide range of methods to measure the impacts of the CSE can be classified considering the following elements (Grieco *et al.*, 2015), as summarized in Figure 11.1:

- Typology of indicator for the assessment: can be quantitative, qualitative and qual/quantitative;
- Typology of impact to assess: on people, environmental, social, economic and holistic;
- Purpose of the measurement: screening, assessment, management, certification, reporting, etc.;
- Industry/domain of activity: general, specific (e.g. accommodation, mobility, food, etc.);
- The user of the measurement: CSE initiatives (for internal evaluation), universities, research institutions, not-for-profit networks and organizations, consulting firms, local institutions, etc.

There is no single evaluation methodology that can fully capture the variety of impacts and complexities of CSE practices/initiatives. Consequently, creative ways to combine different evaluation frameworks, tools and techniques are sought. In the context of this contribution, we propose two approaches: (i) a Multidimensional assessment and (ii) the Theory of Change (Table 11.1).

One option is to focus on the CSE initiative itself as a 'unit of analysis' using a multidimensional assessment methodology. The assessment would therefore be focused on the internal and external dimensions of its potential impact. By way of an example: (i) an internal impact may be related to the ability of the initiative to effectively deliver benefits to users and/or participants (i.e. goods and services at

Table 11.1. Main frameworks of analysis.

	Multidimensional Assessment (Paragraph 4)	Theory of Change (ToC) Framework (Paragraph 5)
Unit of analysis	CSE initiative	Wider organization level of a CSE platform
Data typology	Quantitative	Qualitative + quantitative
Impact typology	Environmental, social, economic, political	Holistic
Purpose of the assessment	Assessment	Reporting
Industry/Domain of activity	Multi-domain	Specific
Geographical level	Mainly local (suggested)	Any
User of the assessment	All (CSE practice, university, not-for-profit organization, etc.)	Organizations, consulting firms

Source: Explanation by the authors based on typology of methodological approach.

lower prices), while (ii) external effects can mostly be associated with the impacts on the "outside" world (i.e. environmental effects). Accordingly, we discuss existing literature in Section 11.3 and then in Section 11.4 we propose a series of practical indicators for the assessment of the wide range of impacts thus revealed.

Another option is to consider the wider organizational level of a CSE platform as the 'unit of analysis'. This means that the CSE platform is analysed as a whole, to allow the organization that manages the platform to assess and report on its impacts. This is the lens of analysis adopted in Section 11.5, leveraging the Theory of Change.

This contribution, therefore, is based on a mixed-methods (MM) evaluation approach. The purpose is to strengthen the reliability of data, the validity of the findings and recommendations, and to broaden and deepen our understanding of impacts assessed – and how these are affected by the broader context within which the CSE initiative operates.

11.3 Mapping the Impact of the Collaborative and Sharing Economy

The review and estimation of the impacts of the Collaborative and Sharing Economy is a relatively new area of research and is still a controversial field of study, with some authors emphasizing a series of positive social, economic, political, environmental values and effects, and others debating its negative implications.

The impacts of CSE have attracted growing interest and research, which explore the topic using different approaches (i.e. considering the perceived impacts, direct and indirect impacts, internal and external areas of impact, potential outcomes, etc.), by addressing the sharing economy and/or the collaborative economy as a whole, or by focussing on different industries (i.e. hospitality, mobility, food, services, etc.).

However, to date, little rigorous research has been done on the quantitative assessment of economic, social, political and environmental impacts of CSE which remain largely unknown and to date unquantifiable. Moreover, a range of important induced effects may be indirectly triggered by the increased number of platform-mediated transactions and/or behavioural changes induced by the collaborative and sharing economy – so-called 'rebound' effects. These are even more obscure and difficult to be identified and quantified.

This section aims briefly to summarize recent findings in the field, with the purpose of suggesting and pointing out the main contributions to existing knowledge on the impact of the CSE. For those who are new to this subject, Table 11.2 contains suggested foundational reading addressing the following issue: how can we analytically conceptualize and empirically assess the various impacts of the Collaborative and Sharing Economy?

One of the most important aspects of the CSE is its **social** value and contribution to the creation or strengthening of social ties and interactions (Frenken, 2017), to the enhancement of social cohesion and the establishment of a sense of community among participants (Parigi *et al.*, 2013; Rosen, 2011; Schor, 2016), including the creation of "digital communities" (Hamari *et al.*, 2016; Reischauer and Mair, 2018; Vaskelainen and Piscicelli, 2018) of collaborative consumption (Botsman and Rogers, 2010; Dillahunt and Malone, 2015; Martin, 2016). In this regard, the role played by trust, reciprocity and sense of belonging to a community have been crucial factors pushing forward the adoption of the collaborative and sharing economy (Belk, 2010; Celata *et al.*, 2017).

Many CSE initiatives – mostly those operating in the food or mobility domain – have also been noted as favouring the generation of positive effects on wellbeing, health (Woodcock *et al.*, 2014), and quality of life (e.g. sharing of healthy food). While other social impacts are related to the creation of (often temporary) employment (De Groen *et al.*, 2017) and the lack of protection of workers' rights (Berger *et al.*, 2017).

The CSE has also been accused of producing a series of negative social impacts, and commentators have raised concerns, identifying it as a source of "selective exclusion" (Benkler, 2004), social, racial and even digital social discrimination

Table 11.2. Suggested reading by area of potential impact.

Social

Botsman and Rogers, 2010, What's Mine Is Yours: The Rise of Collaborative Consumption

Celata *et al.*, 2017, The sharing economy as community marketplace? Trust, reciprocity and belonging in peer-to-peer accommodation platforms

Edelman and Luca, 2014, Digital Discrimination: The Case of Airbnb.com

Frenken, 2017, Putting the sharing economy into perspective

Martin, 2016, The sharing economy: A pathway to sustainability or a nightmarish form of neoliberal capitalism

Parigi *et al.*, 2013, A community of strangers: the dis-embedding of social ties

Schor, 2016, Debating the sharing economy

Environmental

Demailly and Novel, 2014, The sharing economy: make it sustainable

European Commission, 2016a, Environmental potential of the collaborative economy

Frenken, 2017, Political economies and environmental futures for the sharing economy

Heinrichs, 2013, Sharing economy: A potential new pathway to sustainability

Leismann *et al.*, 2013, Collaborative consumption: Towards a resource-saving consumption culture

Martin and Shaheen, 2011, Greenhouse gas emission impacts of car-sharing in North America

Economic

Barron *et al.*, 2018, The Effect of Home-Sharing on House Prices and Rents: Evidence from Airbnb

Guttentag, 2015, Airbnb: disruptive innovation and the rise of an informal tourism accommodation sector

Malhotra and Van Alstyne, 2014, The Dark Side of the Sharing Economy … and How to Lighten It.

Picascia *et al.*, 2019, The airification of cities. Making sense of the impact of peer to peer short term letting on urban functions and economy

Sundararajan, 2016, The Sharing Economy. The End of Employment and the Rise of Crowd-Based Capitalism

Wachsmuth *et al.*, 2017, Airbnb's Impact on Canadian Housing Market.

Zervas *et al.*, 2017, The Rise of the Sharing Economy: Estimating the Impact of Airbnb on the Hotel Industry

(Continued)

Table 11.2. Continued

Political
De Groen *et al.*, 2017, The Impact of the Platform Economy on Job Creation
Edelman and Damien, 2016, Efficiencies and Regulatory Shortcuts: How Should We Regulate Companies like Airbnb and Uber?
Frenken, 2017, Putting the sharing economy into perspective
Katz, 2015, Regulating the sharing economy
Malhotra and Van Alstyne, 2014, The Dark Side of the Sharing Economy … and How to Lighten It.
Wachsmuth and Weisler, 2018, Airbnb and the rent gap: Gentrification through the sharing economy

Source: Elaboration of the authors.

(Edelman and Luca, 2014; Edelman and Damien, 2016; Zukin *et al.*, 2015), and for the commodification of personal assets, sociality, intimacy and identities (Hearn, 2010; Ronzhyn, 2013; Ert *et al.*, 2016). Issues related to consumer protection and safety, privacy and (big) data use and treatment (Koopman *et al.*, 2014) are still open and much debated.

Finally, at a wider urban level, much literature – mainly concerning the hospitality industry – has reported conflict at the community level (e.g. between tourists and residents), and has raised concerns about the ability of CSE practices to reshape the landscape of extant urban conflicts around issues such as gentrification, social and economic segregation, quality of life and coexistence between different communities and groups (Trudelle and Pelletier, 2016; Gutiérrez *et al.*, 2017).

In terms of **environmental** impacts, recent research has generally identified positive impacts derived from the new paradigm of more sustainable consumption and production practices that result in ecological resilience, the reduction of Greenhouse Gases (GHG) (Martin and Shaheen, 2011, 2016; Chen and Kockelman, 2016), waste, and resource usage, reduction of ecological footprint, provision of green infrastructure (Heinrichs, 2013; Leismann *et al.*, 2013), and more sustainable behaviour at various levels (European Commission, 2016b).

Also in terms of environmental impact, rebound effects that are due, for example, to shifts and/or additional consumption of goods and/or services due to a decrease in the market prices should be considered (Denegri-Knott, 2011). As stated by (Plepys and Singh, 2019, pp. 67–68), "secondary increases in consumption of other goods and services when the residual savings from the consumption of primary goods and services are made available" potentially produce a series of potential environmental that the authors elegantly illustrated using a simplified causal loop diagram (CLD).

Nevertheless, despite a shift towards more sustainable consumption models and increased environmental awareness, some authors remain sceptical of highlighting the positive effects of the sharing economy on the environment, in particular on its ability to change behaviour at consumer level (Demailly and Novel, 2014; Frenken, 2017). Moreover, some research has pointed out that platforms (e.g. in the case of food sharing) focus on waste reduction instead of trying to address waste prevention.

In conclusion, as Frenken (2017) argued, thanks to few systematic studies available, "one can—tentatively—conclude that the environmental impacts of sharing are likely to be positive, but possibly much smaller than some claim and hope for" (p. 7).

From an **economic** perspective, the CSE impacts can be examined at different scales and viewpoints. Positive impacts vary from job opportunities brought into the economic systems, occasions for (extra) income for individuals, families and more generally local communities – also improving deprived neighbourhoods (Dillahunt and Malone, 2015), as well as the development of micro-entrepreneurship and start-ups and the economic impetus given to a microcosm of secondary market entrepreneurs supporting the CSE ecosystem (Malhotra and Van Alstyne, 2014).

For the consumer, the CSE has the potential of delivering concrete benefits. The possible decline in the prices of goods and services, cost savings or revenue generation for individuals, allows, in fact, to increase consumer welfare in absolute terms (Frenken, 2017). By contrast some authors have highlighted various negative effects related to increased income inequalities due to the ability of CSE platforms to depose those with less education and lower incomes (Schor, 2016), uneven wealth distribution, the shifting of income and opportunity to better-off households and providers, and a deepening polarization of power in the digital economy and equality of access to the CSE.

In terms of the workforce, on one hand many platforms presented themselves as providers of flexible employment opportunities, but on the other the rights, benefits and fair-pay levels of this emerging sector of freelancers seem to be under threat (Sundararajan, 2016, Murillo *et al.*, 2017; Stabrowski, 2017).

The economic impacts of the CSE at sector- and macro-level are less clear-cut. Some CSE platforms have been associated with fiscal irregularity and tax evasion (Malhotra and Van Alstyne, 2014; Frenken, 2017), lack of transparency and unfair competition in a previously 'fair' market, resulting in their being disruptive for traditional markets and economic sectors (Einav *et al.*, 2015; Zervas *et al.*, 2017; Dogru *et al.*, 2019; Gyódi, 2019). Airbnb, for example, has heralded a vast enlargement in the accommodation capacity of cities, with the potential to displace the traditional hospitality sector (Guttentag, 2015; Zervas *et al.*, 2017), has had a significant impact on housing values, availability and affordability (Lee, 2016;

Barron *et al.*, 2018), and has contributed to the "touristification" of many areas (Wachsmuth *et al.*, 2017; Picascia *et al.*, 2019).

In a similar way to other areas of impact, the CSE has the potential to cause economic rebound effects, i.e. impacts on the wider economy from spending the money saved and earned due to new models of collaborative and sharing production and consumption. In terms of "price effects", for example, improvements in efficiency increases the use of the same product or service (direct rebound) and – often in parallel – net gains and savings increase the demand for other goods and services (indirect rebound) (Warmington-Lundstroma and Laurenti, 2020). Because of the many difficulties in identifying, assessing and quantifying the actual impacts the CSE creates, rebound effects have also been largely overlooked.

When it comes to the **political** impacts, these can be identified by adopting, for example, two different perspectives: (i) the "top-down" viewpoint: when considering the political effects that government policy and its administrative practices (e.g. legislation or regulation) can have on a phenomenon, and (ii) the "bottom-up" perspective: when considering, for example, the opportunity offered by CSE initiatives and/or platforms to increase or transfer various elements of power (resources, capabilities, and positions) to those who do not have them.

In the sphere of "top-down" effects, it is often reported that the exponential growth of the CSE has increased the capacity to bypass regulation and favour tax avoidance because many platforms do not adhere to the market regulations and tax obligations that apply to 'regular businesses' (Malhotra and Van Alstyne, 2014). Legislation regulating the CSE (taxation, registration, licensing, etc.) is therefore a fast-changing and a controversial topic of analysis (Edelman and Damien, 2016; Katz, 2015; Codagnone and Martens, 2016).

From a "bottom-up" perspective, some research suggests more positively the opportunity offered by CSE initiatives and/or platforms to enhance the participation in social movements (e.g. new forms of social and solidarity economy practices), to cultivate the political and social skills necessary for citizenship and activism and a "socially-inclusive form of development" (Frenken, 2017), to represent the preferences and wishes of communities and local people in the policy process, and to reclaim the 'commons' (Schor, 2016) or the "right to the city" (Wachsmuth and Weisler, 2018).

Measurement of the political impacts of CSE initiatives is yet to come, but could represent a practical tool to inform and improve policies at different levels. Despite the growing attention given in recent years to issues related to the most innovative impacts of CSE practices, the debate on possible ways to integrate the lessons learned from these experiences in the structuring of new policies or in the implementation of existing social, economic, environmental and political systems, is still undeveloped and unclear.

Finally, the **rebound effects** of CSE initiatives should also be considered in the definition of an assessment method. Individuals are increasingly using online CSE platforms to share or offer their skills, time, and/or underutilized resources to others who need to access, rent, or borrow these goods and services through bartering, swapping, lending, social exchanging, trading or reselling. While this is expected to produce (for example) positive environmental impacts, it is unclear to what degree the savings or earnings from the platforms might increase resource use. Lending, borrowing, or renting items instead of buying new products may be seen as an efficient way to promote a more circular economy. However, changes in the consumption model might liberate resources – such as raw materials or funding – which might inevitably be injected into the system anyway and thus increase production or consumption in other ways. Therefore, the degree to which resources thus liberated can be used is an important issue, and should not be excluded from any analysis.

11.4 A Multidimensional Assessment of CSE Initiatives: Areas of Impact and Indicators for the Analysis

Adopting an "internal" perspective – that is, considering CSE practices as the "unit of analysis" – from the multidisciplinary discussion among researchers and practitioners, as well from the literature review on the topic and major evidence coming from previous research (Sanna, 2018; Celata and Sanna, 2019), the main areas of impact for which it is possible to derive feasible measurements are: social, economic, political and environmental.[1] For each of these areas, a set of indicators aimed at measuring the potential positive and negative impacts is presented in Table 11.3.

The proposed indicators are mainly non-fiscal: physical (quantity of emissions, quantity of waste saved or produced, etc.) and behavioural (social inclusion, public involvement and participation, environmental awareness, etc.). These have been selected from the wider field as the most relevant and significant for each domain object of the study in our opinion, but clearly the list is indicative and not exhaustive of all the possible direct and indirect, positive or negative impacts that a CSE initiative may produce.

Nevertheless, the set of proposed indicators is diverse and wide, and a case-by-case selection process is therefore needed according to a series of factors:

1. Some indicators have been developed and used in the framework of previous research conducted by Celata and Sanna (2016) in which (i) tools for data collection (e.g. questionnaire), (ii) formulas for calculating the indicators and (iii) indications about the methods for the analysis (e.g. Multicriteria Analysis) are available.

Table 11.3. Areas of impact and indicators for assessment.

Area of Impact	Indicator	Description of the Indicator
Social	**Social cohesion**	To strengthen/reinforce the sense of belonging of a community and the relationships among P/U* within the community itself.
	Human and civil rights	To promote the protection of human and civil rights (e.g. protection of minorities/targeted groups, worker's rights, consumer's rights, data treatment, etc.).
	Social capital	To strengthen social interaction/networking between P/U.
		The creation of new relationships between people who did not previously know each other, which would be unlikely without the initiative relationships.
		The creation/reinforcement of trust.
	Social inclusion and equity	The heterogeneity of P/U in terms of origin, gender, age, social status (internal social inclusion).
		The degree to which the needs of the local community are taken into consideration (external social inclusion).
		Gender equality, measured e.g. by the percentage of women and men fulfilling key roles in the initiative (e.g. founders, leaders, etc.).
	Quality of life and wellbeing	To improve quality of life, to promote leisure and/or health of P/U.
	Accessible learning	To create opportunities for learning and the willingness to make learning resources accessible to P/U.
	Capacity building and knowledge transfer	To improve the ability of a person, group, organization, or system to meet its objectives or to perform better.
		To retain and/or make available skills, knowledge, tools, equipment, and other resources.
	Opportunity creation	The ability to create opportunities for P/U, e.g. to find a job, for social mobility, etc.
		To promote/increase/encourage new forms of micro-volunteering and/or exchange of assets.

(*Continued*)

Table 11.3. Continued

Area of Impact	Indicator	Description of the Indicator
	Social empowerment	The enhancement self and social awareness, self-worth, dignity of P/U.
	Public involvement and participation	The ability to allow people to "do/work together", increasing a community's social fabric and resilience.
		To strengthen/promote/support civic participation, collective action towards a political goal. E.g. the platform helps and/or gives people and individuals the power to influence, and increases individual ability to act.
	Digital divide	The ability to reduce/contrast the "digital divide". (Normally platforms are accessible only to "digital/expert" P/U. By relying only on digital platforms, the digital divide can be exacerbated.)
Environmental	**GHG Emissions**	To reduce greenhouse gas emissions and to improve the quality of the environment.
	Consumption of resources	To reduce resource consumption (water, energy, land, etc.) and/or waste production.
	Waste production	To reduce waste production and/or to promote the reduction of waste.
	Utilization of idle resources	To promote activities aimed at maximizing the utilization of idle resources by sharing commodities such as vehicles, spaces, tools (etc.) with others.
	Environmental awareness	To raise environmental awareness among members/local community/population (e.g. CSE initiative/platform shares/promotes the values of biodiversity and the steps they can take to conserve and use it sustainably).
	Sustainable production	To promote production and/or use of goods/services derived from sustainable sources.
	Corporate Social Responsibility and Environmental Management	Product/service providers can take environmental protection and improvement of societal well-being as a Corporate Social Responsibility (CSR) (e.g. implement/improve their performance and accountability in these areas).

(Continued)

Table 11.3. Continued

Area of Impact	Indicator	Description of the Indicator
	Rebound effects (environmental)	Positive or negative rebound effects e.g. additional consumption of other products or services due to e.g. income effects and/or lower per unit price.
Economic	**Economic impact on P/U***	Economic effect on P/U. Can be (a) positive, (b) negative, and (1) direct or (2) indirect. Can be measured as a variation of (e.g.): (i) income, (ii) expenditure, (iii) debt, (iv) savings.
	Job creation	To create new jobs, directly or indirectly.
	Opportunities for providers	Create new economic opportunities for providers, and growth of new economic sectors (e.g. secondary market supporting the CSE ecosystem).
	Local economic impact	Impact on the local economy (positive: e.g. to revitalize deprived areas, the growth of local economic activities, etc. and/or negative: effects on some local businesses or incumbent business e.g. by displacement). To improve the self-sufficiency of the local community (e.g. local money is spent locally/off-the-grid/self-organization).
	Fiscal impact and public spending	Impact on tax revenues e.g. for local governments. This can refer to any direct or indirect tax paid/avoidance (including licence, registration fees, etc.). Economic interventions and economic growth induced/due to increase in public spending.
	Rebound effects (economic)	Measurement of decline of demand and substitution effect on "traditional sectors", e.g. in some sectors services/products offered by CSE platforms may cause unintended negative cross-sectoral impacts.
Political	**Regulatory framework**	Effects on the regulatory framework e.g. need/demand/introduction for new regulations and ad hoc legislation regulating e.g. taxation, registration, licensing. Effects on different areas of legislation such as labour law, sectoral/specific tax rules, etc.

(Continued)

Table 11.3. Continued

Area of Impact	Indicator	Description of the Indicator
	Policies and instruments	Effect (direct or indirect) on environmental and more sustainable policies and instruments (e.g. integration of principles of sustainable development with policies and programs).
	Political empowerment	Political empowerment and participation of P/U, including in decision-making processes.
	Political mobilization	Political mobilization and increased demand for political change. This involves P/U organising with a common purpose or common understanding to achieve collective goals, social mobilisation, building alliances and coalitions.
	Power	Forms of control of one person or group over others: the alignment of P/U to the interests and ideologies of its founders, volunteers, and users. In this case can be expressed as the 'power' of a CSE initiative or platform to influence its direct surrounding and/or its 'power' to influence meaningful decisions. May be seen as negative.
	Partnership	Establishment of partnership with local government/s and/or NGO/s to e.g. promote principles of sustainable development or biodiversity protection into CSE Platform's programs.

*P/U = For an easy use of the table we would generally refer to "participants and/or users". Nevertheless, a CSE initiative can reach a wide range of people: Members (e.g. founders), participants (e.g. those taking part to its activities), users (e.g. in the case of a platform-based initiative), etc. These different typologies should be carefully defined while developing the indicators and the analysis.
Source: Elaboration of the authors.

methodology for analysis, availability of data and information, specific needs of the research, etc.

11.5 The Application of the Theory of Change for Impact Assessment of CSE Platforms

In order to develop and test a new framework to assess the impact of CSE platforms we decided to apply the Theory of Change (ToC) at an industry level, even though it is generally used at an organizational level, and then to perform an empirical study on CSE platforms operating in the food industry. Specifically, food sharing platforms (e.g. "Too Good To Go" and "Copia") that focus on enabling a digital connection between suppliers and beneficiaries of edible food waste while having social impacts related to reducing waste (Michelini *et al.*, 2018; Ciulli *et al.*, 2020).

The remainder of the section is organized as follows: the first paragraph presents an overview of the theory of change; the second describes the methodology adopted for the considered case study; in the final two paragraphs, the results obtained are presented.

11.5.1 Theory of Change: Overview

The ToC approach seems to have first emerged in the United States late in the 1990s, in the context of improving evaluation theory and practice in the field of community initiatives (Weiss, 1995). Over the years, ToC has become very popular among non-profit sector practitioners, and it has been defined in different ways. A clear and practical definition was provided by Rogers (2014), who states that the ToC "explains how activities are understood to produce a series of results that contribute to achieving the final intended impacts. It can be developed for any level of intervention – an event, a project, a programme, a policy, a strategy or an organization". The ToC is a flexible framework and it can be applied to identifying the data that need to be collected for measuring impact, it can be a framework for reporting, for strategic planning, and monitoring and evaluation. This is one reason ToC is becoming popular in the social impact assessment field of study.

ToC represents a useful starting point for a social impact assessment, as it makes it possible to identify how an organization's social mission will be achieved. A representation of the ToC includes outcome, output, and activities (Clark *et al.*, 2004). Activities are those initiatives that enable the planned output to be achieved; output is the direct and tangible results that come from the activities and help to achieve the output and the outcome is the social change expected in the long-term (Figure 11.1).

The ToC provides a flexible approach to identify how activities are understood to produce a set of results that, in turn, contribute to achieving a final impact (Grieco, 2015; Rogers, 2014), and describes the change an organization aims to make and

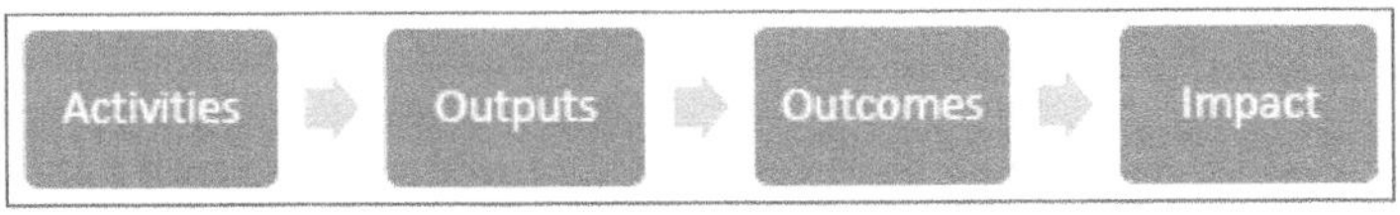

Figure 11.1. Impact value chain.
Source: Adapted from Clark *et al.* (2004).

the steps involved in achieving that goal. In short, the ToC helps an organisation to show how it makes an impact, what it aims to change, and how that change occurs.

Moreover by applying the ToC, it is possible to identify an organisation's main areas of impact and related indicators. Indicators can be *qualitative*, such as positive attitudes and perceptions in consumers, corporate image, organizational climate, and *quantitative*, such as weight/items of food wasted/donated over the years, waste tax reduction, inventory management cost reductions, etc. For a full list of indicators see Michelini *et al.* (2020).

11.5.2 Methodology

To identify the main areas of impact in the food sharing industry, we adopted the methodology developed and tested by Michelini *et al.* (2020). It comprises two focus groups which were held involving different stakeholders: academics, platform managers, institutions (as representatives of policy makers), distributors/suppliers, and non-profit organizations. Each focus group consisted of eight individuals and lasted for approximately 90 minutes. The group was composed of three managers of food sharing platforms, one manager from the distribution sector, one representative of the third sector, and three academics expert in the field of sharing.

Questioning of the focus group was based on the ToC framework and consisted of the following sections: stakeholders, activities, outputs, and outcome (see Table 11.4).

Table 11.4. Focus group questioning.

STAKEHOLDERS
1. What are the main stakeholders of the platform?
2. Does the platform directly or indirectly affect stakeholders? (describe the type of impact)
ACTIVITIES
1. What kind of and how many activities do you carry out?
2. How does their assessment occur?
OUTPUTS: direct and tangible results in the short term
1. What are the main results that will help you achieve the desired changes?
2. Through which indicators do you analyse your short-term performance?
OUTCOMES: The ultimate long-term change in stakeholder life
1. What are the changes you want to generate in the long term?
2. Who benefits from this change?
3. What are your main goals?

Source: Elaboration of the authors.

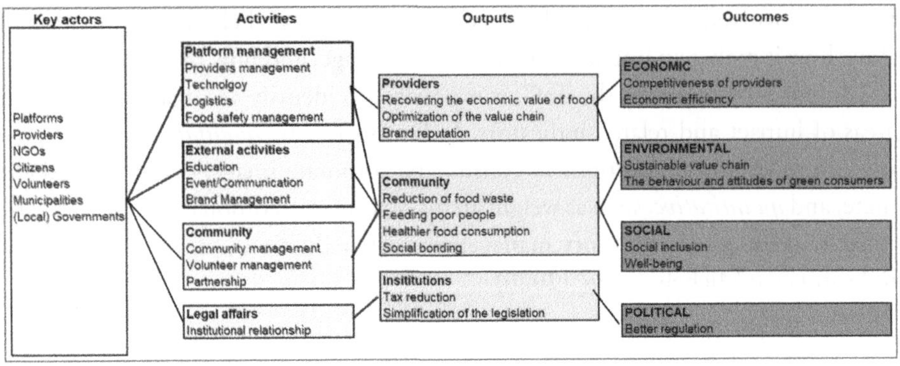

Figure 11.2. ToC: CSE platforms in the food industry.
Source: Elaboration of the authors.

The elements of the ToC framework that drove the focus group were adopted as categories for analysis in the coding process: all the portions of text that were considered as referring to a specific category were recorded under that label. All the elements were then positioned within the map (Figure 11.2), in order to demonstrate the connections between them. To this end, following the ToC methodology, we started by analysing the activities implemented by different types of platform, then we identified the concrete outputs and outcomes that the organizations aim to achieve.

11.5.3 Main Results

The focus groups allowed the identification of the main stakeholders, activities, outputs and outcomes of food sharing platforms. Figure 11.2 summarizes the main results.

Firstly, the main types of **stakeholders** affected directly or indirectly by the platforms were identified: providers of food (such as restaurant, bakeries, supermarkets etc.), not-for-profit organizations that are beneficiaries of food, citizens (involved in peer-to-peer platforms), volunteers wo were generally involved in supporting the logistical process, and finally municipalities and governments.

The first set of **activities** relate to the management of providers, aimed at involving and maintaining suppliers who "feed" the platforms, such as distributors, retailers or restaurants, and technology and data management. More specifically, data analysis is essential for providers that require information which can then be used to improve the supply chain and procurement process in order to prevent waste. A further area of activity pertains to all those actions relating to the establishment of the external presence of the platforms, such as events and communications, and training and education, both of which have emerged as being essential for promoting sustainable behaviours and raising awareness concerning the social good, environmental protection, and safety. An additional key area of activity is related to food management in terms of logistics, and the processes and procedures related to food safety. The focus group identified several activities aimed at establishing and managing relationships with other relevant actors in the ecosystem: volunteers, users/consumers and non-profit organizations. Lastly, institutional relationships play a key role as a mean to carry out lobbying, in an attempt to raise the issue of food waste at a political level, and to influence the development of specific policies to tackle it.

The **outputs** reflect the multiple perspectives of the stakeholders involved in the ecosystems of food sharing platforms. The first set of outputs relates to the providers' perspective, such as recovering the economic value of food, the optimization of the value chain, and the opportunity to enhance the reputation of their brand. A second perspective refers to the community: collecting surplus food is also linked to the opportunity to feed poor people, who would otherwise have no access to this food. More specifically, these outputs concern the distribution of safe food to people, and also the promotion of healthier food consumption. From a community perspective there are other several outputs: the reduction of food waste, the enhancement of neighbourhood relationships (social bonding) and of sustainable attitude and behaviour. A further perspective is that of the government and local authorities. It implies potential externalities that result from the actions of the platforms when they are able to influence legal regulations, particularly in terms of tax reduction and simplification of the legislation on food sharing.

The focus group participants identified four main areas of **outcomes**: economic, environmental, social and political. Economic outcomes are related to the providers; thanks to the platform, businesses were able to increase their economic efficiency and maximize their profits. Environmental outcomes pertain both to providers and to consumers. Providers were able to improve the sustainability of the value chain, while consumers were able to move towards more environmentally responsible attitudes and behaviours. In terms of social outcomes, beneficial sharing practices are seen as a way for consumers to increase their well-being, improve social cohesion, and reduce social distance within society.

Food sharing platform initiatives also have an impact in terms of policy change, working with governments and local authorities to improve legal regulation, thereby maximizing the societal impact they can have.

11.6 Conclusion: Policy Recommendations

The impact assessment of CSE practices and platforms raises many implications at policy level that can be addressed via two perspectives. The first viewpoint refers to *implications derived from the general impact of the CSE*, the second one is related to the *impact assessment methods and tools*. The next two paragraphs will shed light on both of these.

11.6.1 CSE Impact: Policy Implications

In general, the debate around the impact of collaborative and sharing economy is dynamic and multifaceted, with some authors describing its social value and positive effects, and others emphasizing its negative implications. Some scholars have pointed out that while the effects of the CSE are likely to be positive, they are likely to be much smaller than some claim and hope; furthermore due to the uncertainty regarding institutional and technological changes to come, scepticism would be advisable regarding the possibility of providing a realistic assessment of the impact of the sharing economy (Frenken, 2017).

However, analysis of the current academic literature, reports and articles written by practitioners and consultancy firms on the topic allows the identification of the main trends and controversial issues in this field of study. According to Murillo *et al.*, 2017, the main controversial issues can be classified into the following areas: (i) market, (ii) government, (iii) workers, (iv) consumers and (v) environment. In order to provide insights and potential implications for policy makers, we propose a discussion of these most relevant controversial issues.

With regard to the **market**, the CSE poses the matter of whether or not a particular activity can be qualified as a two-sided or multi-sided platform. Two-sided (or more generally multi-sided) markets are roughly defined as markets in which one or several platforms enable interactions between end-users, and try to get the two (or multiple) sides "on board" by appropriately charging each side (Rochet and Tirole, 2004). Its importance concerns competition policy implications, since some economic principles used in competition policy do not hold when markets are two-sided or multi-sided. For example, in two-sided markets, pricing to one side below marginal cost is not a predatory behaviour but it can be a profit maximising strategy. Looking at only one side can lead to a market definition that is too narrow (Codagnone and Martens, 2016; Evans, 2008).

Another important issue refers to the debate between new entrants and incumbents. Incumbents argue that they still face various regulatory burdens that new entrants are evading; these include licensing requirements, price controls, service area requirements, marketing limitations, and technology standards (Koopman *et al.*, 2014).

At a **political (governmental)** level it is often reported that the exponential growth of the CSE has increased the capacity to bypass regulation and favour tax avoidance. Many actors in the sharing economy do not pay taxes, existing outside the "traditional" economy. Furthermore, the headquarters of these platforms may be located in tax havens, making tax collection even more difficult (Malhotra and Van Alstyne, 2014; Klobučar *et al.*, 2016; Frenken, 2017).

To this regard, as indicated by Frenken (2017), according to three possible "platform scenarios" (capitalism, redistribution, cooperative) and related "institutional logics" (market, state, community), governments could adopt new taxation systems, e.g. making effective use of platforms to tax activities that previously were hard to monitor, and/or considering putting new and *ad hoc* taxes on property, revenues, or claiming tax on the profits that platforms make as a function of the volume of transactions in their respective territories.

Regarding the **workforce**, on one hand the CSE has indeed presented itself as a provider of flexible employment opportunities, but on the other hand the rights, benefits, and fair-pay levels of this emerging sector of freelancers seem to be under threat. Many platforms do not guarantee social security or pension rights since most of their workers are not considered employees, but as independent contractors (Murillo *et al.*, 2017; Klobučar *et al.*, 2016; Stabrowski, 2017; Sundararajan, 2016).

From the **consumer's perspective**, Schor (2016) investigated how CSE activity is affecting the distribution of income and opportunity. She found that the CSE increases income inequality amongst the bottom 80% of the population (in terms of wealth distribution), shifting more income and opportunity to better-off households and providers.

The CSE could also have a negative effect on social inclusion, since CSE service providers do not always have to follow the same legislation as traditional service providers, and thus certain groups could find themselves excluded. For example, disabled individuals may not be able to use services such as car sharing, because providers are not required to have vehicles adapted to their needs, and private drivers are not obliged to take them as passengers (Klobučar *et al.*, 2016).

In terms of **environmental** impacts, recent research mainly identifies the positive impacts derived from the new paradigm of more sustainable consumption and production practices. However, industry-specific literature has lighted some concerns, for example in the case of food sharing, one of the main issues is the focus on waste reduction, instead of trying to address waste prevention (Michelini *et al.*, 2020).

In the mobility sector, some studies have argued that if shared cars are cheaper and more readily available, this could prompt more people to use such cars instead of public transport, leading CO_2 emissions to rise (Klobučar *et al.*, 2016). Moreover, as noted in Section 11.3, literature concerning the hospitality industry has reported conflict, e.g. between tourists and locals at a community level, decreases in the quality of neighbourhood life and coexistence, rising accommodation prices, and gentrification (Gutiérrez *et al.*, 2017; Trudelle and Pelletier, 2016; Sundararajan, 2016).

Policy implications	Market	Government	Workers	Consumers	Environment
Improve the policy framework	Ensure fair competition between new entries and the incumbents.	Uniform or similar regulation all over the EU (e.g. on taxation issues).	Review all the labour-market relevant legislation (e.g. to guarantee workers protection, minimum wage, etc.).	Data security and protection of consumer rights.	Reform environmental guidelines.
Digital divide reduction	Improve the taxation and regulatory framework.	To train Public Administration (PA) in order to improve and make uniform the digital literacy and make CSE useful at the governmental level as well.	To improve the digital literacy of the workers and guarantee equal access to CSE services.	To improve the digital literacy of the population through effective educational and training systems.	Advance the social goals of protecting the environment, also implementing and/or improving environmental accountability requirements.
Promote and support inclusiveness	Foster equal access to all the interested categories.			To improve the services inclusiveness (i.e., disabled people).	
Transparent lobby campaign		Encourage the dialogue between platforms and policy makers.			
Foster social entrepreneurship	Encourage the start-up of fair and sustainable oriented platforms.				

Figure 11.3. Policy implications and recommendations.
Source: Elaboration of the authors.

Starting from these premises Figure 11.3 shows in brief the most important implications that should be addressed by policy makers for each of the following areas:

- improve the policy framework;
- reduce the digital divide;
- promote and support inclusiveness;
- make transparent lobbying campaigns;
- foster social entrepreneurship.

11.6.2 CSE Impact Assessment: Policy Implications

Identifying qualitative and quantitative metrics for reporting information to stakeholders is not straightforward, which makes assessing the impact of the CSE complex. This complexity has resulted in the development of many different

Main issues	Possible pitfall	Policy recommendations
Heterogeneity **Domain of activity.** **Business models.** **Stakeholders.** Lack of... **Financial resources.** **Standards.** **Scientific approach.** **Culture.**	**Risk of encouraging:** The development of models that are too complex or too vague. The proliferation of indicators. The development of unrealistic and/or unfeasible models.	Facilitate the creation of an open-source library of indicators and/or easy-to-use toolbox of feasible indicators. Foster the networking of academics, practitioners, and policy makers. Support the development of methodologies to foster impact-investing tools

Figure 11.4. Impact on policy.
Source: Elaboration of the authors.

models aimed at providing guidelines and indicators for such assessments (Grieco, 2015). The ongoing proliferation of models is due to the fact that organizations differ in size, capacity, activities, and focus, and consequently there is no single model that is suitable to assess all of them.

The work carried out so far has allowed us to identify the main issues that have characterized the impact assessment of CSE, and its potential pitfalls, and to provide some policy recommendations that will be useful to improve the quality of impact assessment (see Figure 11.4).

An initial and important issue concerns the "heterogeneity" of domain of activity as CSE practices and/or platforms operate in different industries (such as food, hospitality, mobility, finance, etc.) and can adopt different business models (e.g. for-profit and non-profit). Furthermore, there are many interested parties – such as academics, practitioners, entrepreneurs' regulatory agencies, policy and program developers – that may have different needs and purposes in using impact assessment methodologies.

A second major issue is related to the "lack of…": particularly financial resources, shared knowledge and available data and information; and also to standards, such as shared metrics or indicators, and the lack of a methodological approach that characterizes accounting practices aimed at assessing financial returns and culture among organizations, institutions, entrepreneurs and public administrations. Even if generally-accepted accounting principles exist to aid financial reporting, a comparable standard related to the measurement and communication of social impact does not yet exist because it is difficult to define the concept universally, and related measurement tools often lack the rigour that characterizes scientific approaches aimed at assessing financial returns (Grieco, 2015).

The main pitfalls relate chiefly to the risk of encouraging the development of models that are too complex or too vague, the proliferation of models, approaches, indicators, and the development of unrealistic or inapplicable measurements.

Based on prior analysis, the following **policy recommendations** can be made:

- ✓ Facilitate the creation of an open-source library of methodologies and/or easy-to-use toolbox of feasible indicators: it is recommended to create and make available to researchers, specialists and policymakers an open-access and easy-to-use toolbox of feasible indicators and/or a library of methodologies which would allow them to make accessible and consolidate existing knowledge and available and tested tools, such as, for example, those realised in the framework of a variety of EU funded projects (some of which have been also used and mentioned in this contribution).
- ✓ Foster the interaction of academics, practitioners and policy makers: it is recommended to encourage and support improved communications between academics (from different fields of study), specialists, practitioners and policymakers by creating networks that bring these groups together. The creation of networking opportunities and long-term programmes – such as this COST Action – can play a crucial role in making the existing knowledge accessible and advance.
- ✓ Support the development of methodologies useful to foster impact-investing tools – namely a new and innovative model that leverages market-driven efficiencies – to provide social services such as Social Impact Bonds (SIB). An SIB is a contractual arrangement between an entity with a mandate to promote social welfare (e.g. governments, development banks, and philanthropic organizations) and a private sector investor that will finance social service interventions up-front in exchange for future payouts. The amount of the payout level is linked to the resulting effectiveness of the social service (Bergfeld *et al.*, 2019).

To conclude, taking into account the disruptive nature of many CSE initiatives and platforms, despite the fact that a common and shared metric for the measurement of their wide range impacts does not yet exist, it is suggested that a reinforced interdisciplinary approach based on the networking of academics, practitioners and policy makerspolicymakers, supported by an open-source library of methodologies, and an integrated collaboration between private sector and public entities, could represent a feasible way to assimilate the lessons learned from different fields of study, research and practice, and translate them into the structuring of new and more effective policies.

References

Bardhi, F. and Eckhardt, G.M. (2012). Access-Based Consumption: The Case of Car Sharing. *Journal of Consumer Research, 39*(4): 881–898.

Barron, K., Kung, E., and Proserpio, D. (2018). The Effect of Home-Sharing on House Prices and Rents: Evidence from Airbnb. Retrieved from https://ssrn.com/abstract=3006832

Belk, R. (2010). Sharing. *Journal of Consumer Research, 36*(5): 715–734.

Benkler, Y. (2004). Sharing nicely. *Yale Law Journal, 114*: 273–358.

Bergfeld, N., Klausner, D., and Samel, M. (2019). Improving Social Impact Bonds: Assessing Alternative Financial Models to Scale Pay-for-Success. *Annual Review of Policy Design, 7*(1): 1–20.

Berger, T., Chen, C., and Frey, C.B. (2017). Drivers of Disruption? Estimating the Uber Effect. Retrieved from www.oxfordmartin.ox.ac.uk/downloads/academic/Uber_Drivers_of_Disruption.pdf

Botsman, R. (2013). The Sharing Economy Lacks a Shared Definition, available at https://www.fastcompany.com/3022028/the-sharing-economy-lacks-a-shared-definition

Botsman, R. and Rogers, R. (2010). What's Mine Is Yours: The Rise of Collaborative Consumption. New York: Harper Collins Publishers.

Celata, F. and Sanna, V.S. (2016). Multi-criteria analysis for carbon efficient projects, Deliverable 4.2 of the FP7 TESS project. [Online] www.tess-transition.eu/wp-content/uploads/2016/11/TESS_D4.2_Multi-criteria-analysis-for-carbon-efficient-projects.pdf

Celata, F. and Sanna, V.S. (2019). A multi-dimensional assessment of the environmental and socioeconomic performance of community-based sustainability initiatives in Europe. *Regional Environmental Change, 19*(4): 939–952.

Celata, F., Hendrickson, C.Y., and Sanna, V.S. (2017). The sharing economy as community marketplace? Trust, reciprocity and belonging in peer-to-peer accommodation platforms. *Cambridge Journal of Regions, Economy and Society, 10*(2): 349–363.

Ciulli, F., Kolk, A., and Boe-Lillegraven, S. (2020). Circularity brokers: digital platform organizations and waste recovery in food supply chains. *Journal of Business Ethics, 167*(2), 299–331.

Clark, C., Rosenzweig, W., Long, D., and Olsen, S. (2004). Double bottom line project report: Assessing social impact in double bottom line ventures.

Codagnone, C. and Martens, B. (2016). Scoping the sharing economy: Origins, definitions, impact and regulatory issues. *Digital Economy Working Paper*, Institute for Prospective Technological Studies.

De Groen, W.P., Kilhoffer, Z., Lenaerts, K., and Salez, N. (2017). The Impact of the Platform Economy on Job Creation, in *Regulating the Platform Economy: How to Protect Workers While Promoting Innovation, 52*(6): 345–351.

Dillahunt, T. and Malone, A. (2015). The Promise of the Sharing Economy among Disadvantaged Communities. *Proceedings of the 33rd Annual ACM Conference on Human Factors in Computing Systems*, 2285–2294.

Demailly, D. and Novel, A.S. (2014). The sharing economy: make it sustainable. Study No. 03/14, Institut du développement durable et des relations internationales (IDDRI), Paris, France.

Denegri-Knott, J. (2011). Have it now: Ebay and the acceleration of consumer desire. *European Advances in Consumer Research*, (9): 373–9.

Dogru, T., Mody, M., and Suess, C. (2019). Adding evidence to the debate: Quantifying Airbnb's disruptive impact on ten key hotel markets. *Tourism Management, 72*: 27–38.

Edelman, B.G. and Luca, M. (2014). Digital Discrimination: The Case of Airbnb.com. *Working Paper* 14-054, Harvard Business School.

Edelman, B.G. and Damien, G. (2016). Efficiencies and Regulatory Shortcuts: How Should We Regulate Companies like Airbnb and Uber? *Stanford Technology Law Review 19*: 293–328, Harvard Business School NOM Unit Working Paper No. 16-026.

Einav, L., Farronato, C., and Levin, J. (2015). Peer-to-peer markets. National Bureau of Economic Research NBER *Working Paper Series* 21496, Cambridge, USA.

Ert, E., Fleischer, A., and Magen, N. (2016). Trust and reputation in the sharing economy: The role of personal photos in Airbnb. *Tourism Management*, (55): 62–73.

European Commission (2016a). A European agenda for the collaborative economy, available at https://ec.europa.eu/transparency/regdoc/rep/1/2016/EN/1-2016-356-EN-F1-1.PDF

European Commission (2016b). Environmental potential of the collaborative economy, available at http://trinomics.eu/wp-content/uploads/2018/09/DG-ENV-Collaborative-Economy.pdf

Frenken, K. (2017). Political economies and environmental futures for the sharing economy. *Phil. Trans. R. Soc. A, 375*: 2095.

Frenken, K. and Schor, J. (2017). Putting the sharing economy into perspective. *Environmental Innovation and Societal Transitions*, (23): 3–10.

Grieco, C. (2015). Assessing social impact of social enterprises: Does one size really fit all? Heidelberg: Springer.

Grieco, C., Michelini, L., and Iasevoli, G. (2015). Measuring value creation in social enterprises: A cluster analysis of social impact assessment models. *Nonprofit and Voluntary Sector Quarterly, 44*(6): 1173–1193.

Gutiérrez, J., García-Palomares, J.C., Romanillos, G., and Salas-Olmedo, M.H. (2017). The eruption of Airbnb in tourist cities: Comparing spatial patterns of hotels and peer-to-peer accommodation in Barcelona. *Tourism Management,* (62): 278–291.

Guttentag, D. (2015). Airbnb: disruptive innovation and the rise of an informal tourism accommodation sector. *Current Issues in Tourism, 18*(12): 1192–1217.

Gyódi, K. (2019). Airbnb in European cities: Business as usual or true sharing economy? *Journal of Cleaner Production, 221*: 536–551.

Hamari, J., Sjöklint, M., and Ukkonen, A. (2016). The sharing economy: Why people participate in collaborative consumption. *J. Assoc. Inf. Sci. Technol. 67*: 2047–2059

Heinrichs, H. (2013). Sharing economy: A potential new pathway to sustainability. GAIA Ecol. *Perspect. Sci. Soc., 22*: 228–231.

Katz, V. (2015). Regulating the sharing economy. *Berkeley Technology Law Journal, 30*(4): 1067.

Klobučar, S., Brunelet, N., Dallhammer, E., and Schuh, B. (2016). Urban Impact Assessment. The Sharing Economy. *European Committee of the Regions,* available at https://cor.europa.eu/en/our-work/Documents/Territorial-impact-assessment/sharing-economy.pdf

Koopman, C., Mitchell, M., and Thierer, A. (2014). The Sharing Economy and Consumer Protection Regulation: The Case for Policy Change. Mercatus *Working Paper,* Mercatus Center at George Mason University, Arlington.

Lee, D. (2016). How Airbnb Short-Term Rentals Exacerbate Los Angeles's Affordable Housing Crisis: Analysis and Policy Recommendations. Retrieved from http://blogs.ubc.ca/canadianliteratureparkinson/files/2016/06/How-Airbnb-Short-term-rentals-disrupted.pdf

Leismann, K., Schmitt, M., Rohn, H., and Baedeker, C. (2013). Collaborative consumption: Towards a resource-saving consumption culture. *Resources, 2*: 184–203.

Malhotra, A. and Van Alstyne, M. (2014). The Dark Side of the Sharing Economy … and How to Lighten It. *Communications of the ACM, 57*(11): 24–27.

Martin, C.J. (2016). The sharing economy: A pathway to sustainability or a nightmarish form of neoliberal capitalism? *Ecological Economics, 121*: 149–59.

Martin, E. and Shaheen, S. (2011). Greenhouse gas emission impacts of car-sharing in North America. *IEEE Transactions on Intelligent Transportation Systems 12*(4): 1074–1086.

Martin, E. and Shaheen, S. (2016). Impacts of car2go on vehicle ownership, modal shift, vehicle miles traveled, and greenhouse gas emissions: An analysis of five North American cities, *Working Paper,* Transportation Sustainability Research Center (TSRC), Berkeley.

Michelini, L., Principato, L., and Iasevoli, G. (2018). Understanding food sharing models to tackle sustainability challenges. *Ecological Economics, 145*: 205–217.

Michelini, L., Grieco, C., Ciulli, F., and Di Leo, A. (2020). Uncovering the impact of food sharing platform business models: a theory of change approach. *British Food Journal 122*(5): 1437–1462.

Murillo, D., Buckland, H., and Val, E. (2017). When the sharing economy becomes neoliberalism on steroids: unravelling the controversies. *Technological Forecasting & Social Change, 125*, 66–76.

Parigi, P., State, B., Dakkhallah, D., Corten, R., and Cook, K. (2013). A community of strangers: the dis-embedding of social ties. *PLoS One, 8*: 1–7.

Picascia, S., Romano, A., and Teobaldi, M. (2019). The airification of cities. Making sense of the impact of peer to peer short term letting on urban functions and economy. https://doi.org/10.31235/osf.io/vs8w3

Plepys, A. and Singh, J. (2019). Evaluating the sustainability impacts of the sharing economy using input-output analysis, Chapters, in: Oksana Mont (ed.), *A Research Agenda for Sustainable Consumption Governance*, chapter 5, pp. 66–84, Edward Elgar Publishing.

Reischauer, G. and Mair, J. (2018). How organizations strategically govern online communities: Lessons from the sharing economy. Acad. Manag. Discov.

Rochet, J.C. and Tirole, J. (2004). Two-sided markets: an overview. *Institut d'Economie Industrielle Working Paper.*

Rogers, P. (2014). Theory of Change: Methodological Briefs-Impact Evaluation No. 2 (No. innpub747).

Rosen, D., Lafontaine, P. R., and Hendrickson, B. (2011). CouchSurfing: belonging and trust in a globally cooperative online social network. *New Media & Society, 13*: 981–998.

Ronzhyn, A. (2013). Online identity: construction interpersonal trust and openness through participating in hospitality social networks. *Journal of Education Culture and Society, 1*: 47–56.

Sanna, V.S. (2018). Grassroots initiatives for sustainability transitions: community-wide impacts and economic functioning. Special Issue "Post-Growth Organization". *Management Revue, 29*(4): 349–380.

Schor, J. (2016). Debating the sharing economy. *Journal of Self-Governance and Management Economics, 4*: 7–22.

Slee, T. (2015). What is Yours is Mine: Against the Sharing Economy. OR Books, New York.

Stabrowski, F. (2017). 'People as businesses': Airbnb and urban micro-entrepreneurialism in New York City. *Cambridge Journal of Regions, Economy and Society, 10*: 327–347.

Sundararajan, A. (2016). The Sharing Economy. The End of Employment and the Rise of Crowd-Based Capitalism. MIT Press: Cambridge, UK.

Trudelle, C. and Pelletier, M. (2016). Analysis of urban conflict networks: Theoretical and methodological perspectives. In *Conflicts in the City. Reflections on Urban Unrest*; Del Romero, L., Ed.; Nova Publishers: New York, NY, USA.

Vaskelainen, T. and Piscicelli, L. (2018). Online and Offline Communities in the Sharing Economy. *Sustainability, 10*, 2927.

Wachsmuth, D., Kerrigan, D., Chaney, D., and Shillolo, A. (2017). Airbnb's Impact on Canadian Housing Market. Policy report.

Wachsmuth, D. and Weisler, A. (2018). Airbnb and the rent gap: Gentrification through the sharing economy. *Environment and Planning A: Economy and Space, 50*(6):1147–1170.

Warmington-Lundstroma, J. and Laurenti, R. (2020). Reviewing circular economy rebound effects: The case of online peer-to-peer boat sharing. *Resources, Conservation & Recycling: X*, (5) 100028.

Weiss, C.H. (1995). Nothing as practical as good theory: Exploring theory-based evaluation for comprehensive community initiatives for children and families. In *New Approaches to Evaluating Community Initiatives: Concepts, Methods, and Contexts*, The Aspen Institute, 65–92.

Woodcock, J., Tainio, M., Cheshire, J., O'Brien, O., and Goodman, A. (2014). Health effects of the London bicycle sharing system: health impact modelling study. *BMJ, 348*: g425.

Zervas, G., Proserpio, D., and Byers, J.W. (2017). The Rise of the Sharing Economy: Estimating the Impact of Airbnb on the Hotel Industry. *Journal of Marketing Research, 54*(5): 687–705.

Zukin, S. Lindeman, S., and Hurson, L. (2015). The omnivore's neighborhood? Online restaurant reviews, race, and gentrification. *Journal of Consumer Culture*, 1–21.

DOI: 10.1561/9781680838411.ch12

Chapter 12

Rethinking the Stack: New Narratives for an Era of Collective Intelligences

By David Crombie, Revathi Kollegala and Soenke Zehle

As holistic views of systems design gain ground, the question of how we might turn the "accidental megastructure" of existing infrastructure stacks into consciously designed collective intelligence architectures has emerged as a "matter of concern" for those engaged in designing cooperative and sharing economies (Bratton, 2014).[1] If we are aiming at a "technological sovereignty" facilitated by cooperative approaches rather than framed by the power planetary platforms, future ecosystems for products and services must be based on alternative technology stacks. And as planetary perspectives are beginning to frame more and more of what we do, the sustainability of such technological ecosystems comes into focus. That is why we want to enter the stack – and add stack design to the conversation around cooperative and sharing economies. While one goal of our research is to imagine a "collective intelligence design stack", the focus of this essay is the why-and-how of such a

1. On the distinction between "matters of fact" (assumed to be given) and "matters of concern" (negotiated among and across a multitude of human and non-human actors), see Latour (2004).

use of the metaphor of the stack. And whether the stack is most useful as a mainly technological metaphor – or whether we should think instead about the stack as the conceptual schema of transformation narratives. What follows is an overview of the state of play around "stack design" in the *anticipate* collective intelligence design research network and the presentation of the network's first "collective intelligence design canvas", a concept stack to facilitate a collaborative systems design conversation.

12.1 Integrating "Stack Design" into a Systems Design Conversation

We are neither coders nor engineers but driven by a shared sense that the scope and speed of technological change has literally implicated us – folded us – in the workings of distributed intelligent systems that affect how we relate, speak, work. It is this transformation of our own agency we wish to comprehend, and the emergence of collective intelligences that bring human and non-human agency together in new constellations. While we realize that "stack design" is an abstract design proposition, we want to build on the technical metaphor of the stack as a layered system to facilitate a more holistic systems design conversation – and find out more about the role each of us might play in that process.

Stacks are generally defined as "the set of technologies an organization uses to build a web or mobile application ... a combination of programming languages, frameworks, libraries, patterns, servers, UI/UX solutions, software, and tools used by its developers".[2] A quick look at the stacks used by key players in and across the platform economy suggests that these modularized systems have more in common than their fierce competition for users might suggest. Rather than repeating the usual invocations of global brands (and their market power), we want to shift the focus of our conversation to the stacks that sustain the power of these actors – a first step toward a more comprehensive understanding of the relational infrastructures behind data-driven societies, what we have termed the condition of distribution, and a different way of thinking about the design of (fairer) markets for a more cooperative economy.[3]

Adopting the principle that technology stacks are in principle subject to co-design, we showcase a series of complementary co-design processes to explore how we might best facilitate such a conversation. In these examples, we focus on

2. https://stackshare.io

3. For an exercise to build a prototype of the internet based on the Open Systems Interchange (OSI) stack and a network of Raspberry Pi microcomputers, see Howser (2020).

the role of data – from the data monetization controversy (data-as-right vs data-as-commodity) to "platform" cooperatives adding a digital technology and data governance layer to the much older organizational practices of the cooperative. We do not, however, assume a comprehensive understanding of data governance approaches. Our invitation to join the "stack design" conversation begins with a much simpler idea – the search for a narrative (with a wide range of characters and multiple conflicting plots) that might be able to frame a cross-sectoral systems design conversation in which many actors see a role for themselves.

So before we turn to technology, we define our non-technological point of departure, leaving the question of where *exactly* to draw the line between the technological and the non-technological to the side for now. While we are not yet sure what the scope and structure of such a conversation will be, we do know that we want such a narrative to allow us to comprehend the condition of distribution that serves as infrastructural context for the ways in which we live and work.

12.2 Combining Technological and Non-Technological Design Approaches

Critical observers are still debating whether it makes more sense to approach the situation we find ourselves in as a "technological" or more fundamentally a "mathematical" condition that assumes a translatability of experience into data no longer marked by the worldly remainders – accident, context, history, situation – of such necessarily incomplete translations (Hörl, 2015; Mersch, 2013). Echoing the latter, some even fear a disappearance of the political into the paradigms and protocols of prediction altogether – leaving us with a politics incapable of relating to that which is aleatory, other, incommensurable (Rouvroy, 2020; see also Rouvroy, 2016). It is in part our difficulty in drawing the line between the technological and the non-technological – the *no-longer* as well as the *not-yet* technological – that has encouraged us to engage with what we approach as collective intelligences.

Such an approach owes much to experiences with the experimental cultures of arts-and-technology research, but also a proximity to the performing arts where the staging of interactive multi-plot narratives is an established aesthetic practice sustained by a rich methodological tradition. Integrating "performative" elements not only builds on democracy-theoretical work on the key role of performance, notably Judith Butler's performative theory of assembly reminding us that if we rely only on speech we miss the expressive dimension of physical bodies and fail to attend to the dynamics of exclusion in speech-only assembly formats (Butler, 2015). It also incorporates research by performing artists on "reducing implicit cognitive biases through the performing arts" Arch *et al.* (2020). In this context, choreography or *mise-en-scène* are not only invoked metaphorically, referencing the spaces and stages of aesthetic experimentation only to relegate these to the (irrelevant, irreverent) sidelines of the field of technological transformation. Quite the contrary: in an economy increasingly turning to experience as the final frontier of value creation, the work of imagination and institution (from "experience" designs facilitating complex and collaborative forms of use to the education of the "creative" agency expected to drive such processes) assumes a centrality that is not always easily reconciled with the self-styled exile from the din of the world we have come to associate with the arts, or views of aesthetic autonomy that see in artistic freedom a model for other freedoms. But here we are, creating, measuring, organizing experience across linked lifeworlds whose degree of interconnectedness is yet another symptom of the condition of distribution.

So yes, arts and culture offer us methodological inspiration to better understand how the concepts we use interact with each other, how placing them in different constellations affect their analytical reach, change what they can tell and reveal to us about the world. In our attempt to bring technological and non-technological design processes together, we do not want to simply relegate the arts to the domain of the non-technological but need to highlight those formats that engage the question of technology from within. This does not mean to look only for technology-based media arts practices, but for forms of artistic expression that address one or more of the registers of the shift towards prediction as a ubiquitous system and the rise of scalable collective intelligences embedding us. Among those that interest us the most are the ones engaging with the not-yet, the speculative registers of artistic intuition open to and opening up multiple futures. Which is why we frame our exploration of collective intelligence through the question of anticipation.

12.3 Anticipate! A Research Network on Collective Intelligence Design

Building on future studies, scenario development, and speculative design, the interest in anticipation as a holistic framework continues to grow (Poli, 2019; Miller, 2018). In this context, authors and colleagues initiated *anticipate* – a collective intelligence design research network. Many of the ideas in this essay have been co-developed in the context of conversations across the research network.[4] When the network was initiated, the researchers involved decided to not position the project in the context of a specific academic discipline or research field but maintain a "non-disciplinary" stance. Such a stance is, in turn, necessarily dynamic (tracking

4. https://www.anticipate.network

transformations rather than observing final states) and collective (as there can be no single perspective from which the whole of a process or system comes into analytical view). It is through the "anticipatory arts" of play (role-playing, storytelling, dystopian and utopian thinking in cinema, games, literature) and their what-if-power that we can develop new ways of engaging in fore-sighting activities on all levels of generative engagement – from the individual citizen to the policy maker structuring the different stages of such engagement. In the course of the anticipate network conversation, we have come to embrace a type of play where players change the rules of engagement and exploration as they move through the process through the following principles:

1. Explore Impact. Rather than solely focusing on a single technology (such as artificial intelligence) as a field of applied innovation, the network makes the question of collective intelligence the focus of non-disciplinary inquiry and experiment. This allows us to approach, explore and comprehend the wide-ranging implications and possible impact of machinic intelligences without locking us into the dynamics of technological development.

2. Imagine Innovation. New imaginaries, new narratives, new horizons – if we are to anticipate worlds in which human and non-human actors become part of collective intelligences, we will need all of these. In imagining alternative futures, the network widens the space of innovation. This goes both ways, as we also need to innovate imagination. New technologies change the way we can arrive at concepts, tell stories, foresee futures. This opens up new problem spaces and calls for new conceptual blueprints to ultimately create new instruments for the organization of change.

3. Co-Create Discourses. We can only find the new if we have a language that allows room for the unknown. Otherwise we may not be able to name the new when we encounter it – or miss it altogether. The network critically assesses the terms we have come to use to talk about the new – and creates new terms whenever we think existing terms won't do. The co-creation of new languages is one way to anchor technology design in a broader and more holistic conversation about how we want to live and work.

4. Make Worlds. The distinctions we have established in education and research have served us only so well in building new alliances. Rather than struggling to re-connect what we have come to accept as always already separate – IT, SSH, Arts and Culture – we begin with a multi- and even non-disciplinary view of the systems and worlds of which we are a part. In the context of ecological crisis, we need to have a better sense of how the world exists – its interdependencies, its timescales, its spatialities. If technology is to play a role in addressing this crisis, we need a way of speaking about worldmaking that acknowledges that technologies can play multiple roles, and that our ways of exploring impact must acknowledge the complexity of technological agency.

5. Contextualize Agency. The conditions of change frame our agency – as they change, so do our options for individual and collective action. Awareness of contexts directly translates into new possibilities for action. We need to rethink how we can explore anticipatory assumptions, harnessing structures for mutual learning to meet these challenges. By collaborating with a wide range of actors, we can devise new educational formats to properly assess, scope and tackle more complex and chaotic problems.

Exemplifying a new generation of socio-technical systems, automated decision making is on the rise, with capacities for data analysis and prediction that far exceed the powers of earlier generations of expert systems that simply offered static contextualizations for individual decisions (Algorithm Watch, 2020). While data ethics strategies affirm the centrality of the human as "decision-maker of last resort", these systems (applied cross-sectorally, including automotive, banking/finance, health, human resources) effectively confront us with a sober truth: a hybrid system in which machines will propose and ultimately make better decisions. If we want to resolve this dilemma – that ethics place the human in command, while technological innovation improves the quality of (soon) autonomous assistive systems – we need cooperative decision systems that offer dynamic recontextualisation, design a co-agency model that can comprehend the hybridization of human and machinic agency, and safeguard not only our individual human agency but our ability to co-exist in a new generation of collective intelligences that radically alter our understanding of the ways in which humans and machines collaborate. Building a "stack design" conversations is one way to do that – stressing our embeddedness in socio-technical systems designed to amplify our collective agency and open up new practices of co-creation.

12.4 Data Institutions: Stack Design as a Way to Frame Matters of Concern

One of the ongoing conversations offering points of entry and reference for a "stack design" conversation revolves around data governance. Combining data management and the definition of user rights, stack design appears here as the design of data institutions.[5] Trusts, for example, feature prominently in the EU's data governance

5. For the design vision of one of the major data space industry alliances, see OPENDEI (2021).

framework.[6] For the Open Data Institute, data trusts apply the existing concept of fiduciary duty to data management: "With data trusts, the independent person, group or entity stewarding the data takes on a fiduciary duty. In law, a fiduciary duty is considered the highest level of obligation that one party can owe to another – a fiduciary duty in this context involves stewarding data with impartiality, prudence, transparency and undivided loyalty"[7] Such visions have inspired a wide range of approaches in the search for alternatives to the current system of data extractivism driven by the agendas of a small number of companies that dominate the platform economy.

In "Data Union", digital artisan, educator and feminist hacker Larisa Blazic and her collaborators "established the first International Data Union to represent the rights of individuals, outside the control of commerce or the state".[8] The goal was to "support members in leveraging the value of their data" as they retain their data rights but define the term of its use by others. If there is to be a translation of collective bargaining traditions to the new field of data labor, this means conflict – including the refusal to work ("data strike") if there is no agreement. Recently, the "data union" project was followed-up by a "Data Union Fork: tools for data strike", a workshop to co-create the technical means of such collective self-determination of the ways in which the data we generate throughout our daily lives are used by others.[9] In these design conversations, Blazic featured the use of DECODE OS, part of the toolkit created by her fellow software activists in the DECODE project, a collaborative software and policy research process revolving around citizen-driven strategies for technological sovereignty.[10]

Her work on data unions exemplifies a brand of arts-and-technology research that allies experimental approaches, agile technology development, and the design of distributed systems. While Blazic and DECODE relate their work directly to the technopolitical grassroots, Steamr is part of a current wave of data monetization

---

6. https://ec.europa.eu/digital-single-market/en/european-data-governance

7. https://theodi.org/article/what-is-a-data-trust

8. https://www.e-w-n-s.net/dataunion/dataunion.html

9. https://waag.org/en/event/data-union-fork-tools-data-strike

10. One of the motivations for our research is existing "technological sovereignty" efforts such as DCENT (Decentralised Citizens ENgagement Technologies) that have already engaged in such stack design. Building on a wide array of activist/social movement engagement, DCENT already developed a "collective intelligence framework" that links what we refer to as "stack design" to the "new municipalism" involving citizens in the future of urban development and cross-city cooperation in a broader technopolitics. See D-CENT (2014); on new municipalism, see Barcelona En Comú (2019). A follow-up project, DECODE was "an experimental project to develop practical alternatives to how we use the internet today – four European pilots will show the wider social value that comes with individuals being given the power to take control of their personal data and given the means to share their data differently." See https://decodeproject.eu/what-decode.

start-ups, representing another strand of the current effort to explore the future role of data intermediaries. While both Blazic and Streamr refer to "data unions" as organizational forms to advance shared interests, Streamr explicitly links its proposals to market design strategies.

The team behind Streamr, "the missing real time data protocol for the decentralized web", has made the case that it is through the creation of "data unions" that we will be able to create an alternative data ecosystem (Ronstedt, 2020).[11] The design of an automated data marketplace offers an example of trying to understand how (our) data acts once we release it into a dynamic of self-organized value creation. Whether such marketplaces are a good answer to the question of ethical monetizations of data or whether monetization is in conflict with technological sovereignty and self-determination continues to be a matter of controversy.

The Electronic Frontier Foundation, for example, "strongly opposes data dividends and policies that lay the groundwork for people to think of the monetary value of their data rather than view it as a fundamental right" (Tsukayama, 2020). Both proposed privacy regulation (leaving the determination of the value of data essentially to the companies using the data) and the vast profits of platform economy giants (vast in total but tiny in per-user amounts) suggest that "if where we've been is any indication of where we're going, there won't be much benefit from a data dividend". Because payouts are likely to remain low, creating incentives for those to whom such small amounts matter will affect the most vulnerable populations; and even data that appears to have little value (like the location of the school from which you graduated) might affect the outcome of automated decision systems (such as denying loans to people from the "wrong" neighborhood). EFF also opposes "pay-for-privacy" schemes that offer discounts in return for greater access to data to better target ads. Instead, EFF makes the case that privacy is far from dead – and neither dividends nor discounts should convince us to turn a freedom into a commodity.

Responding to criticism of his polemical "privacy is dead" proposition, Streamr's head of growth has argued that it is precisely the expectation to get involved in market design that will end up weakening privacy: "because it is too difficult to protect what is precious, people end up giving up on all of it and their privacy becomes entirely worthless by default. So why not put a value on it, and ask people to figure out those decisions for themselves?" (Malik, 2020). Being asked to put a value on their data, he claims, will in fact lead many of them to differentiate more, not less, between data to be kept private and data made available to others. He is also convinced that appropriate forms of property will be found to

11. https://swash.io is one of the first use cases demonstrating the data union principle, https://www.clture.io and https://datacy.com are other examples of automated data marketplaces.

facilitate such sharing: "When transferring data as property, Data Unions, who act as mediators of people's data, will likely adopt leasing rights more akin to authorship rights than simplistic property rights".[12] Such sharing can easily co-exist with the use of privacy-protecting tools; and a simple agency matrix that only sees states, markets and individual consumers obscures the key role of civil society intermediaries who – like data unions – protect the interests of their members. Which is why "we're going to need legislation to stop unscrupulous players, and to establish healthy relations between a union's managers and its owners". Data unions offer a way to comprehend how data-based value creation works; it also helps us understand how "the platform economy" could and should be regulated, how user rights (including rights to share their data) might drive the design of technology policy and ultimately "the stack" itself as such regulation is built into the system in the form of depersonalization standards or other ethics-by-design features.[13]

What we take from these two use cases: the question of intermediaries – collectives – identifies an important terrain for "collective intelligence design" as the focus shifts from the individual to the individual-in-context (Ruhaak, 2020).[14] We also welcome the call to reimagine concepts of property and public goods, especially if the emphasis on "selling" data shifts toward the conditions of "sharing" data and the rules and relationships needed to govern such sharing.

Data monetization is one of the areas where key decisions regarding our collective agency (and intelligence) are currently made. Whenever (libertarian) trust in (new) markets competes with (activist) faith in enlightened policy making, our interest in the transformation of agency and value comes into play: what are the assumptions we make regarding users? Is the assumption of vulnerability to the "bribes" of data

---

12. Malik references the work of Savona here. See Savona (2020). Reviewing the ideas of "data labor" and "collective bargaining" promoted by Weyl and others (see Weyl and Posner, 2018, the foundation based on these ideas <https://www.radicalxchange.or>, and the "short manifesto for the data labor movement" by Ibarra *et al.* (2018); for a modest example of a data-dignity-based design approach, also see https://www.microsoft.com/en-us/garage/profiles/trove). Savona concludes that in these proposals, "The notion of labour dignity becomes 'data dignity'. . . . Creating a credible institutional actor that can represent and collectively bargain on behalf of data generators is a necessary, though not sufficient, condition to make this governance model work. . . . Current labour markets issues – such as technological unemployment, skill-biased technical change, and wage polarisation – would simply be reproduced in a data labour market. . . . Advocating for data labour markets to address data value redistribution is an endeavour whose success would depend on an adequate system of collective representation and bargaining." In our view, this turns the question of "data labor" into a question of collective intelligence design.

13. Also see the MyData Global community's comments on the draft version of the EU's Data Governance Act https://mydata.org/2020/11/06/mydata-view-on-the-leaked-eu-data-governance-act-nov-5-2020.

14. "New data governance models – such as data commons, in which groups of people collectively decide on data collection and use – can help us. In other cases, we may rely on intermediaries and data trusts to help us exercise our rights and execute our decisions. Such intermediaries should have a fiduciary duty to make decisions in our best interest, and the scope of their power should be restricted to a specific purpose. Importantly, collective data governance models are needed for non-personal and personal data alike."

dividends and discounts simply an expression of policy paternalism, or is it even more naive to have faith in collective bargaining when the history of organized labor and social movements includes fantastic victories (whether and to what extent the authoritarian economies of China and the Soviet Union qualify as examples of the power of organized labor is a concern whose complexity lies beyond the scope of this essay) but also (especially in Europe and the US) many defeats by those whose agency was amplified to a much greater extent by "market" forces? What would the fora for such negotiations be, and how would those involved arrive at a shared sense of the creation and distribution of value? What are the strengths and limitations of a "data labor" analogy, beyond the references to work and collective bargaining? And what if the vision of organized labor is that of the "wobblies", the International Workers of the World (IWW) refusing a politics of representation? Exploring the radical imaginaries invoked by a "data union" might be one of the tasks for historically-aware research and organizing.

What is intriguing about the data union vision is that machine intelligence (as in automated sharing of data governed by user-defined rules) may be harnessed in new forms of sharing. But to fully take advantage of such possibilities, it makes sense to expand the "data union" vision and return to more comprehensive imaginaries of cooperation and commoning. The horizon here, however, is not the reinvention of markets (through theories of data labor and economic theories of auctioning), but to think "other-than" markets.

Commons have been considered as key elements in the building of cultures of trust: "we should look for ways of making data available for the common interest ... Technology companies would move from being data owners to becoming data stewards" (Shah, 2018). And the idea of user-defined rules for the shared use and maintenance of resource is, after all, one way to define the commons: "Users negotiate their own rules, assign responsibilities and entitlements, and set up monitoring systems to identify and penalize free riders".[15] And yet, market design theorists Posner and Weyl (referenced explicitly by the framers of Streamr's "data union" perspective) continue to recycle the clichéd version of a "tragedy of the commons" (Posner and Weyl, 2018).[16] But as Helfrich et al also note in their analysis, "critiques do not necessarily help us imagine how to remake our institutions and

15. https://www.freefairandalive.org/read-it

16. Garret Hardin's controversial 1968 essay on the "tragedy of the commons" played a key role in popularizing contemporary uses of the commons; what is often forgotten is that Hardin himself revisited this essay and admitted that the omission of the term "unmanaged" in the original publication facilitated multiple misunderstandings. See Hardin (1998). Research on the (historical) diversity of commons management approaches (especially the work of Elinor Ostrom) has been a major contributor to the current renaissance of the commons as conceptual and organizational vision.

build a new world".[17] Which is why, above and beyond conceptual critiques of the marginalization of cooperative and commons-based perspectives in the design of a new generation of data institutions, we turn to experiments that bring such cooperative institutions into being.

12.5 New Collective Action Institutions: Platform Cooperatives

In the platform cooperatives movement, the technology layer of data-driven efforts comes together with a tradition of cooperative value creation. The platform cooperatives movement began in the US, but has already expanded into a strong international network integrating many cooperative traditions into the platform coops conversation.[18] Platform Coops Germany started in Germany in 2020.[19] For us, it is a use case in "translating" a global approach to organizational development by integrating regional cultures and traditions of cooperation with a platform design agenda.[20] More schematically, it is an effort to transform from below existing (technology-centric) innovation narratives to address commons-oriented ideation and value creation and bring existing dynamics around peer-to-peer cultures into a regional innovation context. The project has resonated especially with those who have struggled to open up the rather conservative tradition of German cooperativism, foregrounding the close relationship between more recent sharing economy trends and a much broader historical tradition of cooperativism that is currently experiencing a renaissance. The latter is key because it shapes the context in which sharing economy initiatives exist and expand – an opening of definitions and narrative of innovation, of (public) value, and of collaborative agency and cooperative management.[21]

While cooperativism in Germany has its roots in the agricultural sector, other sectors with a strong cooperative dynamic include banking, energy, food, and

17. https://www.freefairandalive.org/read-it

18. https://platform.coop/who-we-are/people, also see https://coopcycle.org. Driven by art and design students to create an alternative to individualist startup culture, s:coop is an example of "next generation" cooperativism; while start-up culture is all around us, there are fewer efforts to bring younger people into the cooperative ecosystem, see https://scoop.vision.

19. https://platformcoop.de. Platform Coops Germany is supported through the first-ever federal research fund for non-technological innovation, the Innovation Programme for Business Models and Pioneering Solutions (IGP).

20. https://platformcoop.de/, https://www.coopgo.de

21. For a mapping of German sharing and cooperative economy actors, see https://www.i-share-economy.org.

housing (Brummer, 2018). Cooperative approaches to "shared infrastructure" are much older than the platform/sharing economy; like their counterparts across Europe, German farmers have "shared" equipment for over 60 years in machinery associations (Hasenpusch, 2018). The cooperative movement just celebrated the 130th anniversary of the "cooperative law" from 1889. Today, 40 million peopel across Germany are members of a cooperative, the cooperative idea has been accorded UNESCO Intangible Cultural Heritage status in 2016. While it is historically inaccurate to argue that this tradition implies that "Germany invented the sharing economy", not least because the modern cooperative movement has its roots in England (ICA) and sharing is common to all cultures, it is true that the cooperativism pioneers "Raiffeisen and Schulze-Delitzsch ... were ahead of their time in anticipating a sharing economy based more on mutual benefit than on personal or public property" (Reay, 2018).

In order to support the further development of the sharing economy, establishing and strengthening links with the cooperative sector is important, as the new forms of "platform cooperativism" and the assessments of the future potential of the sharing economy are likely to benefit from exploring these wider historical connections with cooperative ownership and use. The German cooperative movement is currently going through a fair amount of soul-searching; organisations such as Igenos e. V., an association of cooperatives' members, are making the case for a bottom-up renewal of cooperative practices hampered by the movement's top-down traditions (Igenos, 2018).[22] The new discussion of "platform cooperatives" offers additional inspiration. Traditional cooperative actors are beginning to engage with the platform cooperativism trend, but the key actors in critically assessing this trend have been civil society organisations like Supermarkt Berlin and rethink coop (a member of the US-based "Platform Cooperativism Consortium") (Bott and Giersberg, 2018). Recent initiatives such as Platform Cooperatives Germany reflect a growing interest in a renaissance of cooperativism in the context of contemporary peer-to-peer cultures and infrastructures. It illustrates the dynamics of creating alternative narratives through processes of organizational development that are, in turn, a key register of systems design. Their attention to the technology layer of cooperative value creation opens up another register of the systems design conversation we wish to engage in.

The focus on platform coops also helps argue that the rise of "data societies" as narratives of change and innovation does not simply call for a corresponding set of "data institutions", but of collective action institutions that facilitate different

22. The report references the work of Günter Ringle on the impact of National-Socialist on the cooperative tradition of democratic selforganization (see Ringle, 2018).

kinds of individual and collective agency. Future *anticipate* research will turn to the insights of collective action research and its attention to long-term change, as we explore the social and cultural institutions framing our actions: "we need to look at institutions for collective action from a long term perspective. First of all, an institution needs time to get in shape, to be modeled according to the needs of those involved, and these institutions change slowly: a (semi-)democratic process for the change of rules requires time-consuming consultation of all the stakeholders involved. Secondly, the success of an institution, once well in place, can to a certain extent be measured by its longevity. By doing this, in combination with an examination of the stimulating and/or threatening factors that these institutions were dealing with we can understand what makes cooperation successful and when it fails. History thus is essential to our understanding of the mechanisms underlying institutions for collective action".[23] Key examples of institutions foregrounded by *anticipate* are **commons** and **cooperatives** – two collective action institutions currently experiencing a renaissance and supported across many communities of change as **integral to the co-creation of shared knowledges and futures**. However, we do not explore them primarily in terms of their institutional design, but explore the role of storytelling in building new collective action institutions – much more than organizational forms, they reflect shared visions and values.

12.6 Creating Narratives of the Future

In her Nobel Lecture, the writer Olga Tokarczuk addressed the quest for new narratives:

> "Today our problem lies – it seems – in the fact that we do not yet have ready narratives not only for the future, but even for a concrete now, for the ultra-rapid transformations of today's world. We lack the language, we lack the points of view, the metaphors, the myths and new fables. Yet we do see frequent attempts to harness rusty, anachronistic narratives that cannot fit the future to imaginaries of the future, no doubt on the assumption that an old something is better than a new nothing, or trying in this way to deal with the limitations of our own horizons. In a word, we lack new ways of telling the story of the world" (Tokarczuk, 2019).

By definition, artistic intuition comprehends the current conjuncture in ways that differ radically from empirical assessments, and such an intuition makes no claim to representativeness in its sweeping diagnosis. Yet wherever we have looked,

23. http://www.collective-action.info/introduction

we find evidence of both the exhaustion of existing narratives – of autonomous agency whose conflictual dynamics of constitution are rarely acknowledged as citizens, people, users are expected to always already *have* such agency; of infinite economic growth as material condition safeguarding the coherence of open societies; of technology-centric innovation struggling to imagine non-technological vectors of change – and a growing interest in new stories of the world. The 2019 report *Understanding our political nature: how to put knowledge and reason at the heart of policymaking*, the first outcome of the European Commission's *Enlightenment 2.0* initiative, makes attention to the role of narrative a key measure of future policy success.[24] And related efforts such as the "European Moments" research project coordinated by Timothy Garton Ash has already documented people's willingness to contribute to such a co-creation of narratives.[25]

Tokarczuk includes the qualifier "it seems" – more than a rhetorical phrase, "it seems" can be interpreted as an attempt to capture that the world is trying to tell us something. To ask ourselves whether there is indeed a "language of things", whether we need to look at the question of representation from the other side – focusing less on the question of the subject, of identity and intention, and much more on the object, its autonomy, its other-than-human temporalities (think nuclear or plastic waste) and spatialities (think pollution particles). Whether or not this involves an epistemological effort to actually "side with the object", we see a useful change in perspective to approach the condition of distribution.[26]

So we imagine a non-human perspective, we inscribe alterity into our relation to the world, no matter what else we do, it seems that we must follow this ethical injunction. Grasping the threat of extinction, activists and the arts alike want us

---

24. https://ec.europa.eu/jrc/en/facts4eufuture/understanding-our-political-nature

25. https://europeanmoments.com

26. For explorations of Walter Benjamin's "primitive" reflections on such a "language of things", see Bracken (2002) as well as Steyerl (2006). As noted by David Cunningham in his "Photography and the Language of Things" series, "it's hard, too, not to see this concept of a language of things as prompted by, and as alluding to, a far more pervasive turn to things and objects in recent theory and art than could be disinterred from the 'weird' mix of pantheism, nominalism and early romanticism to be found in the young Benjamin's essay itself (written when he was 24 years old). Indeed, from the post-Deleuzian vitalism of a so-called new materialism to the generalized 'actants' of Latourian anthropology to the 'withdrawing' but always active objects of object-oriented ontology, speaking things are pretty much everywhere in the humanities and social science today" (<https://www.fotomuseum.ch/en/explore/still-searching/articles/29092_if_things_could_speak>). We share this interest in part because the "algorithmic" object is challenging not only our attempts to translate its manner of "speaking" but the whole idiom of "making the invisible visible" so often invoked to account for the (presumed) power of the arts. It is, in fact, not clear to us at all what the role of the arts might be in exploring the condition of distribution, nor what conceptual idioms might best facilitate such exploration. So by "siding with the object", we try to follow what the arts *do* – and see where such practice-driven research leads us.

to finally listen to the world-as-other.[27] So Gaia is our interlocutor.[28] During an exchange around the question of "the collective" organized by Bruno Latour, the conservationist David Western comments on the frequent invocation of the figure of a self-organizing whole popularized by Lynn Margulis and James Lovelock that "We had better get our metaphors right, because if we don't convince the public about the threats to our planet and mobilize action in the next ten, we're in real trouble. Metaphors matter".[29] While consensus builds that "Gaia" also makes sense scientifically, Western contends that "we have to distinguish between a metaphor which has public valence and galvanizes action, and a metaphor that captures the essence of the planetary process".[30] And Lovelock himself has noted that "Even if in the end Gaia should turn out to be no more than a metaphor, it would still have been worth thinking of the Earth as a living system" (Latour, 2016). So the whole it is; our situation demands that we maintain the question of the whole – of the horizon of our agency and the effects it creates as it interacts with the agency of others, human and non-human – as one layer of our inquiry.

Which brings us, incidentally, back to the question of the stack. GAIA-X, a European initiative to create a federated cloud infrastructure, explicitly references Gaia as figure of the planetary.[31] The project has been accompanied by a collaborative coding effort to create a "sovereign cloud stack", a collection of open source code powering a wide variety of cloud-based services.[32] In this context, "Gaia" has become the cypher of a socio-technological system safeguarding "sovereignty". Whether or not its initiators think "Gaia" beyond the geopolitical conflicts such "sovereignty" is meant to address (the proposal references the need for an alternative to existing authoritarian and libertarian approaches to the internet, read: China and US), or

---

27. Also see the "sustainability turn" in (organized) culture; examples include the *Network of European Museum Organizations* https://www.ne-mo.org/advocacy/our-advocacy-work/museums-and-sustainability.html and *We are Museums* http://www.wearemuseums.com/museumsforclimate/museums-facing-extinction, UNESCO is currently revisiting its *Agenda 2030* strategy to better address Sustainable Development Goals (https://en.unesco.org/news/how-could-science-museums-contribute-achievement-sustainable-development-goals).

28. Initiated by a flagship publication of the natural sciences, the launch of *The Lancet Planetary Health* (which "seeks to be the pre-eminent journal for enquiry into sustainable human civilisations in the Anthropocene") is indicative of the extent to which this conversation is already underway, see https://www.thelancet.com/lanplh/about. The authors of this chapter are active in https://www.planetaryhealthalliance.org.

29. See "Feedback on Day One," p. 131, in Latour *et al.* (2020).

30. ibid. On the state of scientific discussions on Gaia, see Clark (2020). The biological research of Margulis (1938–2011) features prominently in the "Critical Zones" engagement and exhibition project co-curated by Latour, see https://zkm.de/en/exhibition/2020/05/critical-zones.

31. https://www.data-infrastructure.eu

32. http://scs.community

beyond a vision of data-driven value creation we do not yet know, what counts is that a metaphor has been put into play that at least strives to *think the planetary* to engineer a (more) viable system.[33] It now assumes, in Western's terms, "public valence" beyond the spheres of ecopolitics and might facilitate translations between the technological and the non-technological.[34]

What appears like a shift away from the non-technological to the technological is much closer to the original intent than often assumed; Lovelock (inventor of the electron capture detector) has recently suggested that the Gaia theory "is just engineering written very large indeed ... with Gaia you can go out in the world and start measuring things" (Watts, 2020a).[35] What is important in the context of our argument here is that we use "the stack" as an *ecosystemic* metaphor to engage with the relational infrastructures sustaining collective intelligences. And while the term "ecosystem" is commonly used *metaphorically* to describe such infrastructures, it makes sense to speak of the infrastructures for our data-driven lives in ecosystemic terms not least because their operation assumes a wide range of resources.[36] The miniaturization of devices has been made possible by the planetary expansion of the distributed systems that allows these ever-smaller devices to operate, the networks of sourcing, production, use, and disposal are planetary indeed.[37] And so are the stacks that allow these distributed systems to operate – whether or not they reference "Gaia" in their names.

33. Technical specifications are under development, a draft is expected mid-2021.

34. The quickly-expanding list of GAIA-X partners, already ranging from regional cloud providers to the data intelligence company Palantir, suggests that GAIAX is on its way to becoming another example of the "Brussels Effect" of passive externalization of regulatory approaches primarily aimed at actors in Europe's common market, see Bradford (2020).

35. Bruno Latour has hailed Lovelock's invention as one of our age's greatest discoveries: "While Galileo used a telescope to reveal that the Earth is part of an infinite universe, Lovelock used his electron capture detector to reveal that the Earth is completely different from any other planet because it has life. He and [Lynn] Margulis spotted Gaia. Lovelock from space, taking the question as globally as possible;Margulis from bacteria, taking the question from the other end, both realising that Life, capital L, has managed to engineer its own conditions of existence. For me that is the greatest discovery of this period, though it is still not very much accepted by mainstream science." (See Watts, 2020b).

36. Needless to say, the political pathos presenting Europe as a leader in sustainability is not exactly substantiated by many of the EU's policy initiatives, and artists and designers have already criticized the call for a "New European Bauhaus" (NEB) to guide the green transition and called for a new holism: "the transnational character of climate change challenges the idea of borders and national sovereignty, showing us that the only way forward is in thinking of the world as a whole" (https://janvaneyck.info/apply/letter-to-object-to-the-term-new-european-bauhaus). At the same time, the NEB process pioneers a design-driven policy process and has set up a wide engagement process around regenerative economies and planetary perspectives, see https://blogs.ec.europa.eu/eupolicylab/portfolios/social-economy-canvas.

37. See the Good Electronics Networks for reports from all corners of the world of electronics production and disposal. Members include environmental, human rights, labor and research organizations, https://goodelectronics.org/aboutus/members.

In *Novacene,* Lovelock has developed this vision of a cybernetic culture further. Extrapolating the potential of machinic intelligence, he imagines the emergence of cybernetic organisms (cyborgs) cooperating with humans to maintain (through geoengineering and other efforts) Gaia's capacity for self-regulation (Lovelock, 2019). Whether or not readers agree with such an ecomodernist proposal – what inspires us is that Lovelock *speculates* about possible futures as a way to identify paths of action in the present.[38]

12.7 Feeding Forward: Elements of a Collective Intelligence Design Stack

The complexity of new technical infrastructures that cut across national boundaries and call into question established governance framework challenges publics and policy makers observing current acceleration of technological change – autonomous systems, artificial intelligences, decentralised technologies. Many analysts suggest that we stress the role of technology even more. But what we learn from the arts is that it is crucial to explore the impact of such change – and the societal challenges that arise from such change – through the lens of individual and collective experience: what impact do these changes have on the ways in which we live, work, and anticipate individual and collective futures? In the course of these conversations, the authors have created a first draft of a design canvas to explore how the condition of distribution affects how we approach objects, comprehend emerging forms of human-machine agency, co-create value, integrate anticipation into what we do, and work toward an understanding of intelligence that breaks out of the mold of measuring machine intelligence against narrow comprehensions of consciousness (Zehle *et al.*, 2021).

Returning us to the "stories of the world" perspective, such efforts challenge the conventional view of being overwhelmed by complexity. While designing the *technological conditions* that frame local action – around health, justice, and/or climate issues – is not always part of the vision of political action, a wide range of practices of collaborative "infrastructuring" – from autonomous media networks to local

38. Lovelock uses the term "cyborg" in the generic sense of "cybernetic organism" that has accompanied the popularization of cybernetic approaches and was introduced by Manfred E. Clynes and Nathan S. Kline in their 1960 *Astronautics* article "Cyborgs and Space." In the context of science-and-technology studies, the term is more frequently associated with the work of Donna Haraway. For an overview of the impact Haraway's cyborg had expecially on feminist and queer theory, see "cyborg" in the *New Materialism Almanac* https://newmaterialism.eu/almanac/c/cyborg.html. In Haraway's recent work, the cyborg as that which cuts across the boundaries between human and non-human has given way to "new figures of promise such as the coyote, the trickster, companion species, or her recent turn to the chthonic forces of the Earth" (ibid.). For an analysis of Haraway's nuanced engagement with "Gaia" see Clark (2019).

commons – exists (Lyle *et al.*, 2018). We will continue to explore such strategies of dealing with complexity in addressing a broad range of societal challenges, and it is the nexus of agency-in-distributed-systems to which we keep returning. To do so, we need to explore whether (and if so, how) the dominant narratives of change end up decontextualizing (and thus weakening) our individual and collective agency to the extent that the very idea that we could have an impact on the contexts that frame this agency becomes almost unthinkable.

Only a few decades ago, technology assessment was at the center of discussions of how best to address societal challenges. Today, it has narrowed its focus and contributed to a corresponding narrowing of innovation narratives (Sand, 2019). We have found that engaging in a multi-layered "stack design" conversation is much more demanding than embracing metaphors of the whole, and we are far from being able to specify the elements of a "collective intelligence design stack". However, the "stack design" perspective has been particularly useful in locating interdepencies of technological and non-technological transformations and the search for ways to co-design such assemblages to better amplify our individual and collective agency.[39] Guided by the concept stack of our draft "collective intelligence design canvas", we will continue to iterate this canvas to facilitate such engagement with use cases that help us understand these layers, both individually and in their interaction.

Across the cooperative economy, related design proposals have already been made and we are looking to integrate their findings. This includes the work of the MyData Global cooperative, bringing coders, engineers, policy makers, and researchers together, MyData to pioneer a design-oriented "data literacy" conversation to which we would like to contribute.[40] An innovation research agency that has already prioritized collective intelligence in its work, NESTA has created the draft of a "Collective Intelligence Design Playbook".[41] It schematizes a great deal of research into the dynamics of collaboration, citizen science, civic technology, and social movements engaging in technology design processes, a new grant program encourages innovation in this emerging field.[42] And the Systemic Design Association has offered its own approach, based on comparative co-creation research.[43] Exploring these and related efforts to structure design conversations, we have integrated the research issues explored above in a canvas that links six different vectors

39. For a comparative analysis of co-creation approaches with a systemic design focus, see Jones (2018).

40. https://mydata.org

41. https://www.nesta.org.uk/toolkit/collective-intelligence-design-playbook

42. https://www.nesta.org.uk/feature/collective-intelligence-grants

43. https://www.systemicdesigntoolkit.org

influencing how we imagine and approach behavioural, social, and cultural change: object, agency, value, situation, intelligence, and world.

ANTICIPATE CANVAS

< PERSPECTIVE SHIFT	CONCEPT	PERSPECTIVE SHIFT >
MATERIAL / DISCRETE	OBJECT	HYBRID / PROCESSUAL
ACTORS / INDIVIDUAL / DECONTEXTUALIZED	AGENCY	ACTANTS / DISTRIBUTED / ECOSYSTEMIC
COMPETITIVE / COMMENSURABLE	VALUE	CO-CREATIVE / INCOMMENSURABLE
HERE / NOW	SITUATION	EMERGENT / FUTURAL
HUMAN-VS-NON-HUMAN	INTELLIGENCE	COLLECTIVE
CULTURE-VS-NATURE	WORLD	GAIA / PLANETARY

OBJECT: Building Processual Awareness for Regenerative Economies

Anticipate **facilitates the long-term thinking** required for citizen-driven transformation processses, circular economies and "triple bottom line" models of value creation. While cycles of extraction, production, use, and disposal keep raw materials, pollution particles, or waste in motion across space and time, we think of "objects" mainly in their (spatial and temporal) discreteness. It is a real challenge to cognition, perception, and agency to comprehend that a plastic bag thrown into a river and the detection of micro-plastic particles in an unborn infant **involve the same object**. We wish to advance a processual comprehension of (technical and non technical) objects to better address the **ecosystemic effects they continue to produce on spatial and temporal scales that far exceed those of individual human lives** (500+ year lifecycle of plastics, 300.000+ years nuclear waste).

AGENCY: Distributed and Ecosystemic

Anticipate uses co-creation to facilitate future-oriented **visions of collective action taking the full range of human and non-human agency into account**. We employ co-creation methods to better contextualize our agency, attend to the dynamics of sense-making, and bring into view systemic contexts. These include the ecosystems as well as the distributed technological infrastructures we use to communicate, relate, and work, informing and guiding our decisions. Building on such an expanded sense of distributed objects and systems, it integrates non-human – including natural (example: bacteria) and machinic (example: learning systems) – agency in its visions of collective agency. In its effort to expand the space of agency, we invite participants to **mobilize culture and science to reimagine our relationalities**, make visible a wider array of actors, and expand the spatial and temporal scope of human agency – embracing a "futures" spirit to explore even more radical expansions of agency in processes of behavioural, social, and cultural change. Such an expanded sense of agency is all the more urgent as the climate crisis can neither be comprehensively understood nor effectively addressed without **attention to non-human forms of agency.**[44]

As already discussed in the context of the 2015 COP21 and related legal initiatives that have endowed several natural entities in various places around the world with rights and the capacity to be represented in courts, the question of **integrating non-human agency in a new politics of representation** acquires a new urgency in our times of crisis.[45]

In its experimentation with the role of arts and culture in the reimagination of (collective) agency, *anticipate* builds in these experiments, specifically the "Théâtre des Négociations" organized on 2015 and related formats. The argument invoked by the organizers (including Bruno Latour) of the "theatre of negotions", a simulation inspired by Latour's ideas of a "parliament of things" still holds: "As the experimentation with forms of representation is a basic element of theater, it seemed appropriate to work with and within an institution that is dedicated to the production of theater".[46] The project integrated a simulation to simulate alternative political decision making dynamics – a play-based approach that also informs the vision of arts-and-culture oriented *anticipate* research.

VALUE: Co-Creative

Despite the truism that "everything has its price", experience tells us that not all values can be mapped onto each other. This reality severely limits current market-based efforts to "internalize" ecosystemic dynamics through the attribution of commercial value. "Broader European social values and citizens' needs should be at the heart of the new policy narrative. Innovation, and the economic benefits that flow from it, will follow. This is the only way to achieve sustainable wellbeing and concrete solutions for citizens".[47] In the spirit of mission-driven transformation, *anticipate* attends to the complex dynamics of translation between incommensurable values

---

44. For an example of actant/environment-centered design, see Sznel (2020).

45. https://www.eurozine.com/representing-nature/, also see https://unfccc.int/process-and-meetings/the-paris-agreement/the-paris-agreement

46. https://raumlabor.net/le-theatre-des-negociations

47. https://ec.europa.eu/info/publications/101-ideas-future-research-and-innovation-europe_en

and a vision of co-creating value involving the multiple value scales at the heart of collective action institutions such as commons and cooperatives. Its emphasis on complex transformation narratives build capacity to further explore such value co-creation beyond the project, contributing to social and technological innovation across a diversity of communities of change.

SITUATION: Navigating Uncertain Futures

Anticipate's co-creative research formats build futures literacy.[48] The exercise of collective, cooperative, and co-creative agency always occurs in the now, everything we do has the potential to contribute to behavioral, social, and cultural change. The ambition is to integrate a "futures" perspectives into the exercise of such agency.[49] We do so by focusing on co-creation and advancing a definition of creativity that ranges from the artistic and cultural to the technological, linking the daily exercise of citizenship to the longer-term co-creation of viable systems.

INTELLIGENCE: Mobilizing Human + Non-Human Intelligences

To fully harness the power of intelligent (machinic and natural) systems in addressing the challenges of the current transformation to a regenerative economy, we must address – as part of a comprehensive behavioral, social, and cultural change – the framing of our agency both by technological and "natural" systems.

Anticipate is motivated by a broader vision of collective intelligences that integrate human and non-human agency in new assemblages. We believe that we will only harness the power of intelligent systems if we redefine our relationship to non-human agency. In its final consequence, this means that we need to move beyond a vision of "human-centric design" – which is why *anticipate* develops an alternative view of collective agency and intelligence open to non-human agents and intelligences.

Focusing on the central role of language – the way we describe and discuss these systems – in framing the kinds of agencies and intelligences we imagine, the IEEE has recently called for a **"de-anthropomorphizing" of machinic intelligences** to create space for **new visions of collective intelligences** and address the full range of ethical implications of co-designing such intelligences (Kostopoulos, 2011). *Anticipate* explores artistic practices that reimagine citizenship and democracy for an age of autonomous systems and artificial intelligences. We are

48. Our use the term is inspired by UNESCO's futures literacy initiative, https://en.unesco.org/futuresliteracy/about.

49. https://www.lincolninst.edu/sites/default/files/pubfiles/how-use-exploratory-scenario-planning-full.pdf

already beginning to imagine machines as actors in democratic societies, are giving "automated decision-making systems" an ever-expanding role,[50] and the EP (for European Parliament) has already (and passionately) debated whether or not to grant machines independent legal status ("robot personality").[51] Through the collaborative analysis and assessment of embodied experiences across the arts, *anticipate* facilitates our comprehension of contextual agency in embedded systems and human-machine-networks more generally, **couples such technological visions with the distributed non-human agency of ecosystems**, supports the co-creation of new interfaces to these systems, and hence fosters the future-oriented engagement with societal challenges needed to comprehensively analyze and effectively address them.

WORLD: GAIA and Planetary Perspectives

Popularized by the engineer James Lovelock and the biologist Lynn Margulis and recently embraced by sustainability thinkers like Bruno Latour as well as the "earth systems sciences", the figure of "GAIA" invoked in European projects like "GAIA-X" describes a self-regulating systems whose survival hinges not on "environmentalism" – a view that continues to separate us from the world – but on a **comprehensive understanding of our embeddedness** in such a self-regulating system, and of the effects our behaviour has on the capacity of that systems to maintain and repair itself.

There are several ways to read and use this canvas. The left column could be labeled "default / standard" (an understanding of the technical object, agency, and value that is mainstream and considered the unquestioned point of departure in policy papers and innovation narratives), the right column "challenge / what-if" – it schematizes the kind of conversation we would like to contribute to. Another way to read the canvas is to suggest that the key concepts in the middle column oscillate between these poles, so "discrete" and "processual" here refer to the respective ends of a spectrum of possible ways to view each of the concepts in the stack.

Such a concept stack is our first approximation of stack design – before we can even define the stack as operating principle of a system, we want to map it as a conceptual constellation, treating "concepts" as as "boundary" concepts whose analytical reach changes as we change our points of view.[52] The "research briefs" offered

50. https://algorithmwatch.org/en/project/automating-society

51. https://www.europarl.europa.eu/RegData/etudes/STUD/2016/571379/IPOL_STU(2016/571379_EN.pdf

52. This is an experimental adaptation of the idea of boundary objects: "Boundary objects are those objects that are plastic enough to be adaptable across multiple viewpoints, yet maintain continuity of identity." See (Star,

here will be part of the next round of *anticipate* conversations, expanding "stack design" into a holistic form of engagement.

References

Algorithm Watch. (2020). *Automating Society Report 2020.* Available at: https://automatingsociety.algorithmwatch.org

Arch, J.G., Ventura-Gabarró, C., Adamuz, P.L., Calvo, P.G., and Fuentemilla, L. (2020). Reducing Implicit Cognitive biases Through the Performing Arts, *Front. Psychol.*, 17 May 2021. Available at: https://doi.org/10.3389/fpsyg.2021.614816, https://www.frontiersin.org/articles/10.3389/fpsyg.2021.614816/full

Barcelona En Comú. (2019). *Fearless Cities: A Guide to the Global Municipalist Movement.* Oxford: New Internationalist. Available at: http://fearlesscities.com/en/book

Bott, J. and Giersberg, K.-W. (2018). Über die Bedeutung deutscher Genossenschaften im globalisierten Wettbewerb" [On the significance of German cooperatives in a globalized competitive economy], *ADG Scientific Center for Research and Cooperation.* Available at: https://www.adgonline.de/forschung/whitepaper-genossenschaften-im-globalisierten-wettbewerb; https://supermarkt-berlin.net, https://www.rethinkcoop.de

Bracken, C. (2002). The Language of Things: Walter Benjamin's Primitive Thought, *Semiotica, 138*1/4, 321–349.

Bradford, A. (2020). The Brussels Effect: How the European Union Rules the World. Oxford: Oxford University Press.

Bratton, B.H. (2014). The Stack. On Software and Sovereignty. Cambridge, MA: MIT Press.

Brummer, V. (2018). Of Expertise, Social Capital, and Democracy: Assessing the Organizational Governance and Decision-making in German Renewable Energy Cooperatives. Available at: https://doi.org/10.1016/j.erss.2017.09.039

Butler, J. (2015). Toward a Performative Theory of Assembly. Cambridge: Harvard University Press.

1989). Star has suggested to replace the "Turing test" with a "Durkheim" test – an approach that resonates with our interest in ecological perspectives: "The test of intelligence of a distributed system is necessarily an ecological one. . . . In order to understand the acceptance and use of a machine in and by a community, that community must be actively present as it evolves" (ibid. 41). The idea of boundary objects has seen a variety of adaptations, most recently as "boundary objects in sustainability transitions," see Franco-Torres *et al.* (2020).

Clark, B. (2019). Gaia as Cyborg. Available at: https://www.gaian.systems/research/gaia-as-cyborg.

Clark, B. (2020). Gaian Systems: Lynn Margulis, Neocybernetics, and the End of the Anthropocene. Minneapolis, MN: Minnesota University Press.

D-CENT. (2014). *Collective Intelligence Framework in Networked Social Movements.* Barcelona: D-Cent. Available at: https://dcentproject.eu/wp-content/uploads/2015/09/collective_intelligence_framework.pdf

Franco-Torres, M., Rogers, B.C., and Ugarelli, R.M. (2020). A Framework to Explain the Role of Boundary Objects in Sustainability Transitions, *Environmental Innovation and Societal Transitions*, *36*, 34–48.

Hardin, G. (1998). Extensions of the 'Tragedy of the Commons', *Science*, New Volume, *280*(5364), 682–3.

Hasenpusch, M. (2018). Die Dritte Bauernbefreiung [The third liberation of peasants], *Spectrum*, 12–16. Available at: https://www.deutsche-leasing.com/site/DL.com/get/documents_E-466730824/cms/downloads/Unternehmen/spectrum/DL_SPECTRUM_56_web1_Einzel.pdf

Howser, G. (2020). Computer Networks and the Internet: A Hands-On Approach. Springer.

Hörl, E. (2015). The Technological Condition, transl. Anthony Enns, *Parrhesia*, *22*, 1–15. Available at: https://www.parrhesiajournal.org/parrhesia22/parrhesia22_horl.pdf

Ibarra, I.A., Goff, L., Hernández, D.J., Lanier, J., and Glen Weyl, E. (2018). Should We Treat Data as Labor? Moving beyond "Free", *AEA Papers and Proceedings*, *108* (May 2018). Available at: https://www.aeaweb.org/articles?id=10.1257/pandp.20181003

Igenos. (2018). Alt Aber Sexy: Unsere Rechtsform Genossenschaft – Darum verdient der dritte Weg eine zweite Chance!, Diskussionspapier [Old but sexy: Our legal form coopertive – that is why the third way needs a second chance!]. Available at: https://www.igenos.de/wp-content/uploads/2018/07/igenos-AP-1.4-13.07.2018-1.pdf

Jones, P. (2018). Contexts of Co-creation: Designing with System Stakeholders. In Jones, P., Kijima, K. (Eds.), *Systemic Design: Theory, Methods, and Practice*, 3–52. Tokyo: Springer, Japan.

Kostopoulos, L. (2021). *Decoupling Human Characteristics from Algorithmic Capabilities*, IEEE. Available at: https://standards.ieee.org/initiatives/artificialintelligence-systems/decoupling-human-characteristics.html

Latour, B. (2004). Why Has Critique Run out of Steam? From Matters of Fact to Matters of Concern. *Critical Inquiry*, *30*(Winter 2004), 225–48.

Latour, B. (2016). Why Gaia is not a God of Totality, *Theory, Culture & Society*, *34*(2–3), 61–81, 72. doi:10.1177/0263276416652700

Latour, B., Schaffer, S., and Gagliardi, P. (Eds.). (2020). A Book of the Body Politic: Connecting Biology, Politics and Social Theory. *San Giorgio Dialogue 2017.* Venice: Fondazione Giorgio Cini.

Lovelock, J. (2019). *Novacene: The Coming Age of Hyperintelligence.* London: Penguin.

Lyle, P., Sciannamblo, M., and Teli, M. (2018). Fostering Commonfare. Infrastructuring Autonomous Social Collaboration, In *Proceedings of the 2018 CHI Conference on Human Factors in Computing Systems* (*CHI'18*). Association for Computing Machinery, New York, NY, USA, Paper 452, 1–12. DOI: https://doi.org/10.1145/3173574.3174026

Malik, S. (2020). You Can't Sell Your Kidneys! Responding to Objections to Data Ownership, *Streamr Blog* (2020-07-02). Available at: https://blog.streamr.network/objections-to-monetising-your-data

Mersch, D. (2013). Ordo ab chao – Order from Noise. Berlin: Diaphanes.

Miller, R. (Ed.). (2018). Transforming the Future: Anticipation in the 21st Century. New York, NY: Routledge.

OPENDEI. (2021). *Design Principles for Data Spaces*, Position Paper #1 (April 2021). Available at: https://design-principles-for-data-spaces.org

Poli, R. (Ed.). (2019). Handbook of Anticipation: Theoretical and Applied Aspects of the Use of Future in Decision Making. Springer.

Posner, E. and Weyl, G. (2018). Radical Markets. Princeton: Princeton University Press, 44. Available at: http://assets.press.princeton.edu/chapters/s11222.pdf

Reay, D. (2018). Very Cooperative: How Germany Invented the Sharing Economy, *Handelsblatt* (2018-08-04), Today. Available at: https://www.handelsblatt.com/today/finance/very-co-operative-how-germany-invented-the-sharing-economy/23581748.html

Ringle, G. (2018). Verfremdung der Genossenschaften im Nationalsozialismus: Gemeinnutzvorrang und Führerprinzip, *Wismar Discussion Papers* 01, Hochschule Wismar, Wismar Business School. Available at: https://ideas.repec.org/p/zbw/hswwdp/012018.html.

Ronstedt, M. (2020). It's Time to Build Data Unions, *Streamr Blog* (2020-10-05). Available at: https://blog.streamr.network/its-time-to-build-data-unions

Rouvroy, A. (2016). 'Of Data and Men'. Fundamental Rights and Freedoms in a World of Big Data. Report to the Council of Europe's Directorate General of Human Rights and Rule of Law. Available at: https://rm.coe.int/16806a6020

Rouvroy, A. (2020). Algorithmic Governmentality and the Death of Politics, *Green European Journal.* Available at: https://www.greeneuropeanjournal.eu/algorithmic-governmentality-and-the-death-of-politics

Ruhaak, A. (2020). Surveillance Capitalism Wasn't Built by Powerful Companies Alone, *Center for International Governance Innovation* (2020-11-18). Available at: https://www.cigionline.org/articles/surveillance-capitalism-wasnt-built-powerful-companies-alone

Sand, M. (2019). On 'Not Having a Future', *Futures*, *107*, 98–106.

Savona, M. (2020). Governance Models for Redistribution of Data Value, *VOX* (2020-01-10). Available at: https://voxeu.org/article/governance-models-redistribution-data-value

Shah, H. (2018). Use Our Personal Data for the Common Good, *Nature* (2018-03-28). Available at: https://www.nature.com/articles/d41586-018-03912-z

Star, S.L. (1989). The Structure of Ill-Structured Solutions: Boundary Objects and Heterogeneous Distributed Problem Solving. In Hubs, M. and Gasser, L. (Eds.), *Readings in Distributed Artificial Intelligence* II, 37–54, 37. Menlo Park, CA: Morgan Kaufmann.

Steyerl, H. (2006). The Language of Things, *transversal* 06. Available at: https://transversal.at/transversal/0606/steyerl/en

Sznel, M. (2020). Tools for Environment-centered Designers: Actant Mapping Canvas. Available at: https://uxdesign.cc/tools-for-environment-centered-designers-actant-mapping-canvas-a495df19750e

Tokarczuk, O. (2019). Nobel Lecture, Nobel Laureate in Literature 2018. Available at: https://www.nobelprize.org/uploads/2019/12/tokarczuk-lecture-english.pdf

Tsukayama, H. (2020). Why Getting Paid for Your Data Is a Bad Deal, *EFF* (2020-10-26). Available at: https://www.eff.org/deeplinks/2020/10/why-getting-paid-your-data-bad-deal

Watts, J. (2020a). James Lovelock: 'The biosphere and I are both in the last 1% of our lives', *The Guardian* (2020-07-18). Available at: https://www.theguardian.com/environment/2020/jul/18/james-lovelock-the-biosphere-and-i-are-both-in-the-last-1-per-cent-of-our-lives.

Watts, J. (2020b). "Bruno Latour: 'This is a global catastrophe that has come from within', *The Guardian* (2020-06-06). Available at: https://www.theguardian.com/world/2020/jun/06/bruno-latour-coronavirus-gaia-hypothesis-climate-crisis

Weyl, E.G. and Posner, E.A. (2018). Radical Markets: Uprooting Capitalism and Democracy for a Just Society. Princeton, NJ: Princeton University Press.

Zehle, S., Kollegala, R., and Crombie, D. (2021). Co-creating Value with the Cooperative Turn: Exploring Human-Machinic Agencies Through a Collective Intelligence Design Canvas. In Ahram, T., Taiar, R., and Groff, F. (Eds.). *Human Interaction, Emerging Technologies and Future Applications IV. IHIET-AI 2021. Advances in Intelligent Systems and Computing*, vol. 1378. Springer, Cham. Available at: https://doi.org/10.1007/978-3-030-74009-2_20

DOI: 10.1561/9781680838411.ch13

Chapter 13

Mapping the Landscape of Sharing and Cooperativism for Design Research and Practice

By Özge Subaşı, Anton Fedosov and Oliver Bates

Emerging studies of local cooperatives, their sharing practices, and the use of platforms for cooperation call for specific designs and design guidelines to support the endurance and growth of a community-oriented collaborative economy. These efforts also indicate that design has the potential to shape cooperative engagements. However, to-date, only a few design resources are tailored for exploring and further developing design insights from empirical and conceptual research on sharing and cooperativism. To bridge this gap, we report on an international workshop that included a diverse group of scholars, designers, and activists. During the workshop, we aimed to unpack the role of design regarding sharing and cooperativism. Through the synthesis of workshop outcomes, we present new insights pointing towards the development of an ecosystemic approach in design for cooperativism.

We call on designers to (1) proactively adopt design goals that focus on ecosystemic design and tools; (2) be inclusive, equitable, and justice-oriented to ensure solidarity and collectivism; and (3) rethink terms such as currency and data while designing for cooperativism in their projects. In this chapter, we conceptualise and discuss the key ideas from the workshop highlighting potential implications for design research and practice.

13.1 Introduction

Sharing as a form of social exchange existed long before the invention of online sharing platforms (Belk, 2010; Cook *et al.*, 2013). The sharing economy – following a model of sharing idle capacities such as time, lodgings, cars and skills – has been fast-growing, incorporating many technology-centric and technology-driven designs and platforms (Botsman and Rogers, 2010; Belk, 2018). Recent studies on several sharing platforms indicate a number of design challenges associated with the emerging platforms such as discrimination of people of colour as entrepreneurs e.g., as Airbnb flat hosts (Edelman and Luca, 2014), inequality, distrust and safety concerns for disadvantaged populations (Dillahunt and Malone, 2015), the insecurity and inequality created via on-demand work practices (Alkhatib *et al.*, 2017; Dillahunt *et al.*, 2018; Dumančić *et al.*, this volume), the high emotional labour required to deliver a service (Lutz *et al.*, 2018).

As an alternative, sharing systems and digital services with a community orientation hold a promise to incorporate new ideas beyond existing financial and monetary platforms, such as creating micro-enterprises and collaborative currencies (Carroll and Bellotti, 2015), activating local neighbourhoods (Fedosov *et al.*, 2021a) and fostering community resilience (Light and Miskelly, 2015). There has been a significant amount of prior research in media and communication scholarship (e.g., John, 2017; Kennedy, 2016) as well as in (social) computing research (e.g., Fedosov, 2020) around the alternative versions of sharing beyond just repurposing idle capacities. The key factors to the idealised social perspectives of the sharing (economy) services lie in their relation to human's reciprocal capacity as untapped resources (John, 2017), and their potential for social change (Bellotti *et al.*, 2014).

The collective, and local aspects of sharing are essential and valuable qualities for the future of sharing cultures (e.g., Light and Miskelly, 2015). To minimise the inequalities, and build more sustainable futures, we arranged our workshop around **the social benefits that emerge over time from local sharing initiatives (Light and Miskelly, 2015) rather than concentrating on the idle capacity, and its re-use to form new monetary economies** (Light and Miskelly, 2019).

Inspired from the earlier work on commons, a resource shared by a group of people, Ostrom introduced an early set of design recommendations for shared resource systems (Ostrom, 1990). Ostrom's design principles were about defining boundaries, matching rules and conditions to community needs, giving the members the right to modify the established rules, handling conflict resolution, providing options for monitoring and sanctioning, and attending to the other governance activities (Hess and Ostrom, 2007). In their work on designing technologies for valuing social cooperation as a common good, Bassetti *et al.* (2019) emphasize the role of design for nurturing sharing and autonomous cooperation, cooperation that can point to social change, which are not yet disciplined by the capital (ibid.). They suggest that a platform is a space in which social cooperation emerges autonomously, and to achieve this, designers should create environments for common resources and common values to flow. In turn, Nardi (2019) calls for designers to take a stance and shoulder responsibility in various efforts within post-growth economies, which offers a contrasting narrative to mainstream platform capitalism.

More recently, Botero *et al.* (2020) questions the role of participatory designers in contributing to commons-based forms of organising in the following ways: (a) to design better infrastructures and vocabularies for commoning practices, (b) to connect the vocabularies to the other transformative movements such as feminism, environmentalism, indigenous movements, and (c) to deal with **the contradictions that arise from cultivating commons in capitalist and individualistic cultures**. With the advent of digitalization, platform cooperativism (Scholz, 2016) describes how to organise sharing in the context of community by distributing ownership, establishing democratic governance, and reinvigorating solidarity (Scholz and Schneider, 2017). By the same token, Lampinen *et al.* (2018a) outline how the more traditional forms of organizing (e.g., member-owned cooperatives) deal with the challenges of digitalization.

In this chapter, we concentrate on **the role of design in building sustainable systems for cooperativism**. As an example, a small body of work on sharing and cooperativism within the Human-Computer Interaction (HCI) scholarship shows the potential for new systems, platforms, and ecologies to harness social and ecological sustainability, as well as engender trust in platforms and communities through design (Dillahunt *et al.*, 2018; Glöss *et al.*, 2016; Katrini, 2018; Lampinen *et al.*, 2013; Light and Miskelly, 2019; Fedosov *et al.*, 2021a). This recent turn in design sets the expectation that designers take responsibility by asking who will benefit from the output, and how to address the problems of ecology and inequality caused by profligate development of contemporary consumption practice (Nardi, 2019). Moreover, designers and design decisions have potential in ensuring sustainable development of non-profit resource sharing arrangements (e.g., see Fedosov *et al.*, 2018, 2021a). We aim to foster such a discussion by presenting the outcomes of a design workshop (Subasi *et al.*, 2020) focused on **unpacking decisions made in the design of platforms (e.g., regarding governance structures, platform ownership, privacy strategies for membership) and their relevance in the context of sharing and cooperativism.**

13.2 The Workshop

We conducted a design workshop at the NordiCHI 2020,[1] the main Nordic forum for Human Computer Interaction research, with an aim to engage with sharing and cooperativism. By placing the existing design propositions with alternative economic models at the centre of our inquiry (Light, 2019), we drew upon three key aspects of sharing and cooperativism from the related work: (1) local and relational assets of sharing (Light and Miskelly, 2019); (2) global principles of sharing cultures (Katrini, 2018); and (3) designing beyond markets (Lampinen and Brown, 2017). The aim of this workshop was to illustrate the common threads relating to the question: *how can designers contribute to devising mechanisms for cooperativism and sharing?*

13.2.1 Background

Our work built upon previous design workshops and community efforts to foster collaboration within HCI. The specific examples include: researcher organised panels on the sharing economy (e.g., Lampinen *et al.*, 2015), formulated design statements regarding sustainable sharing in local communities (Malmborg *et al.*, 2015), examinations of power aspects and asymmetries around digitally-mediated labour (Lampinen *et al.*, 2018b), and suggested research agendas around themes of "Collaboration" and "Work" with respect to sharing economy platforms (Lampinen *et al.*, 2016). Fedosov *et al.* (2019a) developed upon these themes by considering the design of platform cooperatives as a beneficial form of the sharing economy. They organised a special interest group at the CHI 2019, a premier international conference of HCI, to provide multidisciplinary perspectives on challenges faced by platform co-ops and opportunities that accompany them. Their outcomes informed our composition of the critical prompts to the participants of the workshop.

In our design workshop, we employed the Sharing Economy Design Cards (Fedosov *et al.*, 2019b) to facilitate hands-on activities where participants were prompted to critically reflect on the existing platforms. Example questions were: Who benefits from platforms? Who is harmed by the platform? (see Design Justice

1. https://nordichi2020.org/

Principles: Costanza-Chock, 2018; N.A., 2018). Our expectations for the workshop were twofold: (1) to share our collective experiences on cooperativism, and (2) to reflect on how cooperativism can help in the design of platforms and technology through imagining how the alternative economic models may work alongside the dominant system of platform capitalism.

13.2.2 Setup and Stages

The workshop tools and materials included video conferencing, several scaffolded Miro boards with vector design elements and prompts, and digital workshop materials such as post-it notes, pen, paper, as well as shared textual documents for note-taking. The 4-hour long workshop was conducted with the help of three facilitators, three note-takers and a technical assistant alongside 15 participants.

We distributed the call for participation through mailing lists, former workshops (Fedosov *et al.*, 2019a; Korsgaard *et al.*, 2020), and multiple community channels such as the Sharing and Caring COST Action, the Sustainable HCI community, the CoTech network, and the LIMITS workshop series.[2] We received contributions from scholars, designers, practitioners, and activists, covering topics such as: platforms, politics, design, governance, infrastructures, markets, data privacy, and fairness.

We asked participants to suggest one platform and to prepare a 1-minute presentation including a critical question related to the themes of the workshop. After receiving the final submissions, we grouped the submissions to the following three descriptive themes: (1) "Platforms, coops, governance", (2) "Models, markets and sustainability", and (3) "Design". Our workshop consisted of an introductory session, a set of hands-on activities on the whiteboard, and a synthesis session. We prepared a running slide setting from the participants' presentations and the virtual environment for the design workshop with Miro boards and virtual prompts. We prepared boards with three main activities: (1) Onboarding, (2) Deconstructing a platform, and (3) Towards Design Justice (Figure 13.1). For the details of the workshop's schedule, see Subasi *et al.* (2020), and the corresponding website[3] for the accepted position papers. Below we briefly outline the three hands-on activities on Miro.

Stage I: Onboarding. This stage ensured that all participants were familiar with practicalities and with the workshop's thematic orientation. We then split participants into three breakout groups relating to their position papers. In each

2. http://sharingandcaring.eu/about-sharing-and-caring; https://groups.google.com/forum/#!forum/sustainable-chi; https://www.coops.tech; https://computingwithinlimits.org

3. https://sharingcoopnordichi2020.wordpress.com

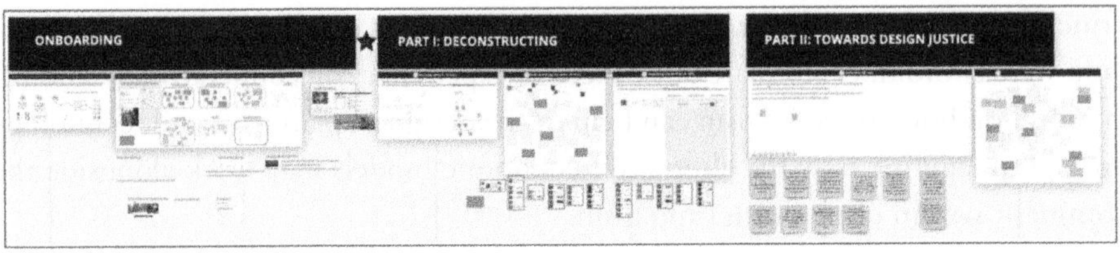

Figure 13.1. Overview of a Miro board used in the design workshop.

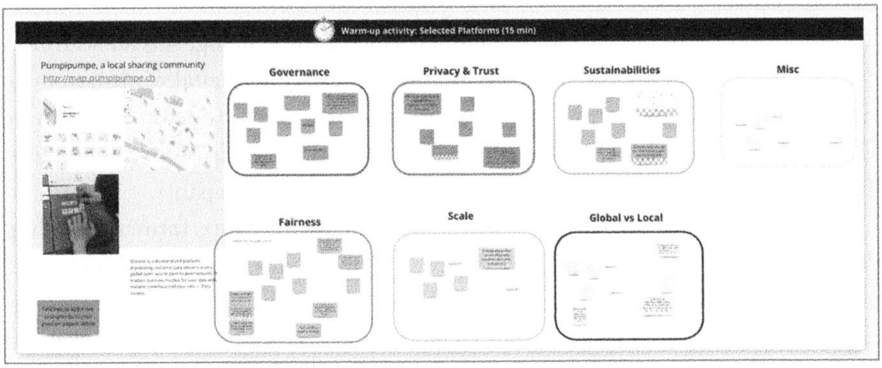

Figure 13.2. Onboarding stage: Categories for initiating discussions.

group facilitators scaffolded examples of the related platforms based on the participants' suggestions. This stage also included a warm-up activity: to sort pre-seeded platforms to construct categories of Governance, Privacy & Trust, Sustainability, Fairness, Scale, Global vs. Local, and Miscellaneous (Figure 13.2) to remind the participants about these categories. We adapted these categories from the outcomes of a previous CHI-community gathering on this topic.[4]

Stage II: Deconstructing. The deconstructing stage helped the participants to concentrate on the specific characteristics of the reviewed platforms from the author's position papers. In each group participants nominated a platform (or a group of platforms), analysed the interaction mechanisms and interfaces of the suggested platforms, deconstructing underlying principles that lead to the core platforms' features. The deconstructing board drew on the Sharing Economy Design Sprint methodology (e.g., Fedosov *et al.*, 2019b) and included three distinct areas to support understanding, mapping and improving activities of the sprint (see Figure 13.1). We also integrated the card-based toolkit for the sharing economy (Fedosov *et al.*, 2019b) to facilitate generative design activities (see Figure 13.3).

Stage III: Towards Design Justice. The last stage of the workshop aimed to reflect on the topics synthesized from the previous stage. The board included several

4. https://medium.com/acm-chi/talking-cooperativism-and-human-computer-interaction-c3df2929b4b4

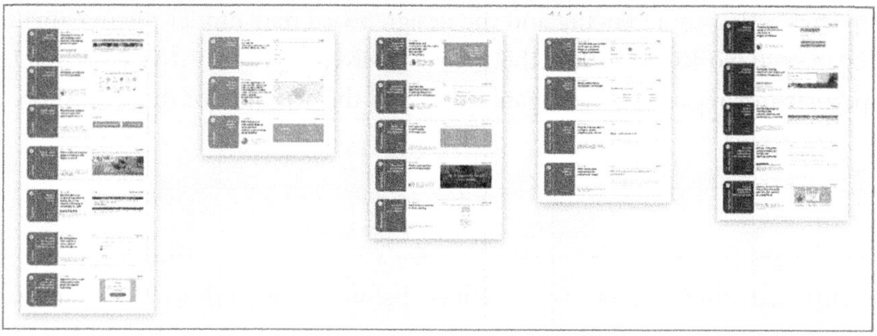

Figure 13.3. Sharing Economy Design Cards toolkit (Fedosov *et al.*, 2019b) were integrated to the boards before the design workshop.

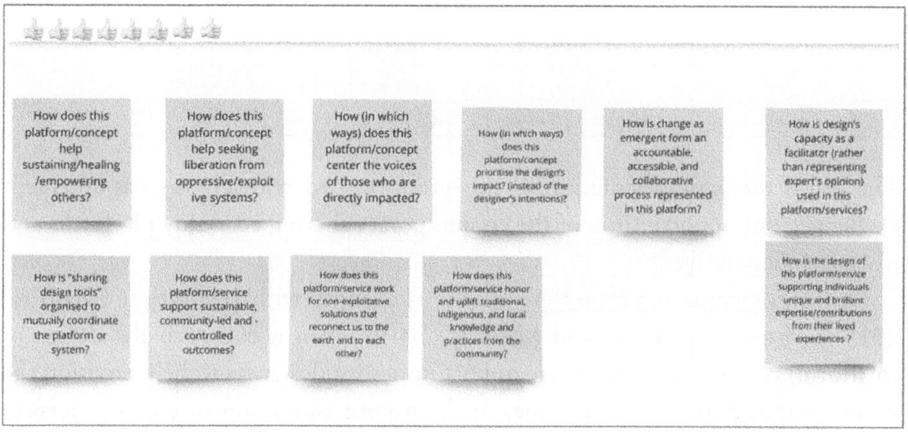

Figure 13.4. The questions from Design Justice Network integrated to the boards as virtual sticky notes.

design justice-related questions on the post-its notes (Costanza-Chock, 2018; N.A., 2018). Participants were asked to elaborate on the most relevant aspects pertinent to the themes of our workshop in relation to the selected platforms at hand (see Figure 13.4).

13.2.3 Limitations

The design workshop was held remotely, using multiple tools. We carefully designed an inclusive procedure and conducted a dry run to ensure a comfortable experience for our participants. The tools selected, their accessibility, the "time zone" factor (we had participants from 6 different time zones), and the duration of the workshop may have caused exclusion for some. During the workshop, the individual interests, the familiarity to tools and other unknown factors resulted in different levels of engagement. The moderation, the constant swapping between group chat, video conferencing and the design board may hinder the maximal efficiency in such workshops. Nonetheless, we received highly positive feedback from the participants regarding the organization and the activities of the workshop.

13.3 Results

As mentioned, three themes were defined before the workshop according to the content of the submitted papers: (1) "Platforms, coops, governance", (2) "Models, markets and sustainability", and (3) "Design". Below we present the results from the hands-on session for each theme. The reader should note that, we use the original language and definitions of the participants with a minimum interpretation of the terms, such as "capitalism", "middle-class", "feminist movements" and so on.

13.3.1 Platforms, Coops, Governance

The broad thematic orientation of this group was to critically discuss emergent challenges and opportunities of the contemporary sharing economy platforms and platform cooperatives.

Our participants have emphasised the challenges with replication (Manzini, 2015) and meshing (Light and Miskelly, 2019) when it comes to the design of (platform) cooperatives. There are opportunities for designers to help envision the potential trajectories when it comes to the scaling aspirations of coops to answer questions like "*What levels of scale are appropriate and desired for coops?*", "*What are the boundaries of scale?*" Furthermore, designers may play a key role in creating reusable design materials (e.g., templates, worksheets, how-to guides), design patterns as components that can be adapted across various instances of platform coops in their cultural, linguistic, and geographical contexts. For example, the Mobility Factory (http://www.themobilityfactory.eu), a European cooperative enterprise, develops software for electrical car-sharing cooperatives, which is currently used by 11 coops from four countries. One prominent example of use is Som Mobilitat (https://www.sommobilitat.coop), a Catalan non-profit electric car-sharing cooperative with over 1600 registered members. While Som Mobilitat deliberately aimed at developing their activity only regionally in Catalonia, they helped other cooperatives adapt their model in different regions in Spain.

Our participants further elaborated on fairness in design for such cooperative platforms: "*How can designers ensure the participation of women in spaces of governance and as users of the service?*". Currently, only a very few women use the Som Mobilitat platform and none are in a position of power to influence decision-making at the design phases of the new features/products. This renders questions around diversity, inclusion, fairness, and participation critical at the early stages of establishing a coop. What is more, in the efforts of supporting various non-profit sharing economy initiatives to be more sustainable and resilient, the outputs of the future designs can also aim to balance between customization and abstraction required to make them appealing to other initiatives with helping them to govern their shared resources beyond the original context of use. In other words, a car sharing platform developed in Belgium may be reused, adapted, and customised by a tool sharing coop in Switzerland.

The participants have also added to the importance of adapting existing approaches to the design (e.g., the Google Design Sprint) to the needs of the community. For example, for distributed cooperatives such as Karrot (https://karrot.world/), a food sharing platform, it remains an open challenge (to run generative activities across different locations). Karrot serves groups of people who want to save food from being wasted and aim at sharing it instead, allowing these groups to organise efficiently, on a local and voluntary basis. One platform's designer emphasised that adopting the sprint in designing governance features and processes of the platform takes significantly more time and resources than for profit-driven platforms. Hence, there is a need for new design methods and approaches, which embrace the volunteering nature and the context of the coop's activity. One may ask: *"How does a design approach specifically support local groups to be more sustainable and resilient?", "How can we build sustainable models for communities with small numbers of participants?"*. We thus envision three salient avenues for design research and practice:

- developing new design methods taking into consideration time-limited engagement in the community with different stakeholders and levels of engagement,
- devising techniques to attain to longer turn-around design cycles; and
- addressing the need for defining models of governance and ownership early on, while incorporating different stakeholders.

A few participants elaborated that open-source tools for community governance already exist. For example, one participant has developed CommunityRule (https://communityrule.info), an online tool to facilitate democratic governance by utilizing pre-defined and customizable templates based on common practices of informal online communities. Furthermore, other computational tools, like PolicyKit (Zhang *et al.*, 2020), have been recently devised by scholars in computing, drawing on established theoretical frameworks for describing governance arrangements, e.g. from early design guidelines of Ostrom's Institutional Analysis and Development on commons (1990), which suggests the right of community members to formulate their own rules. Still, those visual tools used specific domain terminology, as well as often require technical expertise to set them up and to author policies. Drawing upon the concept of "technological sovereignty" and using a cooperative mindset, future technological coop ecosystems may also be based on alternative technology stacks. One example that we have examined within this workshop was DAOstack (https://daostack.io/), an open-source software stack designed to support a global collaborative network. The stack can be used to build organizations for any kind of collective work, and it also contains tools to link these organizations together, so as the network grows, all its member organizations are strengthened. The designers of online governance tools need to create usable and useful wrappers around emerging powerful computational toolkits for easy adoption and use in the community's practices. **By allowing flexibility of governance, tool design would help accommodating circumstances where initiatives may change their rules based on learning from each other**.

A better understanding of the potential value of the distributed technologies can prompt various monetization opportunities that are in line with their values (e.g., the immutability of data) and principles (e.g., of autonomy), with the caveat that some of those general-purpose distributed technologies may intensify requirements on access when it comes to scaling (e.g., music streaming services over blockchain may be prone to latency). Along these lines we have examined Streamr (https://streamr.network/), a decentralised platform distributing real-time data streams over a global open-source peer-to-peer network. It enables business models for user data with scalable crowdsourced data sets – so-called – *data unions*.[5]

Finally, exploring further the governance considerations in platforms' design, the workshop participants have discussed the lack of appropriate governance of the current sharing economy platforms' owners to ensure adequate "parity between those that use them and those that build them" (Jamieson and Wilson, 2020). Specifically, the participants introduced a conceptual framework for data trusts, "a legal structure that provides independent stewardship of data" to address "platform data tax", the exchange of service for personal data (Jamieson and Wilson, 2020). Similarly, MyData Global, a non-profit initiative, puts this in practice with the purpose to "empower individuals by improving their right to self-determination regarding their personal data".[6] Design for transparency and equitability becomes central in the context of neoliberal capitalism. The specific questions designers may engage with include *"At what stages of interaction with the platform data trust designers should particularly give voice to their users?"*, *"What are we centring when it comes to user*

---

5. https://blog.streamr.network/its-time-to-build-data-unions/

6. https://mydata.org

experience interacting with data trusts", "*Is it reasonable to pay for data handling?*", "*Who is to be held accountable for data processing and handling in the sharing economy?*", and, finally, "*How (local) platform (coops) can have fair access to the instruments to handle 'data trust' containers given the limited resources they may possess?*" Last but not least, designers of future platform cooperatives may reconsider centring on the cooperative model itself but, rather, focus on supporting users' everyday experiences in the communities.

13.3.2 Models, Markets and Sustainability

The initial discussion from this group covered a range of topics: digitisation making sharing harder (on purpose); the non-monetary value that communities add; the role of markets in the sharing economy; and cooperation between precarious workers and shrinking local economies. The discussion moved onto **how non-monetary value is added by communities, and the difficulties in challenging dominant market views of community created, non-monetary value. Community organizations who choose to not participate in capitalist economies are often taken less seriously, presenting barriers to scaling up such initiatives and alternative models**; "*if you're not a business you're not taken seriously in the first place*". The discussion turned to a tension between a growing precarious workforce and the spaces that they find themselves working in. Community owned cooperatives are a valuable way for workers and local organizations to come together and resist forms of precarity whilst challenging dominant economic and technological systems and pressures. This led participants asking, "*What is non-capitalist technology?*", "*What's the model of scaling that is not capitalist?*" A brief discussion followed on how open source models can be capitalist, the need for a non-capitalist culture to make space for models without transactional relationships, the co-option of non-capitalist ideas (e.g., the sharing economy), and using capital to fund non-capitalist models. The discussion concluded on a key point, ***those looking to build non-capitalist platforms can learn from "feminist movements"*** *as it was noted that these movements (alongside others) have a rich history of successful activism and fighting for change.*

During the design activity participants worked through their nominated platforms: Meetup, Karrot and Peerby (https://www.peerby.com/). Peerby, a Dutch-based peer-to-peer platform for lending and borrowing household items, was seen as a "classic sharing platform" and selected as the platform to focus on. Parallels were drawn between Peerby and Library of Things (https://www.libraryofthings.co.uk/) prompting a reflection that users come for the community value aspect (e.g., "sharing of stuff", reduced waste and clutter) whilst the platforms typically become involved in sharing communities for monetisation and profit. The group asked a range of critical questions and reflections highlighting the lack of community, the expensive per-day pricing model, the lack of price regulation, and questioned the true overheads for lenders. Peerby's user base is local, Dutch speaking users who want to share tools and a range of consumer goods to other users who want to borrow things, as well as platform investors. Users (borrowers and sharers) of the platform were seen to be motivated to use it because of proximity to other users and to declutter or reduce purchasing behaviours. In terms of key privacy and trust issues there were concerns identified around whether items would be returned in working order, when there is a dispute who does the platform side with, and insurance. The simple interaction flow was seen as a positive aspect of the platform's user experience. It was however criticised for only being in Dutch, being geographically limited (note that the previous iterations of the platform design had multilingual interface and supported users across Europe (Stofberg and Bridoux, 2019)), having too many questions in order to sign up, and requiring a Facebook account to gain access.

During the "improve the service" task (a subcategory of Stage II: Deconstructing), the following note was written by a participant, summarising the feelings of the group: "*We know this platform isn't going to have technological solutions to social or community problems*". All participants were in agreement that the platform was profiteering on something that could be low tech, low cost and better oriented to building communities and social value. The gaps identified were building consumer trust in the sustainability of local business, trusting lenders and borrowers, creating trust through community boundaries, getting to know the community in meaningful ways, understanding the needs of the community (not just those who can afford to use the platform), and the affordability for those who need to borrow things. The potential improvements described by the group were:

- including sustainability ratings of commodities on the platform
- ratings for borrowers and lenders to help build community and trust
- connecting platforms to existing trusted online communities
- building, creating, and supporting community institutions – helping community members get to know each other
- spending time understanding and designing for the needs of the community
- regulating the market to ensure that lending is affordable for the user base

In Stage III participants highlighted a range of issues that cooperative design can help consider models, markets and sustainability when designing platforms. Platforms like Peerby are designed in such a way that they leave out those who have little or no disposable income, who perhaps have different needs for accessing shared commodities. In terms of sustainability, Peerby does not consider the indirect impacts of their platform on manufacturers or local businesses who sell goods and items that are shared by the platform. Whilst there are net gains regarding environmental externalities due to less waste and less manufacturing of products, voices that care about sustainability are not considered in the platform. The value of sharing, cooperation and building communities is missing. This means that the additional value of borrowing or lending a commodity, such as learning about the items' quirks or history, aren't considered in design, although deemed highly beneficial for community building (Fedosov *et al.*, 2018). Whilst a tool like Peerby might encourage participation in sharing networks, the group wondered whether the platform could encourage the "middle-class" to think about helping others and community building, shifting them away from the focus of making money on their commodities that are currently unused. The final reflection from the group was that *platforms and services like Peerby tend to focus on commercialising the (materialistic) sharing of commodities, whilst not designing for the individuals' unique expertise and contributions surrounding both the commodities themselves or user's capacity to build communities and share expertise (cf. Library of Things).*

The group discussed the role of design justice when designing platforms recognising the need for the monetary transactions as insurance on items to ensure that they came back in the same condition or were returned at all, and that this could be overcome if a community institution was developed to teach, share costs, and share value without the transactional overheads. During this discussion, the participants highlighted:

- Platforms like Peerby could be utilised as a **stepping stone for the "middle-class" to participate more equitably in the sharing economy**.
- Ensuring platforms are governed by the community itself is key as third parties lack the shared needs, values and concerns of communities particularly where sharing commodities have greater social value than monetary value.
- Regulation is needed to combat platform providers also manipulating the prices.
- Not everything needs to be lent through a formal or binding agreement.
- Platforms such as Peerby are utilised by those who have power, commodities and want to see quick profits. These platforms could be seen as a distraction from the profound change that "we" need.
- Capitalism was seen at fault for distorting the true social impact of sharing and cooperative communities.

13.3.3 Design

The focus of the design breakout group was to discuss the emerging design related opportunities for platform cooperatives. Initial discussions were about **the role of**

design in decentralization, and the move from capitalist to cooperative design. The group chose to focus on deconstructing two platforms: Commonfare (https://commonfare.net/), a platform for everyone to share experiences, foster ventures, and connect with people who want to support each other and Karrot, a food sharing platform. The outcome of the group trying to answer two key questions *"What does it address?"* and *"Who is it for?"* was a visualisation of common and diverse points, rather than two separated evaluations (see Figure 13.5). While it was clear that both platforms had similar goals of "long term sustainability", "overcoming poverty" or "developing and maintaining a community", upon closer analysis of the interfaces it was revealed that the two platforms have diverse strategies over anonymity, non-hierarchical conflict resolution, and communication to other platforms. The two most important themes discussed for design were extended from the design goals of the Sharing Economy Design Cards: (1) "Building a scalable common space" and (2) "Defining the design rules of trust across different cultures".

For the *scaling and design* task, several questions posed by the participants were left unanswered such as *"How a systemic change can be pushed forward?"* For a systemic change, open-source was found to be the most prominent strategy among existing approaches to the sharing economy. A major drawback is the funding model. The participants valued an inclusive approach for "people at risk" and the efforts for "a better cooperative future" such as a democratic governance, and social solidarity. The wish for ending poverty and ending social exclusion were repeated for both platforms. Further design questions discussed were: *"How to ensure protecting anonymity as the highest priority?", "How to create systems that collect as little data as possible?" and "How to deal with distrust of institutions?"* **An option for an anonymous sign up, reducing the amount of data collected, or similar requirements were mentioned as design goals that can possibly be realized easily**, as they do not need much contribution from the platform users. The participants heavily critiqued the issue of defining the value, e.g., a currency. Here, it would be important to motivate a transition in how people deal with money. A transparent system can be built that motivates its members towards a new understanding of money and currencies. Previous examples of community currencies (e.g., from Commonfare) can be taken as a departure point to explore the meaning of currency at scale.

The second theme concerned *defining and designing for trust across different cultures.* It was inspired from the Karrot's interface, the participants thought of different models for deconstructing authority. For instance, "*Who decides whether you can join a group and where do we get this information?*" Is it a good model to refer to community peers or contacts via existing community members (e.g., to have a confirmation from two other members to be able to sign in?), and what if a prospective community member does not have a prior contact? Who is the authority to confirm such applications? Shouldn't this be mentioned in the interface? Another key

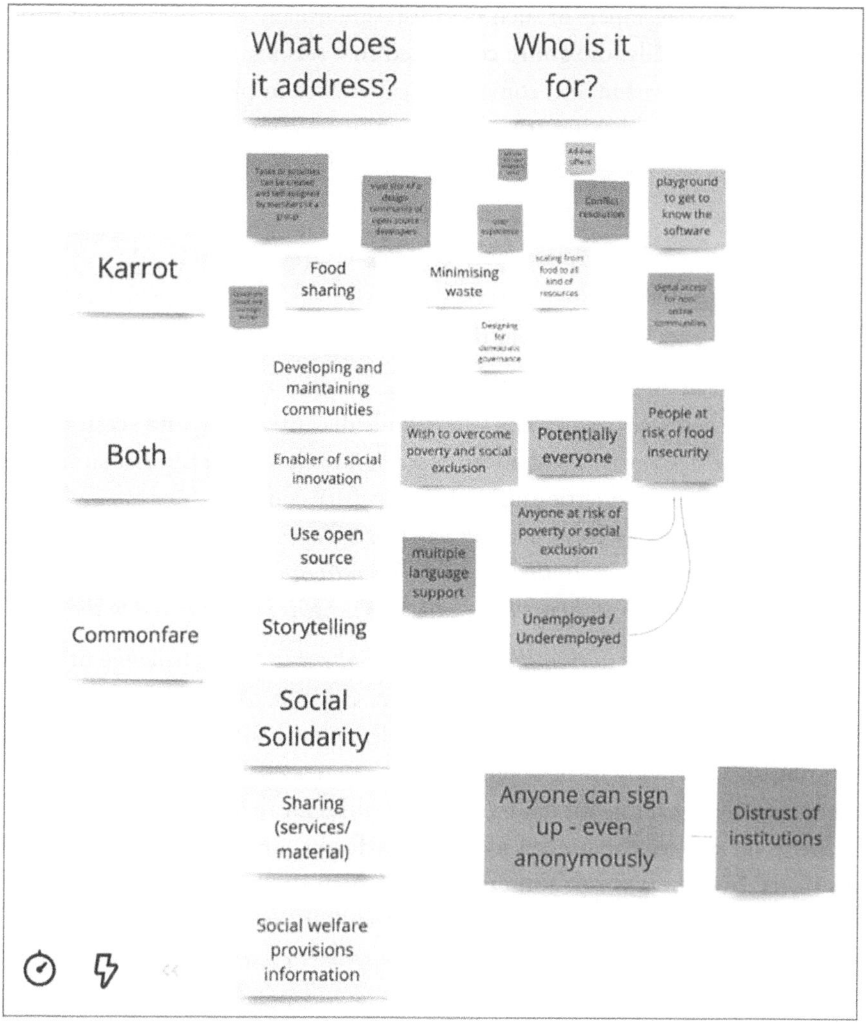

Figure 13.5. A section from the Miro board of breakout group "Design". The comparison of two platforms from the same domain (food sharing) against their target audience and main goal.

issue was forced digital exclusion. "*How do the platforms address non digital users?*" Here, participants pointed to different cultures, and the potential harm that can be caused for local and remote communities, especially outside of Europe. The participants provided instances of the mechanisms that accommodate **hybrid ways of addressing inclusion such as integrating an option of reaching out by phone or holding physical meetings**. Based on the example from Karrot, the participants regarded the efforts to emphasize the social aspects of a community.

From the perspective of platform workers (in this case developers), a vivid community is defined as a place where frustration and love are lived together.

For example, the developers of such platforms are usually very relaxed and patient with each other. Finally, the group concluded that seeking reciprocity in the work environment can be a fun and convenient experience in platform design and can contribute to the establishing trust relationships both in the community and in their supporting digital services.

13.4 Design Challenges and Key Considerations

The design challenges yielded from the whole session are summarised as follows:

Challenges in decentralising and scaling organization platforms. The participants mentioned several challenges, which can occur while **decentralising and scaling** a platform. These include the issues of interpersonal and systems trust and organizational problems. The research questions (RQs) in this space could be: *How can designs contribute to the scaling and decentralizing practices of new business, organizational and meta-organizational models in sharing and cooperativism? Are the process, effects, and interventions sustainable and fair? How can these insights be transferred to the designs?*

Lack of care mechanisms within the platform membership. One challenge of sharing and cooperativism is relying on the volunteers as key persons. There is a need for volunteer involvement, but also their knowledge, plus their efforts in tracking, documenting and archiving of new and emerging practices. However, the volunteers cannot have a central role for taking responsibilities for the whole change. **As coop platforms do not see themselves as work platforms, or work organizations; the volunteers' burden was not something for which the platforms were prepared for**. RQs here can be: *How should care be organised? Where should coop members turn to when something breaks? How can this be transferred to interfaces without conflicting with the inflation of volunteer work, or concerns around anonymity?*

Challenges for accommodating fluid and changing ideals. While exploring food sharing efforts of two different platforms (foodsharing.de, Lebensmittelretten.de), we concluded about the overarching idealism (solving food poverty) and its fluidity. Both **individuals and communities can change or revise their ideals**. At the same time, the practices may bring new ideals that may contrast with the starting point. For example, making food sharing platforms more visible and accountable, is an ideal that is shared by the majority of members but that is, in practice, contrasting with the initial ideals. RQs here can be: *How can platforms reflect the changing ideals of the community? How can platforms support the capturing of the change?*

Cooperativism can become an identity. The idea of associating self with cooperative acts as a part of one's identity was repeatedly brought to the table. This can be addressed in platform designs. However, systems should be built in a way that they support cooperations between people who do not support each other's self-identification or each other's ideas, but they can still share and cooperate. RQs: *How can we develop platforms that avoid normalising designs and thus can favour users' self-identification? How do we ensure that people aren't being "left out"?*

The need for a transition from capitalistic to cooperative platform. Critical positioning of elements such as currencies and the bigger infrastructures such as the open-source movement. Things that "look and smell like money" are difficult for non-capitalistic mindsets. Designers need to find ways of **building new communications that are framed around the related values, and not around the values of capitalistic systems**. While participants saw a great value in using open-source toolkits, several issues were raised as obstacles such as how can these solutions support design of a currency, something that only has a meaning in a closed system defined by fixed values. RQs: *How do we redesign data-driven systems to amplify our collective agency? How can we encourage the design of non-capitalist technologies alongside/in the context of capitalist markets? How can communities who add great non-monetary value (precisely because they don't have a business model) be prioritised in design? How can alternative currencies relate to markets / business models? How can trust be transferred when moving from one community to another? How could these work at scale?*

Our workshop highlighted several opportunities for design researchers and practitioners engaging with cooperativism. We offer three key considerations for designing the next generation services and interactions for sharing and cooperativism: (1) designing and evaluating with a focus on an inclusive ecosystemic approach; (2) resisting financial precarity, by prioritizing convenience, care and fun ourselves; and (3) designing beyond capitalistic systems.

13.4.1 Designers and Design Tools with a Focus on an Inclusive Ecosystemic Approach

Designers are accountable for what they choose to design politically (Nardi, 2019). The designers of sharing economy cooperatives and collectives should clearly emphasise social benefits for their local communities, where a service is envisioned to operate. Prior work on sharing economy pointed to the importance of strengthening the connection of ecological aspects and environmental sustainability (Dillahunt *et al.*, 2017), and to the importance of sharing relational assets that offer an ecology (Light and Miskelly, 2019). Similarly, our participants agreed that the new platforms (that aspire to scale or to create a new business model for cooperativism) should embrace the holistic goal of cooperativism, contributing and promoting an ecosystem in which that platform can survive. Our workshop showed that in the realm of cooperativism and sharing, the designers need to adapt to new forms of governing and ownership and consider new work models with a longer turn around for the design cycles.

Inclusion and equity have been core themes across groups repeatedly in user experience, service and systemic levels concerning class, access, cultures and geographies. Previous work illustrated hazardous results, if no inclusive measures are taken. The two classical examples are discrimination of people of colour as Airbnb flat hosts (Edelman and Luca, 2014), the insecurity and inequality created via on-demand work practices (Alkhatib *et al.*, 2017; Dillahunt *et al.*, 2018). Designers can help redistributing the position of power to influence decision-making at the design phases, and they can work explicitly not to exclude the most underprivileged ones (e.g., the ones who can benefit from lending platforms, but do not have anything to offer in return). Thus, **designers can employ an ecosystems-focused cooperativism stance to support inclusion, equity and justice. They can cooperate with each other to build actionable and open design systems and for valuing good practices.**

In practical terms, emerging approaches to design such as DesignOps[7] and Atomic Design[8] applying to the commons mindset can be seen as the vehicles to accommodate those efforts. Specifically, designers and designs can play a role for integrating our cooperative agenda into the future systems to make the future more inclusive and equitable (e.g., through creating transparent documentation of the process and reusable materials, and by offering services for multiple geographies via easy replication).

When designing with a social justice orientation, conflicts are not problems, but they should be taken as healthy signs that the project is tackling questions worth discussing (Dombrowski *et al.*, 2016). Many conflicts that were discussed in our workshop (e.g., around scaling and trust, the fluidity of the ideas, or cooperativism-as-identity) can be seen as examples of the healthy development process of justice-oriented sharing and cooperative platforms. Utilising the Design justice principles (Costanza-Chock, 2018) and the Sharing Economy Design Cards (Fedosov *et al.*, 2019b) as design tools was instrumental in focusing on the most affected, marginalised stakeholders, and rethinking designs. By asking justice-oriented questions for defining possible harms and benefits from the beginning, designers can go beyond seeking ecosystemic help and start designing their ecosystems for strengthening co-ownership, democratic governance, and solidarity (Scholz, 2016). **Designers can help coops adapt their systems in other localities and contexts of use, and help designing systems that are for both digital**

7. https://www.designbetter.co/designops-handbook

8. https://atomicdesign.bradfrost.com

and non-digital communities (like meetup groups and options to attend with a landline phone number). Yet, the process of transferring UX artifacts (e.g., designs) from one project/context to the other comes with its own challenges (Fedosov *et al.*, 2021b). Platforms can make their costs transparent and apply cost strategies that realistically include all socio-economic classes. **Designers can help create balance between customization and appeal for underprivileged coops so that they get more help from the ones that are already well off.**

13.4.2 Resisting Financial Precarity by Prioritising Convenience, Care and Fun Ourselves

To avoid financial precarity, the ecosystem coops may want to inhabit is a non-capitalist, e.g., a pro-social market of coops. **The aim is to design without transactional, monetary relationships in mind.** Still, the discussed ecosystem is a system in the middle of capitalist structures. One conflicting point is how to integrate a non-capitalistic idea into the dominant monetary system. Previous work on the relationship of commoning and co-design (Bassetti *et al.*, 2019) illustrated the importance of design outputs that embody and facilitate social values as a key to avoid the dominant, market-based and individualistic model of platform capitalism. Instead of individuals set against companies, regulated by punishment rules, the designs for cooperativism can seek for a language that emphasises solidarity and collectivism.

Our participants valued a new paradigm of convenience, care and fun in community experiences with cooperative platforms. One example is the environment where the coop workers will thrive. Rather than mechanisms that lead to burden of a few, the spaces for coops should enable mutual fun experiences. **Designers can embed fun, care and convenience to their platforms as mechanisms to reciprocate the hard work of volunteerism with a sense of solidarity.**

13.4.3 Designing for Non-capitalistic Systems: Rethinking the "Currency" and "Data"

In the process of imagining new markets, two key terms with constantly changing meanings were "currency" and "data". While the way several platforms define a currency, or how they handle data may change, an ecosystemic approach would push forward a design understanding for making these two aspects as understandable as possible. A language should be developed for cooperative platforms, enabling community members to have active ownership of their data and how their transactions are handled. **If a service requires a payment, or if data needs to be collected, the system should make this graspable to the end-users.**

While local, regional and national public funds are still seen as the main sources for seeding new and novel projects as such, a business ecosystem with similar values, and an understanding of the possibilities in the market are needed to ensure long term sustainability. Although there are certain conflicts on the financial models, instead of pushing members to choose one over the other, more importantly, integrating a rationale for why and how the price strategy is defined seems to be crucial to ensure service sustainability over time. For example, in the case of Peerby, participants questioned why the service is so expensive. Without a rationale on that, the scepticism may grow. Such **high entry costs may hinder equal access to the platforms from members of different economic classes.**

Currency and their exchange are human made, as is the transaction or conservation of data. Carroll and Bellotti (2015) offered an overview of currency and exchange paradigms from history, explaining details of their reciprocal and market capacities. We can re-imagine them integrating the power of data, and the values of cooperativism. The new ecosystems should also anticipate, account for, and, ideally, provide instruments to alleviate the high emotional labour (Lutz *et al.*, 2018) and the organizational effort (Luckner *et al.*, 2015) required to start and to deliver a service. The ecosystemic values should not ignore or underestimate novelty and social impact, **interactions and activities**, which often happen in small numbers in local communities on the ground (Fedosov *et al.*, 2021a). These can be counted as **even more valuable assets than scaling in numbers to support endurance and resilience of community-oriented sharing (economy) initiatives** over time.

13.5 Conclusion

In this chapter, we mapped the emergent design space of sharing and cooperativism. We discussed the ecosystemic approach to the design of future post-growth sharing platforms and services. We emphasised the importance of inclusive design strategies in devising new sharing platforms and re-imagining the existing ones. We deem that design efforts towards solidarity and collectivism can open new opportunities for platform cooperatives and the other models of the non-profit sharing economies, which aims to create value within the local communities they operate. We also believe that future research can critically revisit the contemporary interpretations of markets, ownership, and data, to ensure that voices of those that are on the fringes are also adequately pronounced in the scholarly conversations going forward. To support this, we have synthesised the findings of our workshop, presenting three key opportunities for designers of platforms to consider: (1) focusing on ecosystemic design and tools; (2) adopting aims that are inclusive, equitable and justice-oriented to ensure solidarity and collectivism; and (3) rethinking terms such as currency and data while designing for cooperativism. Ultimately, we hope that this work will inspire future design research and practice to develop long-lived techniques and methods for knowledge transfer to the communities on the ground with the view of creating better sharing cultures in the environment we aspire.

Acknowledgements

We are grateful to our workshop participants for their contribution to the workshop. This work was supported by the COST Action CA16121 From Sharing to Caring, Forschungskredit of the University of Zurich grant no. FK-20-021, and the EPSRC grant no. EP/S027726/1.

References

Alkhatib, A., Bernstein, M.S., and Levi, M. (2017). 'Examining Crowd Work and Gig Work Through The Historical Lens of Piecework', in *Proceedings of the 2017 CHI Conference on Human Factors in Computing Systems*. New York, NY, USA: Association for Computing Machinery (CHI '17), pp. 4599–4616.

Bassetti, C., Sciannamblo, Lyle, M., Teli, M., De Paoli, S., and De Angeli, A. (2019). Co-designing for common values: creating hybrid spaces to nurture autonomous cooperation. *CoDesign*, *15*(3), 256–271.

Belk, R. (2010). Sharing. *Journal of Consumer Research*, *36*(5), pp. 715–734.

Belk, R. (2018). Foreword: the sharing economy. The rise of the sharing economy: Exploring the challenges and opportunities of collaborative consumption, pp. 9–12.

Bellotti, V.M.E., Cambridge, S., Hoy, K., Shih, P.C., Handalian, L.R., Han, K., and Carroll, J.M. (2014). 'Towards community-centered support for peer-to-peer service exchange', *Proceedings of the 32nd annual ACM conference on Human factors in computing systems – CHI '14*, pp. 2975–2984.

Bødker, S., Lewkowicz, M., and Boden, A. (2020). 'What's in a Word? Platforms Supporting the Platform Economy', in *Proceedings of the 11th Nordic Conference on Human-Computer Interaction: Shaping Experiences, Shaping Society*. New York, NY, USA: Association for Computing Machinery (NordiCHI '20).

Botsman, R. and Rogers, R. (2010). What's mine is yours. The rise of collaborative consumption.

Botero, A., Marttila, S., Poderi, G., Saad-Sulonen, J., Seravalli, Teli, M., and Frederick, M.C.A. (2020). Commoning Design and Designing Commons. In *Proceedings of the 16th Participatory Design Conference 2020 – Participation(s)*

Otherwise – Vol. 2 (PDC '20: Vol. 2), June 15–20, 2020, Manizales, Colombia. ACM, New York, NY, USA, 3 pages.

Carroll, J.M. and Bellotti, V. (2015). 'Creating Value Together: The Emerging Design Space of Peer-to-Peer Currency and Exchange', *Proceedings of the 18th ACM Conference on Computer Supported Cooperative Work & Social Computing – CSCW '15*, pp. 1500–1510.

Cook, K.S., Cheshire, C., Rice, E.R.W., and Nakagawa, S. (2013). Social exchange theory. In J. DeLamater and A. Ward (Eds.), Handbook of social psychology. Dordrecht: Springer, pp. 61–88.

Costanza-Chock, S. (2018). Design Justice: towards an intersectional feminist framework for design theory and practice. *Proceedings of the Design Research Society.*

Dillahunt, T.R. and Malone, A.R. (2015). 'The promise of the sharing economy among disadvantaged communities', *Conference on Human Factors in Computing Systems – Proceedings*, 2015–April, pp. 2285–2294.

Dillahunt, T.R., Wang, X., Wheeler, E., Cheng, H.F., Hecht, B., and Zhu, H. (2017). 'The Sharing Economy in Computing: A Systematic Literature Review', *Proc. ACM Hum.-Comput. Interact.* New York, NY, USA: Association for Computing Machinery, 1(CSCW).

Dillahunt, T.R., Lam, J., Lu, A., and Wheeler, E. (2018). Designing Future Employment Applications for Underserved Job Seekers: A Speed Dating Study. In *Proceedings of the 2018 Designing Interactive Systems Conference (Hong Kong, China) (DIS' 18).* Association for Computing Machinery, New York, NY, USA, 33–44.

Dombrowski, L., Harmon, E., and Fox, S. (2016). Social Justice-Oriented Interaction Design: Outlining Key Design Strategies and Commitments. In *Proceedings of the 2016 ACM Conference on Designing Interactive Systems (DIS '16).* Association for Computing Machinery, New York, NY, USA, 656–671.

Edelman, B.G. and Luca, M. (2014). 'Digital Discrimination: The Case of Airbnb.com', {SSRN} Electronic Journal. Elsevier {BV}.

Fedosov, A. (2020). Supporting the Design of Technology-Mediated Sharing Practices. Carl Grossmann Verlag. Berlin/Bern.

Fedosov, A., Odom, W., Langheinrich, M., and Wakkary, R. (2018). Roaming Objects: Encoding Digital Histories of Use into Shared Objects and Tools, *Proceedings of the 2018 on Designing Interactive Systems Conference 2018 – DIS '18*, pp. 1141–1153, New York, New York, USA: ACM Press.

Fedosov, A., Lampinen, A., Dillahunt, T., Light, A., and Cheshire, C. (2019a). Cooperativism and Human-Computer Interaction. In *Extended Abstracts of the 2019 CHI Conference on Human Factors in Computing Systems (Glasgow, Scotland UK) (CHI EA'19).* Association for Computing Machinery, New York, NY, USA, 1–4.

Fedosov, A., Kitazaki, M., Odom, W., and Langheinrich. M. (2019b). Sharing Economy Design Cards. In *Proceedings of the 2019 CHI Conference on Human Factors in Computing Systems (Glasgow, Scotland UK) (CHI '19).* Association for Computing Machinery, New York, NY, USA, 1–14.

Fedosov, A., Lampinen, A., Odom, W., and Huang, E.M. (2021a). A Dozen Stickers on a Mailbox: Physical Encounters and Digital Interactions in a Local Sharing Community. *Proceedings of ACM Human-Computer Interaction*, Vol. 4, No. CSCW3. CSCW '20, 23 pages. ACM Press, Article 240.

Fedosov, A., Boos, D., Schmidt-Rauch, S., Ojala, J., and Lewkowicz, M. (2021b). Challenges of transferring UX designs and insights across products and services. In *Proceedings of 19th European Conference on Computer-Supported Cooperative Work.* European Society for Socially Embedded Technologies (EUSSET).

Glöss, M., McGregor, M., and Brown, B. (2016). Designing for Labour: Uber and the On-Demand Mobile Workforce. In *Proceedings of the 2016 CHI Conference on Human Factors in Computing Systems (San Jose, California, USA) (CHI '16).* Association for Computing Machinery, New York, NY, USA, 1632–1643.

Hess, C. and E. Ostrom (Eds.) (2007). Understanding Knowledge as Commons. Cambridge, Ma: The MIT Press.

Interaction Design and Architecture(s) Journal, Peer-to-Peer Exchange and the Sharing Economy: Analysis, Designs, and Implications (2015), N.24; 63–80.

John, N.A. (2017). The age of sharing. John Wiley & Sons.

Jamieson, D. and Wilson, R. (2020). "A Fair(er) Share for All? Digital Platform Governance in the Sharing Economy Era", In *Workshop proceedings of NordiCHI 2020 on Sharing & Cooperativism: Designing For Economies*, Accessed on November 30, 2020: https://sharingcoopnordichi2020.files.wordpress.com/2020/10/112-a-fairer_share_for_all_digital-platform_governance_in_the_sharing_economy_era.pdf

Katrini, E. (2018). 'Sharing Culture: On definitions, values, and emergence'. *The Sociological Review. {SAGE} Publications, 66*(2), pp. 425–446.

Kennedy, J. (2016). Conceptual boundaries of sharing. *Information, Communication & Society, 19*(4), pp. 461–474.

Korsgaard, H., Lewkowicz, M., Boden, A., Avram, G., and Bødker, S. (2020). June. Studying Technical Mechanisms for Supporting Sharing Communities. In *ECSCW 2020: The 18th European Conference on Computer-Supported Cooperative Work.*

Lampinen, A., Lehtinen, V., Cheshire, C., and Suhonen, E. (2013). Indebtedness and Reciprocity in Local Online Exchange. In *Proceedings of the 2013 Conference on Computer Supported Cooperative Work (San Antonio, Texas, USA) (CSCW '13).* Association for Computing Machinery, New York, NY, USA, 661–672.

Lampinen, A., Bellotti, V., Monroy-Hernández, A., Cheshire, C., and Samuel, A. (2015). Studying the "Sharing Economy": Perspectives to Peer-to-Peer Exchange. In *Proceedings of the 18th ACM Conference Companion on Computer Supported Cooperative Work & Social Computing (CSCW '15 Companion).* ACM, New York, NY, USA, 117–121.

Lampinen, A., Bellotti, V., Cheshire, C., and Gray, M. (2016). CSCW and the Sharing Economy: The Future of Platforms as Sites of Work. Collaboration and Trust. In *Proceedings of the 19th ACM Conference on Computer Supported Cooperative Work and Social Computing Companion (CSCW '16 Companion).* ACM, New York, NY, USA, 491–497.

Lampinen, A. and Brown, B. (2017). Market Design for HCI: Successes and Failures of Peer-to-Peer Exchange Platforms. In *Proceedings of the 2017 CHI Conference on Human Factors in Computing Systems (Denver, Colorado, USA) (CHI '17).* Association for Computing Machinery, New York, NY, USA, 4331–4343.

Lampinen, A., McGregor, M., Comber, R., and Brown, B. (2018a). Member-Owned Alternatives: Exploring Participatory Forms of Organising with Cooperatives. Proc. ACM Hum.-Comput. Interact. 2, CSCW, Article 100 (November 2018), 19 pages.

Lampinen, A., Lutz, C., Newlands, G., Light, A., and Immorlica, N. (2018b). Power Struggles in the Digital Economy: Platforms, Workers, and Markets. In *Proceedings of the 21st ACM Conference Companion on Computer Supported Cooperative Work & Social Computing (CSCW '18 Companion).* ACM.

Light, A. (2019). Designing the economics of the sharing economy: towards sustainable management. Edward Elgar Publishing, Cheltenham, UK.

Light, A. and Miskelly, C. (2015). 'Sharing economy vs sharing cultures? Designing for social, economic and environmental good'. *Interaction Design and Architecture(s), 24*(1), pp. 49–62.

Light, A. and Miskelly, C. (2019). 'Platforms, Scales and Networks: Meshing a Local Sustainable Sharing Economy', *Computer Supported Cooperative Work (CSCW). Springer Science and Business Media {LLC}, 28*(3–4), pp. 591–626.

Lutz, C., Newlands, G., and Fieseler, C. (2018). Emotional labor in the sharing economy. In *Proceedings of the 51st Hawaii International Conference on System Sciences.*

Luckner, N., Fitzpatrick, G., Werner, K., and Subasi, Ö. (2015). Setting up and Running a Sharing Service: an Organizational Perspective.

Malmborg, L., Light, A., Fitzpatrick, G., Bellotti, V., and Brereton, M. (2015). Designing for Sharing in Local Communities. In *Proceedings of the 33rd Annual ACM Conference Extended Abstracts on Human Factors in Computing Systems (CHI EA '15).* ACM, New York, NY, USA, 2357–2360.

Manzini, E. (2015). Design, when everybody designs: An introduction to design for social innovation. MIT press.

Nardi, B. (2019). Design in the Age of Climate Change. *She Ji: The Journal of Design, Economics, and Innovation* 5(1), 5–14.

N.A., Design Justice Network Principles. (2018). Available at: https://designjustice.org/read-the-principles (accessed:22.11.2020).

Ostrom, E. (1990). Governing the commons-The evolution of institutions for collective actions. Political economy of institutions and decisions.

Scholz, T. (2016). Platform cooperativism. Challenging the corporate sharing economy. New York, NY: Rosa Luxemburg Foundation.

Scholz, T. and Schneider, N. (2017). Ours to Hack and to Own: The Rise of Platform Cooperativism, a New Vision for the Future of Work and a Fairer Internet. OR Books.

Stofberg, N. and Bridoux, F. (2019). Consumers' choice among peer-to-peer sharing platforms: The other side of the coin. *Psychology & Marketing, 36*(12), pp. 1176–1195.

Simonsen, J. and Robertson, T. eds. (2012). Routledge international handbook of participatory design. Routledge.

Subasi, Ö., Fedosov, A., Bates, O., Lampinen, A., and Light, A. (2020). Sharing & Cooperativism: Designing For Economies. In *Proceedings of the 11th Nordic Conference on Human-Computer Interaction: Shaping Experiences, Shaping Society (NordiCHI '20).* Association for Computing Machinery, New York, NY, USA, Article 135, 1–3.

Zhang, A.X., Hugh, G., and Bernstein, M.S. (2020). October. PolicyKit: Building Governance in Online Communities. In *Proceedings of the 33rd Annual ACM Symposium on User Interface Software and Technology* (pp. 365–378).

Index

About the Editors

Maurizio Teli is Associate Professor of Techno-Anthropology and Participation at the Department of Planning, Aalborg University, Denmark. He works at the intersection of participatory design, the analysis of socio-economic inequalities, and critical theories. He investigates how to co-design technologies contributing to commoning practices, with a particular focus on collaborations with grassroots initiatives. He serves in the editorial board and program committee of international journals and conferences in interaction design, co- design, and participatory design. He is now chairing the Center for Sustainability and Digital Transformation at the Technical Faculty of IT and Design at Aalborg University.

Chiara Bassetti is Senior Assistant Professor at the Department of Sociology and Social Research of the University of Trento, and a researcher at the Institute of Cognitive Sciences and Technologies of the Italian National Research Council (CNR). An ethnographer and ethnomethodologist, her main research focus rests on social interaction, cultural phenomena and collaborative practices. Workplace studies, computer supported collaborative work, and socio-technical systems are areas of interest. Further research concerns cooperation, grassroots initiatives and socio-economic innovation. Since 2012, she has been working in interdisciplinary teams, contributing to the design of artifacts such as digital platforms and complementary currencies. She is Co-Editor in Chief of Etnografia e Ricerca Qualitativa – the Italian journal of Ethnography and Qualitative Research.

Contributing Authors

Ana Aleksić
Faculty of Economics and Business, University of Zagreb, Zagreb, Croatia

Julijana Angelovska
University of Tourism and Management in Skopje, Macedonia

Oliver Bates
School of Computing and Communications, Lancaster University, Lancaster, UK

Chiara Bassetti
Italian National Research Council (CNR) and University of Trento, Italy

Aniko Bernat
TÁRKI Social Research Institute, Budapest, Hungary

Ivana Načinović Braje
PhD Faculty of Economics and Business, University of Zagreb

Anita Čeh Časni
Faculty of Economics and Business, University of Zagreb, Zagreb, Croatia

David Crombie
Center for Research and Innovation, HKU, Utrecht, The Netherlands

Veronica Cruciani
Troyes University of Technology, France
LIST3N/Tech-CICO

Vera Diogo
School of Higher Education – Polytechnic Institute of Porto, Porto, Portugal

Kosjenka Dumančić
Faculty of Economics and Business, University of Zagreb, Zagreb, Croatia

Anton Fedosov
Department of Informatics, University of Zurich, Zurich, Switzerland

Hannelore Goyens
U Hasselt, Faculty of Architecture and Arts, Hasselt, Belgium

Liesbeth Huybrechts
U Hasselt, Faculty of Architecture and Arts, Hasselt, Belgium

Eglantina Hysa
Department of Economics, Epoka University, Tirana, Albania

Sindiola Koka
Department of Business Administration, Epoka University, Tirana, Albania

Revathi Kollegala
K8, Saarbrücken, Germany

Alba Kruja (Demneri)
Department of Business Administration, Epoka University, Tirana, Albania

Airi Lampinen
Department of Computer and Systems Sciences, Stockholm University, Sweden

Justin Larner
Social Balance Consultancy, Manchester, UK

Myriam Lewkowicz
Troyes University of Technology, France
LIST3N/Tech-CICO

Ann Light
K3, Malmö University, Sweden/Engineering and Informatics, University of Sussex, Brighton, United Kingdom

Christoph Lutz
BI Norwegian Business School, Oslo, Norway

Filip Majetić
Institute of Social Sciences Ivo Pilar, Zagreb, Croatia

Laura Michelini
LUMSA University Rome, Italy

Rodrigo Perez-Vega
Kent Business School, University of Kent

Chiara Rossitto
Department of Computer and Systems Sciences, Stockholm University, Sweden

Özge Subaşı
Department of Media and Visual Arts, Koç University, Istanbul

Venere Stefania Sanna
Centre for Politics and International Studies (CeSPI) and Department of Methods and Models for Economics, Territory and Finance (MEMOTEF) Sapienza University, Rome, Italy

Maurizio Teli
Aalborg University, Denmark

Penny Travlou
ESALA, Edinburgh College of Art, The University of Edinburgh, Scotland

Egle Vaiciukynaite
Kaunas University of Technology, Kaunas, Lithuania

Soenke Zehle
xm:lab – Experimental Media Lab, HBKsaar, Saarbrücken, Germany

Printed and bound by CPI Group (UK) Ltd, Croydon, CR0 4YY

15/09/2026

14962365-0003